The HDL Handbook
BIOLOGICAL FUNCTIONS AND CLINICAL IMPLICATIONS

The HDL Handbook
BIOLOGICAL FUNCTIONS AND CLINICAL IMPLICATIONS

Second Edition

Edited by

TSUGIKAZU KOMODA

Amsterdam • Boston • Heidelberg • London
New York • Oxford • Paris • San Diego
San Francisco • Singapore • Sydney • Tokyo
Academic Press is an imprint of Elsevier

Academic Press is an imprint of Elsevier
32 Jamestown Road, London NW1 7BY, UK
225 Wyman Street, Waltham, MA 02451, USA
525 B Street, Suite 1800, San Diego, CA 92101-4495, USA

First edition 2010

Notice
No responsibility is assumed by the publisher for any injury and/or damage to persons or
property as a matter of products liability, negligence or otherwise, or from any use or
operation of any methods, products, instructions or ideas contained in the material herein.
Because of rapid advances in the medical sciences, in particular, independent verification of
diagnoses and drug dosages should be made.

British Library Cataloguing-in-Publication Data
A catalogue record for this book is available from the British Library

Library of Congress Cataloging-in-Publication Data
A catalog record for this book is available from the Library of Congress

For information on all Academic Press publications
visit our website at elsevierdirect.com

ISBN: 978-0-12-407867-3

Typeset by TNQ Books and Journals
www.tnq.co.in

Printed and bound by CPI Group (UK) Ltd, Croydon, CR0 4YY
Transferred to digital print 2013

Working together
to grow libraries in
developing countries

www.elsevier.com • www.bookaid.org

CONTENTS

9. HDL Apoprotein Mimetic Peptides as Anti-Inflammatory Molecules   221

Godfrey S. Getz, Catherine A. Reardon

10. Oxidized High-Density Lipoprotein: Friend or Foe   247

Toshiyuki Matsunaga, Akira Hara, Tsugikazu Komoda

11. Current Aspects of Paraoxonase-1 Research   273

Mike Mackness, Bharti Mackness

When I started to write *The HDL Handbook: Biological Function and Clinical Implications*, Second Edition, Professor Akihiro Inazu of Kanazawa University in Kanazawa nominated some suitable contributors. Of course, he is also a respectable contributor to this book. In addition, David H. Alpers, Professor at William B Kountz Hospital, Washington University, revised this second edition very well. Therefore, I would like to thank him for his excellent revision of this book.

As you know, this book includes up-to-date progress in HDL research. The content of this book is a precise collaboration and collection of the work from all of the contributors, who, despite all being busy, carefully wrote their chapters. Therefore, you will be fascinated by the plethora of new advances highlighted by these HDL researchers. The present *The HDL Handbook: Biological Function and Clinical Implications*, Second Edition, is a useful tool not only for basic researchers in institutions or pharmaceutical companies, but also practical physicians.

Furthermore, since this book is small, it is portable. However, the content of *The HDL Handbook* contains the current advances of HDL molecules.

Finally, I believe that this book provides a unique perspective of the HDL field.

Tsugikazu Komoda, MD

CONTRIBUTORS

Lusana Ahsan
Lipoprotein Metabolism Section, Cardiovascular and Pulmonary Branch, National Heart, Lung and Blood Institute, NIH, Bethesda, MD, USA

Marcelo J. Amar
Lipoprotein Metabolism Section, Cardiovascular and Pulmonary Branch, National Heart, Lung and Blood Institute, NIH, Bethesda, MD, USA

Bela F. Asztalos
Lipid Metabolism Laboratory, Tufts University, Boston, MA, USA

Makoto Ayaori
Division of Anti-aging and Vascular Medicine, Department of Internal Medicine, National Defense Medical College, Tokorozawa, Japan

Lita Freeman
Lipoprotein Metabolism Section, Cardiovascular and Pulmonary Branch, National Heart, Lung and Blood Institute, NIH, Bethesda, MD, USA

Godfrey S. Getz
Department of Pathology, The University of Chicago, Chicago, IL, USA

Scott M. Gordon
National Heart, Lung and Blood Institute, National Institutes of Health, Bethesda, MD, USA

Maryse Guerin
UPMC Université Pierre et Marie Curie, Hôpital de la Pitié, Paris, France

Akira Hara
Laboratory of Biochemistry, Gifu Pharmaceutical University, Gifu, Japan

Katsunori Ikewaki
Division of Anti-aging and Vascular Medicine, Department of Internal Medicine, National Defense Medical College, Tokorozawa, Japan

Akihiro Inazu
Department of Clinical Laboratory Science, School of Health Sciences, Kanazawa University, Kanazawa, Japan

Brian Ishida
Lipid Metabolism Laboratory, Tufts University, Boston, MA, USA

Xian-Cheng Jiang
Department of Cell Biology, State University of New York, Downstate Medical Center, Brooklyn, NY, USA and Molecular and Cellular Cardiology Program, VA New York Harbor Healthcare System, Brooklyn, NY, USA

Tsugikazu Komoda
Department of Urology, Toho University School of Medicine, Tokyo, Japan

Zhiqiang Li
Department of Cell Biology, State University of New York, Downstate Medical Center, Brooklyn, NY, USA and Molecular and Cellular Cardiology Program, VA New York Harbor Healthcare System, Brooklyn, NY, USA

Bharti Mackness
AVDA Princip D'Espanya No 152, Miami Playa, Tarragona, Spain

Mike Mackness
AVDA Princip D'Espanya No 152, Miami Playa, Tarragona, Spain

Akira Matsunaga
Department of Laboratory Medicine, Faculty of Medicine, Fukuoka University, Fukuoka, Japan

Toshiyuki Matsunaga
Laboratory of Biochemistry, Gifu Pharmaceutical University, Gifu, Japan

Alice F. Ossoli
Lipoprotein Metabolism Section, Cardiovascular and Pulmonary Branch, National Heart, Lung and Blood Institute, NIH, Bethesda, MD, USA

Catherine A. Reardon
Department of Pathology, The University of Chicago, Chicago, IL, USA

Alan T. Remaley
Lipoprotein Metabolism Section, Cardiovascular and Pulmonary Branch, National Heart, Lung and Blood Institute, NIH, Bethesda, MD, USA

Robert D. Shamburek
Lipoprotein Metabolism Section, Cardiovascular and Pulmonary Branch, National Heart, Lung and Blood Institute, NIH, Bethesda, MD, USA

Mari Tabuchi
Department of Chemistry, Faculty of Science, Rikkyo University, Tokyo, Japan

Mariko Tani
Lipid Metabolism Laboratory, Tufts University, Boston, MA, USA

Boris Vaisman
Lipoprotein Metabolism Section, Cardiovascular and Pulmonary Branch, National Heart, Lung and Blood Institute, NIH, Bethesda, MD, USA

Elise F. Villard
UPMC Université Pierre et Marie Curie, Hôpital de la Pitié, Paris, France

Amirfarbod Yazdanyar
Department of Cell Biology, State University of New York, Downstate Medical Center, Brooklyn, NY, USA

2d: two-dimensional
2DIGE: two-dimensional gel electrophoresis
aa: amino acid
AAPH: 2,2'-azo-bis(2-amidinopropane) dihydrochloride
ABC: ATP-binding cassette transporter
ABCA1: ATP-binding cassette transporter A1
ABCG1: ATP-binding cassette transporter G1
ACAT: Acyl-CoA:Cholesterol Acyltransferase
ACS: acute coronary syndrome
AHLs: N-acyl homoserine lactones
Aib: aminoisobutyric acid
AKR: aldo-keto reductase
AMP: adenosine monophosphate
Apo: apolipoprotein
apoA-I: apolipoprotein A-I
apoB: apolipoprotein B
BLp: apoB-containing lipoprotein particles
BMI: body mass index
BPI: bactericidal/permeability-increasing protein
C: cholesterol
CAD: coronary artery diseases
cAMP: cyclic AMP
CE: cholesteryl ester
CETP: cholesteryl ester transfer protein
CHD: coronary heart disease
cIMT: carotid intima-media-thickness
CKD: chronic kidney disease
CM: chylomicron
CPE: ceramide phosphatidylethanolamine
CRE: cAMP responsive element
CRP: C-reactive protein
CVD: cardiovascular diseases
DAMPs: damage-associated molecular patterns
DMPC: dimyristoylphosphatidylcholine
DPH-PC: 2-(3-(diphenylhexatrienyl)propanoyl)-1-hexadecanoyl-*sn*-glycero-3-phospholcholine
ecNOS: endothelial constitutive nitric oxide synthase
eNOS: endothelial nitric oxide synthase
EL: endothelial lipase
ELISA: enzyme-linked immunosorbent assay
ER: endoplasmic reticulum
FC: free cholesterol
FCR: fractional catabolic rate

FED: fish-eye disease
FFA: free fatty acid
FLD: familial LCAT deficiency
FMD: flow-mediated dilation
GST: glutathione transferase
GWAS: genome-wide association study
Hb: hemoglobin
HcyTL: Homocysteine thiolactone
HDL-C: high-density of lipoprotein cholesterol
HDL: high-density lipoprotein
HDMECs: human dermal microvascular endothelial cells
HFHC: high-fat, high-cholesterol
HL: hepatic lipase
HNE: 4-hydroxy-2-nonenal
HPLC: high-performance liquid chromatography
Hpr: haptoglobin-related protein
HTGL: hepatic triglyceride lipase
HUVEC: human umbilical vein endothelial cells
ICAM: intracellular adhesion molecule
IDL: intermediate-density lipoprotein
IHD: ischemic heart disease
KO: knockout
LBP: lipopolysaccharide-binding protein
LC: liquid chromatography
LCAT: lecithin:cholesterol acyltransferase
LDL-C: low density lipoprotein-cholesterol
LDL: low-density lipoprotein
LDLR: LDL receptor
LP: lipoprotein
LPL: lipoprotein lipase
LpPLA$_2$: lipoprotein-associated phospholipase A$_2$
LPS: lipopolysaccharide
LpX: Lipoprotein X
LXR: liver X receptor
LXRE: LXR-responsive elements
lyso-PC: lysophosphatidylcholine
MALDI-MS: matrix-assisted laser desorption ionization mass spectrometry
MCA: monocyte chemotactic assay
MCP: monocyte chemoattractant protein
MetO: methionine sulfoxide
MI: myocardial infarction
MPO: myeloperoxidase
MS: mass spectrometry
NADPH: nicotinamide adenine dinucleotide phosphate
NEFA: nonesterified fatty acids
NF-κB: nuclear factor-kappa B
NO: nitric oxide

oxHDL: oxidized HDL
oxLDL: oxidized LDL
PAD: peripheral arterial disease
PAGE: polyacrylamide gel electrophoresis
PC: phosphatidylcholine
PL: phospholipid
PLA_2: phospholipase A_2
PLTP: phospholipid transfer protein
PON: paraoxonase
PON1: paraoxonase1
POPC: palmitoyl oleoyl phosphatidylcholine
PPAR: peroxisome proliferator-activated receptor
PS: phosphatidylserine
PWV: pulse wave velocity
QS: quorum sensing
RA: rheumatoid arthritis
RAR: retinoic acid receptor
RCT: reverse cholesterol transport
rLCAT: recombinant human LCAT
ROS: reactive oxygen species
rPLTP: recombinant PLTP
RXR: retinoid X receptor
S1P: sphingosine-1-phosphates
SAA: serum amyloid A
SM: sphingomyelin
SMS: sphingomyelin synthase
SMSr: sphingomyelin synthase–related protein
SNP: single-nucleotide polymorphisms
SPT: serine palmitoyltransferase
SR-BI: scavenger receptor class B type I
SREBP: sterol regulatory-element binding protein
TBARS: thiobarbituric acid-reactive substances
TG: triglyceride
TGRL: triglyceride-rich lipoprotein
TLF: trypanosome lytic factor
TRL: triglyceride-rich lipoprotein
TRL: TG-rich lipoproteins
VA-HIT: Veterans Administration HDL Intervention Trial
VCAM: vascular cell adhesion molecule
VLDL: very low-density lipoproteins
WT: wild-type

CHAPTER 1

Introduction of HDL Molecules, Past and Brief Future

Mari Tabuchi*, Tsugikazu Komoda**

*Department of Chemistry, Faculty of Science, Rikkyo University, Tokyo, Japan **Department of Urology, Toho University School of Medicine, Tokyo, Japan

Contents

Abstract

Our understanding of high-density lipoprotein cholesterol (HDL-C) and its relationship to coronary heart disease (CHD) has changed dramatically over the past half century. The initial discovery of the protective role of HDL-C was made by Gofman in the mid-1950s. In the mid-1970s, Glomset's pioneering studies on reverse cholesterol transport (RCT) and Gordon's Framingham study supported the concept that HDL-C was a "good" lipoprotein, since there was a strong inverse correlation between the levels of serum HDL-C and the subsequent risk of atherosclerosis. Furthermore, in the late-1980s, a decreased concentration of serum HDL-C emerged as one of the major risk factors for coronary artery disease. However, recent studies in the early 2000s have shown that high HDL-C levels do not necessarily have anti-atherogenic effects. In patients with cholesterol ester transfer protein (CETP) deficiency, high HDL-C levels have no anti-atherogenic effect. Furthermore, drugs that inhibit CETP elevate HDL-C levels but do not decrease the risk of cardiac events. Thus, the traditional anti-atherogenic role of HDL-C has been questioned. There is evidence that HDL-C molecules have additional atheroprotective roles beyond bulk cholesterol removal from cells through RCT. Other studies suggest that the widely used diagnostic measurement of HDL-C may have a limitation and that the qualitative measurement of HDL-C molecules may be important in the near future.

1. INTRODUCTION

The high–density lipoprotein (HDL) molecule consists of particles that are heterogeneous in size, density, and composition. Recently, views on the relevance and utility of HDL have been changing dramatically. For over a half century, plasma HDL-cholesterol (HDL-C) was believed to play an important role in the protection against atherosclerosis. HDL-C was termed "good cholesterol" and thought to be a negative risk factor for atherosclerosis. However, an important discovery has been made recently. Although a deficiency or low-level of HDL-C causes premature atherosclerosis, some conditions where there is elevated HDL-C have been found to increase the risk of coronary atherosclerosis. Thus, HDL-C can sometimes become "bad cholesterol" and increase the risk of atherosclerosis. This chapter gives a brief introduction on the history of HDL-C molecules and suggests future directions for investigation.

1.1. HDL-C as "Good Cholesterol"

HDL was first isolated from horse serum in 1929 [1] and from human serum in 1949 [2]. By 1950 total cholesterol and LDL-C took center stage, although the role of HDL-C was unclear [3]. Several novel techniques for lipoprotein separation were developed, including ultracentrifugation [4,5] and electrophoresis [6], that allowed evaluation of the role of HDL-C in atherosclerosis.

The initial discovery of the protective role of HDL-C was made by Gofman at the Donner Laboratory of the University of California at Berkeley in 1956 [7]. He separated HDL-C subclasses using an ultracentrifuge and demonstrated an inverse association between HDL-C levels and coronary heart disease (CHD). However, his work was not widely accepted, because other researchers were unable to replicate his results, and total cholesterol and low-density lipoprotein cholesterol (LDL-C) were the primary focus at that time. Subsequently, many researchers have investigated the role of HDL-C in CHD.

In the 1960s, Glomset [8] investigated the mechanism of the plasma cholesterol esterification reaction and hypothesized that HDL-C plays a role in the transport of cholesterol from peripheral tissues to the liver that was termed "reverse cholesterol transport (RCT)".

In 1975, Miller found that plasma HDL-C was unrelated to the plasma concentration of total cholesterol and other lipoproteins [9]. This finding supported the "RCT" hypothesis. He also showed that a reduction of plasma HDL-C concentration accelerated the development of atherosclerosis, which supported the protective role of HDL-C against CHD. The anti-atherogenic role of HDL-C was proposed to be a result of its role in RCT—removal of excess cholesterol from peripheral tissues to the liver for re-use.

Berg also investigated the relationship between serum HDL-C levels and atherosclerotic heart disease [10]. The mean HDL-C concentration was significantly lower in men with CHD in comparison with those without CHD. These results were in accordance with the hypothesis that high levels of HDL-C protected against CHD to some extent. Gordon also confirmed that HDL-C played a protective role against CHD in the Framingham study in 1977 [11]. In that study, there was an inverse association between HDL-C levels and the incidence of CHD.

2. HDL MAY BE AN INDEPENDENT PREDICTOR OF CHD

The anti-atherogenic property of HDL-C was gradually accepted in the mid-1970s, since there was a strong inverse correlation between the levels of serum HDL-C and the subsequent risk of atherosclerosis. Furthermore, in the late-1980s, a decreased concentration of serum HDL-C emerged as one of the major risk factors for CHD.

Several studies including the Framingham heart study found that low HDL-C levels, widely prevalent in countries with Western diets, was an independent predictor for cardiovascular disease (CVD) risk [11–14]. This risk was true even in the presence of low LDL-C levels [15]. These findings increased the attention paid to HDL-C as a secondary prevention target for CVD risk [16,17]. Moreover, some researchers were convinced that HDL-C might be an even stronger risk factor for CHD than LDL-C. Conventionally, cardiovascular prevention strategies emphasize therapeutic reductions in LDL-C [18,19]. However, increasing attention had been focused on HDL-C as a secondary prevention target to address residual CVD risk [16,17].

3. HIGH LEVELS OF HDL SOMETIMES CAUSE ATHEROSCLEROSIS: THE EXAMPLE OF CETP DEFICIENCY

Some recent studies have shown that high HDL-C levels do not necessarily have anti-atherogenic effects, although the risk for atherogenesis is lower in the many patients with high HDL-C. Plasma Cholesterol Ester Transfer Protein (CETP) deficiency was originally reported in Japanese siblings with hyperalphalipoproteinemia [20]. Increased HDL-C is sometimes clustered in families. The CETP deficiencies were associated with high HDL-C levels and relatively low LDL-C levels (mean levels of 164 mg/dL and 77 mg/dL, respectively) [21]. Many studies showed that a low CETP mass and increased HDL-C was associated with a decreased risk of CHD [22–24]. In a Kochi

Prefectural Institute of Public Health's cross-sectional survey, no cases of CHD were found in 300 subjects with HDL-C > 100 mg/dL [25]. A meta-analysis of multiple clinical studies also supported the concept that lower CETP levels might have an anti-atherogenic effect [26].

However, epidemiological studies in Japanese Americans living in Hawaii and Japanese in the Omagari area have shown a relatively increased incidence of CHD in CETP deficiency [27], and homozygous CETP deficiency in some cases was thought to be pro-atherogenic [28]. Thus, high–HDL levels in some patients do not appear to have an anti–atherogenic effect.

4. IS HDL-C REALLY "GOOD CHOLESTEROL"?

Some clinical treatments, such as niacin [29], torcetrapib [30], and dalcetrapib [31], do not reduce cardiac risk, although HDL-C and/or LDL-C levels are successfully regulated. Statins effectively lower LDL-cholesterol levels, but residual cardiovascular risk remains [29]. Inhibition of CETP with torcetrapib successfully elevated HDL-C levels, but unexpectedly increased cardiovascular morbidity and mortality [30]. Dalcetrapib, another inhibitor of CETP, raised HDL-C levels in patients hospitalized with acute coronary syndrome, but this change failed to translate into a reduction in cardiovascular events [31]. There was no association between gene variants exclusively related to HDL-C concentration and myocardial infarction [32]. These results indicate that HDL-C might not have a causal role in preventing CHD. Although many early clinical studies showed evidence for a beneficial effect of HDL-C on CHD, the cardioprotective effect of increasing levels of HDL-C has been questioned.

5. QUANTITATIVE AND QUALITATIVE MEASUREMENT OF HDL-C MOLECULES

It is thought that HDL-C molecules have additional atheroprotective roles beyond bulk cholesterol removal from cells through RCT. The anti-inflammatory and anti-oxidant activities of HDL-C molecules have recently attracted considerable attention [33,34]. These activities of HDL-C are associated with protection from cardiovascular disease. One of the promising mechanisms was reported as "response to injury" by Ross in 1993 [35].

There are currently seven different direct homogenous HDL-C assays [36] that use chemical precipitation methods with reagents such as dextran sulfate instead of physical ultracentrifugation. In a recent report, the error between the seven direct assays was higher than 12%, resulting in inaccurate HDL-C measurement and subsequent CVD risk estimated by LDL-C [37]. Recently, results of long-term follow-up using an analytical ultracentrifuge have demonstrated that the subfractions HDL2 and HDL3 are independently related to CHD risk [38]. CETP-deficient patients have very large HDL-C particles [39], and CHD patients often have small discoidal HDL-C particles [40]. Large HDL-C particles increase with weight loss, niacin, certain statins, and CETP inhibitors [41–46]. These findings suggest that the widely used diagnostic measurement of HDL-C may have a limitation, and that the measurement of size and shape of HDL-C molecules could be more important than previously thought [47,48]. Thus, the definition should be changed from quantitative measurement of HDL-C to qualitative evaluation of HDL-C subfractions [35,49–51]. However, since HDL-C consists of polydisperse particles that are heterogeneous in size, density, shape, and composition, separations require complicated methods, such as electrophoresis [50] or chromatography [52], and are not available for routine clinical use.

One of the promising methods for the measurement of HDL-C molecules is a technique that uses bioformulated-fiber matrix electrophoresis [53]. High- and low-density fractions of HDL-C were well separated by this method. Although a higher density is generally believed to correlate with a smaller particle size, this separation contradicted this belief. Direct microscopic observation of fractioned HDL-C confirmed the lack of a relationship between density and size. This technique may be useful for the diagnosis of dyslipidemia in the future.

6. CONCLUSION

In conclusion, according to the many studies including clinical and epidemiological investigations over more than half a century, the effect of HDL-C on CVD may be more complicated than previously thought. These important new findings may be useful in estimating the risk of CHD in the near future.

REFERENCES

[1] Macheboeuf M. Recherches sur les phosphoaminolipides et les sterids du serum et du plasma sanguins. II Etude physiochimique de la fraction proteidique la plus riche en phospholipids et in sterides. Bull Soc Chim Biol 1929;11:485–503.
[2] Gofman JW, Lindgren FT, Elliott H. Ultracentrifugal studies of lipoproteins of human serum. J Biol Chem 1949;179:973–9.
[3] Warnick GR. High-Density Lipoproteins: The neglected stepchildren whose importance as a risk factor continues to be defined. Clin Chem 2008;54(5):923–4.

[4] Havel RJ, Eder HA, Bragdon JH. The distribution and chemical composition of ultra-centrifugally separated lipoproteins in human serum. J Clin Invest 1955;34:1345–53.

[5] Redgrave TG, Roberts DCK, West CE. Separation of plasma lipoproteins by density-gradient ultracentrifugation. Anal Biochem 1975;65:42–9.

[6] Emes AV, Latner AL, Rahbani-Nobar M, Tan BHT. The separation of plasma lipoproteins using gel electrofocusing and polyacrylamide gradient gel electrophoresis. Clinica Chimica Acta 1976;71(2):293–301.

[7] Gofman JW, Delalla O, Flazier F, Freeman NK, Lindgren FT, Nichols AV, et al. The serum lipoprotein transport system in health, metabolic disorders, atherosclerosis and coronary heart disease [Reprint of a 1956 Plasma paper]. J Clin Lipidol 2007;1(2): 104–41.

[8] Glomset JA. The plasma lecithin: cholesterol acyltransferase reaction. J Lipid Res 1969;9:155–67.

[9] Miller GJ, Miller NE. Plasma-high-density-lipoprotein concentration and development of ischæmic heart-disease. Lancet 1975;1:16–9.

[10] Berg K, Børresen A-L, Dahlén G. Serum-high-density-lipoprotein and atherosclerotic heart-disease. Lancet 1976;307(7958):499–501.

[11] Gordon T, Castelli WP, Hjortland MC, Kannel WB, Dawber TR. High density lipoprotein as a protective factor against coronary heart disease. The Framingham Study. Am J Med 1977;62:707–14.

[12] Castelli WP. Cholesterol and lipids in the risk of coronary artery disease: the Framingham Heart Study. Can J Cardiol 1988;4(Suppl. A):5A–10A.

[13] Gordon D, Rifkind BM. Current concepts: high-density lipoproteins—the clinical implications of recent studies. N Engl J Med 1989;321:1311–5.

[14] Wilson PWF, Abbott RD, Castelli WP. High density lipoprotein cholesterol and mortality: the Framingham Heart Study. Arterioscler 1988;8:737–41.

[15] Barter P, Gotto AM, LaRosa JC, Maroni J, Szarek M, Grundy SM, et al. Treating to New Targets Investigators. HDL cholesterol, very low levels of LDL cholesterol, and cardiovascular events. N Engl J Med 2007;357:1301–10.

[16] Gotto AM, Brinton EA. Assessing low levels of high-density lipoprotein cholesterol as a risk factor in coronary heart disease: a working group report and update. J Am Coll Cardiol 2004;43:717–24.

[17] Fruchart JC, Sacks F, Hermans MP, Assmann G, Brown V, Chapman J, et al. Residual Risk Reduction Initiative (R3I). Executive statement, the Residual Risk Reduction Initiative: a call to action to reduce residual vascular risk in dyslipidemic patient. Diab Vasc Dis Res 2008;5:319–35.

[18] Expert Panel on Detection, Evaluation, and Treatment of High Blood Cholesterol in Adults. Executive Summary of the Third Report of the National Cholesterol Education Program (NCEP) Expert Panel on Detection, Evaluation, and Treatment of High Blood Cholesterol in Adults (Adult Treatment Panel III). JAMA 2001;285:2486–97.

[19] Graham I, Atar D, Borch-Johnsen K, Boysen G, Burell G, Cifkova R, et al. European Society of Cardiology (ESC); European Association for Cardiovascular Prevention and Rehabilitation (EACPR); Council on Cardiovascular Nursing; European Association for Study of Diabetes (EASD); International Diabetes Federation Europe (IDF-Europe); European Stroke Initiative (EUSI); International Society of Behavioural Medicine (ISBM); European Society of Hypertension (ESH); European Society of General Practice/Family Medicine (ESGP/FM/WONCA); European Heart Network (EHN). European guidelines on cardiovascular disease prevention in clinical practice: executive summary. Eur J Cardiovasc Preven Rehab 2007;14(Suppl 2):E1–40.

[20] Koizumi J, Mabuchi H, Yoshimura A, Michishita I, Takeda M, Itoh H, et al. Deficiency of serum cholesteryl-ester transfer activity in patients with familial hyperalphalipopro-teinaemia. Atheroscler 1985;58:175–86.

[21] Inazu A, Brown ML, Hesler CB, Agellon LB, Koizumi J, Takata K, et al. Increased high-density lipoprotein levels caused by a common cholesteryl-ester transfer protein gene mutation. N Engl J Med 1990;323(18):1234–8.

[22] Lu H, Inazu A, Moriyama Y, Higashikata T, Kawashiri MA, Yu W, et al. Haplotype analyses of cholesteryl ester transfer protein gene promoter: a clue to an unsolved mystery of TaqIB polymorphism. J Mol Med 2003;81(4):246–55.

[23] Takata M, Inazu A, Katsuda S, Miwa K, Kawashiri M, Nohara A, et al. CETP (cholesteryl ester transfer protein) promoter -1337 C>T polymorphism protects against coronary atherosclerosis in Japanese patients with heterozygous familial hypercholesterolaemia. Clin Sci (Lond) 2006;111(5):325–31.

[24] Curb JD, Abbott RD, Rodriguez BL, Masaki K, Chen R, Sharp DS, et al. A prospective study of HDL-C and cholesteryl ester transfer protein gene mutations and the risk of coronary heart disease in the elderly. J Lipid Res 2004;45:948–53.

[25] Moriyama Y, Okamura T, Inazu A, Doi M, Iso H, Mouri Y, et al. A low prevalence of coronary heart disease in subjects with increased high-density lipoprotein cholesterol levels including those with plasma cholesteryl ester transfer protein deficiency. Prev Med 1998;27(5Pt1):659–67.

[26] Thompson A, Angelantonio ED, Sarwar N, Erqou S, Saleheen D, Dullaart RPF, et al. Association of cholesteryl ester transfer protein genotypes with CETP mass and activity, lipid levels, and coronary risk. JAMA 2008;299:2777–88.

[27] Nagano M, Yamashita S, Hirano K, Takano M, Maruyama T, Ishihara M, et al. Molecular mechanisms of cholesteryl ester transfer protein deficiency in Japanese. J Atheroscler Thromb 2004;11:110–21.

[28] Nagano M, Nakamura M, Kobayashi N, Kamata J, Hiramori K. Effort angina in a middle-aged woman with abnormally high levels of serum high-density lipoprotein cholesterol. Circ J 2005;69:609–12.

[29] The AIM-HIGH Investigators. Niacin in patients with low HDL cholesterol levels receiving intensive statin therapy. N Engl J Med 2011;365:2255–67.

[30] Barter PJ, Caulfield M, Eriksson M, Grundy SM, Kastelein J-JP, Michel KBrewer B for the ILLUMINATE Investigators, et al. Effects of torcetrapib in patients at high risk for coronary events. N Engl J Med 2007;357:2109–22.

[31] Schwartz GG, Olsson AG, Ballantyne CM, Barter PJ, Holme IM, Kallend D, et al. dal-OUTCOMES Committees and Investigators, et al. Ration ale and design of the dal-OUTCOMES trial: efficacy and safety of dalcetrapib in patients with recent acute coronary syndrome. Am Heart J 2009;158:896–901.

[32] Voight BF, Peloso GM, Orho-Melander M, Frikke-Schmidt R, Barbalic M, Jensen MK, et al. Plasma HDL cholesterol and risk of myocardial infarction: a Mendelian randomisation study. Lancet 2012;380(9841):572–80.

[33] Brewer HB. HDL metabolism and the role of HDL in the treatment of high-risk patients with cardiovascular disease. Curr Cardiol Rep 2007;9:486–92.

[34] Rosenson RS. Functional assessment of HDL: moving beyond static measures for risk assessment. Cardiovasc Drugs Ther 2010;24:71–5.

[35] Ross R. The pathogenesis of atherosclerosis: a perspective for the 1990s. Nature 1993;362:801–9.

[36] Robert S, Rosenson H, Brewer Jr B, Chapman JM, Fazio S, Hussain M, et al. HDL measures, particle heterogeneity, proposed nomenclature, and relation to atherosclerotic cardiovascular events. Clin Chem 2011;57(3):392–410.

[37] Miller WG, Myers GL, Sakurabayashi I, Bachman LM, Caudill SP, Dziekonski A, et al. Seven direct methods for measuring HDL and LDL cholesterol compared with ultracentrifugation reference measurement procedures. Clin Chem 2010;56:977–86.

[38] Williams PT, Feldman DE. Prospective study of coronary heart disease vs HDL2, HDL3, and other lipoproteins in Gofman's Livermore Cohort. Atherosclerosis 2011;214:196–202.

[39] Asztalos BF, Horvath KV, Kajinami K, Nartsupha C, Cox CE, Batista M, et al. Apo composition of HDL in cholesteryl ester transfer protein deficiency. J Lipid Res 2004;45:448–55.

[40] Asztalos BF, Cupples LA, Demissie S, Horvath KV, Cox CE, Batista MC, et al. High density lipoprotein subpopulation profile and coronary heart disease prevalence in male participants in the Framingham Offspring Study. Arterioscler Thromb Vasc Biol 2004;24:2181–7.

[41] Brousseau ME, Diffenderfer MR, Millar JS, Nartsupha C, Asztalos BF, Welty FK, et al. Effects of cholesteryl ester transfer protein inhibition on high-density lipoprotein subspecies, apolipoprotein A-I metabolism, and fecal sterol excretion. Arterioscler Thromb Vasc Biol 2005;25:1057–64.

[42] Lamon-Fava S, Diffenderfer MR, Barrett PHR, Buchsbaum A, Nyaku M, Horvath K, et al. Extended-release niacin alters the metabolism of plasma apolipoprotein (apo) A-I and apoB-containing lipoproteins. Arterioscler Thromb Vasc Biol 2008;28:1672–8.

[43] Asztalos BF, Swarbrick MM, Schaefer EJ, Dallal GE, Horvath KV, Ai M, et al. Effects of weight loss, induced by gastric bypass surgery, on HDL remodeling in obese women. J Lipid Res 2010;51:2405–12.

[44] Asztalos BF, LeMaulf F, Dallal GE, Stein E, Jones PH, Horvath KV, et al. Comparison of the effects of high doses of rosuvastatin versus atorvastatin on the subpopulations of high density lipoproteins. Am J Cardiol 2007;99:681–5.

[45] Lamon-Fava S, Diffenderfer MR, Barrett PH, Buchsbaum A, Matthan NR, Lichtenstein AH, Dolnikowski GG, et al. Effects of different doses of atorvastatin on human apolipoprotein B-100, B-48, and A-I metabolism. J Lipid Res 2007;48:1746–53.

[46] Asztalos BF, Horvath KV, McNamara JR, Roheim PS, Rubenstein JJ, Schaefer EJ. Comparing the effects of five different statins on the HDL subpopulation profiles of coronary heart disease patients. Atherosclerosis 2002;164:361–9.

[47] Asztalos BF, Collins D, Cupples LA, Demissie S, Horvath KV, Bloomfield HE, et al. Value of high density lipoprotein (HDL) subpopulations in predicting recurrent cardiovascular events in the Veterans Affairs HDL Intervention Trial. Arterioscler Thromb Vasc Biol 2005;25:2185–91.

[48] Asztalos BF, Batista M, Horvath KV, Cox CE, Dallal GE, Morse JS, et al. Change in alpha 1 HDL concentration predicts progression in coronary artery stenosis. Arterioscler Thromb Vasc Biol 2003;23:847–52.

[49] Kontush A, Chapman MJ. Functionally defective high-density lipoprotein: a new therapeutic target at the crossroads of dyslipidemia, inflammation, and atherosclerosis. Pharmacol Rev 2006;58:342–74.

[50] Asztalos BF, Sloop CH, Wong L, Roheim PS. Two-dimensional electrophoresis of plasma lipoproteins: recognition of new apoA-I containing subpopulations. Biochim Biophys Acta 1993;1169:291–300.

[51] Superko HR, Pendyala L, Williams PT, Momary KM, King SB, Garrett BC. High-density lipoprotein subclasses and their relationship to cardiovascular disease. J Clin Lipidol 2012;6(6):496–523.

[52] Okazaki M, Usui S, Ishigami M, Sakai N, Nakamura T, Matsuzawa Y, et al. Identification of unique lipoprotein subclasses for visceral obesity by component analysis of cholesterol profile in HPLC. Arterioscler Thromb Vasc Biol 2005;25:578–84.

[53] Tabuchi M, Seo M, Inoue T, Ikeda T, Kogure A, Inoue I, et al. Geometrical separation method for lipoproteins using bioformulated-fiber matrix electrophoresis: size of high-density lipoprotein does not reflect its density. Anal Chem 2011;83(3):1131–6.

CHAPTER 2

Apolipoprotein A-I Mutations and Clinical Evaluation

Akira Matsunaga

Department of Laboratory Medicine, Faculty of Medicine, Fukuoka University, Fukuoka, Japan

Contents

Abstract

Apolipoprotein (apo) A-I accounts for 70% of the total protein in high-density lipoprotein (HDL) and plays a key role in HDL biogenesis and function. Analyses of the apoA-I amino acid sequence have revealed that most of its 243 amino acid residues are grouped into amphipathic α-helices of 11 or 22 amino acids in length. Since the hydrophobic C-terminal domain (residues 190–243) of the human apoA-I molecule is critical for lipid binding, deletion of this segment reduces the ability of the protein to solubilize lipid. Sixteen different types of homozygous, compound heterozygous, and heterozygous apoA-I deficiencies, including large deficiency, inversion, frameshift, and nonsense mutations, have been reported. Most of the missense mutations causing low HDL cholesterolemia are associated with alterations to amino acids 143–187, forming α-helices 6–7 of apoA-I. Mutations in this region are accompanied by activation failure of lecithin:cholesterol acyltransferase (LCAT). Other point mutations leading to low HDL cholesterolemia include ApoA-I(S36A), (K107del), (R173C)$_{Milano}$, (L178P), and (E235del)$_{Nichinan}$. Twenty-one mutations that cause amyloidosis have been reported. The hereditary amyloidogenic mutations are clustered within amino acids 26–107 and 154–178 of apoA-I. ApoA-I (R151C)$_{Paris}$ and ApoA-I (R173C)$_{Milano}$ are rare cysteine variants that can form dimers. ApoA-I$_{Milano}$, in particular, has been proven to exert anti-atherogenic effects in animal studies and small clinical trials. Although there are difficulties associated with its formulation, clinical applications are expected.

1. INTRODUCTION

Plasma HDL (hydrated density, 1.063–1.21 g/mL) comprises a heterogeneous group of small discoid and spherical particles (7–12 nm in diameter) that differ in density, size, and electrophoretic mobility. The HDL apolipoproteins include apoA-I and apoA-II, as well as apoA-IV, apoA-V, apoC-I, apoC-II, apoC-III, apoD, apoE, apoJ, apoL, and apoM. The main protein component of HDL is apoA-I, a 28-kDa protein that contains 10 amphipathic α-helical domains of 11 or 22 amino acids each. Approximately 70% of total plasma HDL protein is apoA-I (present in normolipidemic human plasma at ~130 mg/dL), and it is located in essentially all HDL particles. The second most abundant protein is apoA-II (two 77-amino acid molecules forming a homodimer in humans with a disulfide-link at Cys6), which comprises 15–20% of total plasma HDL protein. However, this component is not present in all HDL particles. In human plasma, about 25% of apoA-I is present in HDL particles containing only apoA-I (LpA-I), and the remaining HDL particles contain both apoA-I and apoA-II (LpA-I+A-II), typically in a molar ratio of 1–2:1 [1]. The heterogeneity in HDL size and structure is intimately related to the amphipathic highly dynamic helical structure of apoA-I. HDL particles are constantly being remodeled as they transport cholesterol between cells and other lipoproteins. HDL particles are produced as small (< 8 nm in diameter), lipid-poor (lipid content: < 30%), discoid particles made up of apolipoproteins (mainly apoA-I), which carry small amounts of lipid, mainly phospholipids, and free cholesterol (see Figure 2.1). These particles represent an excellent substrate for lecithin:cholesterol acyltransferase (LCAT), the enzyme that generates most of the cholesteryl esters present in plasma. ApoA-I mutations can cause plasma HDL deficiency, as well as LCAT deficiency, and Tangier disease is caused by mutations of the gene encoding ATP-binding cassette transporter A1 (ABCA1).

2. PHYSIOLOGICAL FUNCTIONS OF apoA-I

As the quantitatively predominant component of HDL, apoA-I is crucial for HDL formation. Reverse cholesterol transport (RCT) is a multistep process resulting in the net movement of cholesterol from peripheral tissues back to the liver via lipoproteins. At the first rate-limiting step of RCT, apoA-I interacts with ABCA1 on the cell surface to form discoid nascent HDL (Figure 2.1). The primary substrate for ABCA1 is lipid-free monomolecular

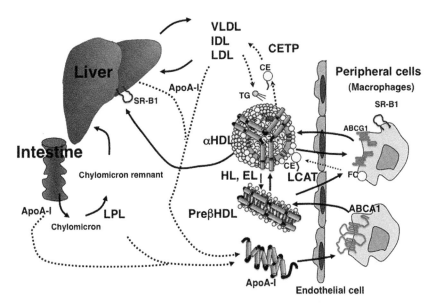

Figure 2.1 *Reverse Cholesterol Transport (RCT) and the Role of High-density Lipoprotein (HDL) and ApoA-I.* RCT is initiated from arterial macrophages by the interaction of lipid-free apoA-I with cellular ABCA1 to generate nascent HDL particles. ABCG1 adds additional cholesterol to HDL, and lecithin:cholesterol acyltransferase (LCAT) esterifies HDL-cholesterol to generate mature spherical HDL particles.

apoA-I. The C-terminal domains of apoA-I, which are characterized by the greatest affinity for lipids, seem to play a crucial role in the interaction between apoA-I and ABCA1 [2–4]. Studies on synthetic peptides that mimic apolipoproteins have indicated that the interaction between apoA-I and ABCA1 is not sequence-specific, and instead the amphipathic helices of apoA-I were identified as the key structural motifs [5,6]. ApoA-I directly interacts with residues located in two extracellular loops of ABCA1 [7]. A three-step model of the interaction between ABCA1 and apoA-I proposed by Vedhachalam and colleagues [8] explains the contribution of the interaction of apoA-I with ABCA1 and membrane lipids, as well as the heterogeneous nature of the nascent HDL particles. The binding of apoA-I by ABCA1 would induce signaling responses that would stabilize ABCA1 and enhance phospholipid translocase activity. This would lead to compression of the phospholipid molecules in the exofacial leaflet of the membrane, which would bulge in the direction of the extracellular space. Consequently, more apoA-I would bind to the membrane protrusions. Step 3 would involve membrane microsolubilization and the creation of discoid nascent

HDL particles. This stage would limit the rate of the entire ABCA1/apoA-I interaction. Variations in the lipid composition of the membrane curvatures created in different membrane environments would cause heterogeneity of the nascent HDLs, particularly in terms of the free cholesterol content. The discoid nascent HDL particles are thought to comprise a cholesterol-containing phospholipid bilayer, with two copies of apoA-I wrapped around the perimeter in a largely α-helical antiparallel "double-belt" conformation (Figure 2.1). ApoA-I on a nascent HDL particle activates LCAT whose apolar products, cholesterol esters, move from the surface to the core of the particle, converting it to a mature core containing spherical HDL (Figure 2.1). ApoA-I is also required to activate LCAT and mediate the interactions of HDL with cell surface receptors, such as scavenger receptor B1 or plasma membrane transporters, including ABCG1 (Figure 2.1) [9,10]. In mice and rabbits, knockout of the *ApoA1* gene causes HDL deficiency, and conversely, transgenic overexpression of *ApoA1* increases HDL cholesterol (HDL-C) in a gene dose-dependent manner. In susceptible animals with an atherogenic lipoprotein profile, atherosclerosis is enhanced by apoA-I deficiency and decreased by transgenic overexpression of apoA-I. Human subjects with apoA-I deficiency and apoA-I-deficient mice fail to form normal HDL particles [11].

3. GENOMIC AND PROTEIN STRUCTURE OF apoA-I

The genes coding *APOA1*, *APOC3*, *APOA4*, and *APOA5* are clustered within a DNA segment on human chromosome 11q23-q24 (Figure 2.2) [12,13]. The *APOA1* gene contains four exons and encodes a 267-amino acid peptide chain including an 18-amino acid prepeptide and a 6-amino acid propeptide (Figure 2.2) [14]. Mature apoA-I in plasma is a 243-amino acid peptide primarily synthesized in the liver and intestine. ApoA-I is a 28-kDa single polypeptide that lacks glycosylation or disulfide linkages. Exon 3 of the *APOA1* gene encodes the N-terminal region of mature apoA-I (residues 1–43), and the remaining regions are encoded by exon 4. The common lipid-associating motif in apoA-I is the amphipathic α-helix. Based on X-ray crystallography and computer modeling, most of the 243 amino acid residues of apoA-I are grouped into amphipathic α-helices of 11 or 22 amino acids in length [15,16] that embrace the carbon chains of several phospholipid molecules like a belt. Analyses of the amino acid sequence have revealed that residues 1–43 form a class G* amphipathic α-helix, while residues 44–243 contain a series of 22-mer

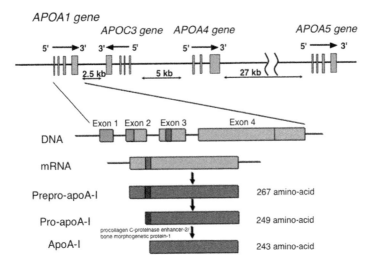

Figure 2.2 *Scheme of the* **APOA1** *Gene and the* **APOA1-APOC3-APOA4-APOA5** *Gene Cluster Localized on Human Chromosome 11.* ApoA-I is synthesized in hepatocytes and enterocytes as a preprotein. The 18-amino acid prepeptide is cleaved prior to secretion, and pro-apoA-I is secreted and cleaved in blood by procollagen C-proteinase enhancer-2 (PCPE2) and bone morphogenetic protein-1.

and 11-mer class A and class Y amphipathic α-helices [17]. The class A helix is a major lipid-binding motif of apolipoproteins, and is characterized by the location of positively charged residues near the hydrophilic/hydrophobic interface and negatively charged residues clustered at the center of the polar face. Class G★ and class Y helices are proposed to have reduced lipid affinity. The class G★ helix is distinguished by a random radial arrangement of positively charged and negatively charged amino acid residues in the polar face, while the class Y helix is characterized by the presence of three clusters of positively charged amino acids in the polar face forming a Y pattern. The repeating amphipathic α-helical segments are critical for the ability of exchangeable apolipoproteins to interact with phospholipids and stabilize HDL particles. Lipid-free or lipid-poor apoA-I molecules remove cholesterol from peripheral cells and transport it in the form of HDL back to the liver for excretion. Lipid-free apoA-I is folded into two structural domains, comprising an N-terminal part (residues 1–189) forming a four-helix bundle and a discrete C-terminal part (residues 190–243), which is predominantly involved in lipid–protein interactions [18]. Mutations between residues 1 and 100 are associated with amyloid formation, whereas mutations within the central region of residues 140–170 are mostly associated with defective activation of LCAT. Few natural mutations have been found within the disordered C-terminal domain (residues 180–243).

4. ApoA-I DEFICIENCIES WITH LARGE DEFICIENCY, INVERSION, NONSENSE, OR FRAMESHIFT MUTATIONS

In total, 16 different types of homozygous, compound heterozygous, and heterozygous ApoA-I deficiencies, including large deficiency, inversion, frameshift, or nonsense mutations, have been reported (Table 2.1) [19–36].

The probands of *APOA1/APOC3* deficiency were two sisters aged 29 and 31 years, who had skin and tendon xanthomas, corneal opacities, and severe coronary artery disease (CAD) [19]. The rearrangement in these probands consisted of a DNA inversion involving portions of the 3′ ends of the *APOA1* and *APOC3* genes, including the DNA region between these genes [37]. Their plasma levels of HDL-C were 4 and 7 mg/dL (0.10 and 0.18 mmol/L), respectively, and only traces of apoA-I were detected in whole plasma. The proband of *APOA1/APOC3/APOA4* deficiency was a 45-year-old white female (Table 2.1) [20]. The defect in this proband was complete deletion of the *APOA1, APOC3*, and *APOA4* gene complex on chromosome 11 [38]. She had severe CAD and corneal opacities, but no hepatosplenomegaly or xanthomas were detected. Her plasma level of HDL-C was 1 mg/dL. She died after coronary bypass surgery.

The proband of Q-2X was a 34-year-old Caucasian female who had xanthelasma, Achilles tendon xanthomas, bilateral cataracts, retinal detachments, mild midline cerebellar ataxia, and asymmetric bilateral neurosensory hearing loss (Table 2.1) [22]. Her plasma HDL-C was 2.3 mg/dL. This proband and four of her 11 siblings had a homozygous mutation of Q-2X, severe deficiency of HDL (< 3.9 mg/dL), and undetectable plasma apoA-I. At 38 years of age, a homozygous sister had premature CAD. An 11-year-old girl of Turkish descent was the proband of Q5fsX30 [23]. She was presented to her pediatrician for unusual yellow-orange xanthomas of several joints at 7 years of age. She had no corneal opacities, no enlarged tonsils, and no splenomegaly. Her plasma HDL-C was 2 mg/dL. Another proband of heterozygous Q5fsX30 was a 10-year-old Japanese girl [24]. Her plasma HDL-C and apoA-I concentrations were 27 and 76 mg/dL, respectively. The proband of a compound heterozygote of Q5fsX6 and L141R$_{Pisa}$ was a 47-year-old Italian man [25]. He was referred to his physician for corneal opacities and complete HDL deficiency (0 mg/dL). He underwent

Table 2.1 Apolipoprotein A-I Deficiencies with Large Deficiency, Inversion, Nonsense, or Frameshift Mutations

	Mutation	Protein Sequence Change*	Genomic Change**	Allele	Age	Sex	Ethnicity	CAD	Characteristics	Reference
1	APOA1/APOC3 gene inversion			Homozygote	29	F	American	+	Two sisters had skin and tendon xanthomas, and corneal opacities	(19)
2	APOA1/ APOC3/APOA4 gene deletion			Homozygote	31	F	American	+		
				Homozygote	45	F	American	+	She died of CAD at 45 years of age	(20)
3	Q-8fsX10	p.Q17fsX10	c.49–50insC	Heterozygote	39	M	Italian	−	Q-8fsX10 mutations were found in five carriers	(22)
4	Q-2X	p.Q22X	c.67C>T	Homozygote	34	F	Canadian	+	She had retinopathy, cataracts, spinocerebellar ataxia, and tendon xanthomas	(22)
5	Q5fsX30	p.Q29fsX30	c.79–86insC	Homozygote	11	F	Turkish	−	She had planar xanthomas	(23)
6	Q5fsX6	p.Q5fsX6	c.85delC	Heterozygote	10	F	Japanese	−		(24)
				Compound***	47	M	Italian	+	Another was apoA-I L141R$_{Psa}$. He had corneal opacities	(25)
7	W8X	p.W32X	c.96G>A	Homozygote	39	M	Japanese	−	He had corneal opacities	(26)
8	R10X	p.R34X	c.100C>T	Homozygote	35	F	Iraqi	−	She had corneal opacities and periorbital xanthelasmas	(27)

Continued

Table 2.1 Apolipoprotein A-I Deficiencies with Large Deficiency, Inversion, Nonsense, or Frameshift Mutations—cont'd

	Mutation	Protein Sequence Change*	Genomic Change**	Allele	Age	Sex	Ethnicity	CAD	Characteristics	Reference
9	Q32X	p.Q56X	c.166C>T	Homozygote	31	F	Italian	–	She had periorbital xanthelasmas	(28)
10	Q84X	p.Q108X	c.322C>T	Homozygote	60	F	Japanese	+	She had planar xanthomas	(29)
				Compound***	13	M	Japanese	–	Another was a missense mutation in the TATA box	(30)
11	E136X	p.E160X	c.478G>T	Heterozygote			French Canadian	+	E136X mutations were found in 17 carriers	(31)
12	Q138fsX40Tomioka	p.Q162fsX40	c.485delAA	Homozygote	64	M	Japanese	+	Two brother had corneal opacities	(32)
13	A154fsX24Shinbashi	p.A178fsX24	c.532delGC	Homozygote	51	F	Japanese	+	She had corneal opacities and planar xanthomas	(33)
14	H162fsX46Sasebo	p.H186fsX46	c.556ins23nt	Homozygote	50	F	Japanese	–	She had corneal opacities, and skin and tendon xanthomas	(34)
15	N184fsX16	p.N208fsX16	c.624delC	Homozygote	69	F	Japanese	–	She had corneal opacities and corneal rings	(35)
16	T202fsX27	p.T226fsX27	c.678delG	Homozygote	42	M	German	–	He had corneal opacities	(36)

*"p." for a protein sequence, and the translation initiator methionine is numbered as +1.
**"c." for a coding DNA sequence, and nucleotide 1 is the A of the ATG–translation initiation codon.
***Compound, compound heterozygote; CAD, coronary artery disease.

three-vessel coronary bypass surgery. His plasma apoA-I level was 3 mg/dL. The patient with W8X was a 39-year-old Japanese male (Table 2.1) [26]. He had corneal opacities, but no CAD or xanthomas. His plasma HDL-C and apoA-I levels were 6 and < 3 mg/dL, respectively. The proband of R10X was a 35-year-old Iraqi woman who presented with xanthelasmas and xanthomas, but showed only minimal changes on cardiovascular examinations and no clinical symptoms. However, her 37-year-old brother with R10X was diagnosed with myocardial infarction at 35 years of age. Their plasma HDL-C and apoA-I levels were 0.4–2.7 mg/dL (0.01–0.07 mmol/L) and 0–2 mg/dL, respectively [27]. The proband of Q32X was a 31-year-old female of Italian origin [28]. She had bilateral periorbital xanthelasmas, but did not suffer from clinical signs of CAD. The proband of Q84X was a 60-year-old Japanese woman (Table 2.1) [29]. She had planar xanthomas at 18 years of age and CAD symptoms started at 52 years of age. She had no measurable plasma apoA-I. Another proband of Q84X was a 12-year-old Japanese boy, who was a compound heterozygote with a point mutation at nucleotide position −27 that changed ATAAATA of the putative *APOA1* gene TATA box signal sequence to ATACATA (Table 2.1) [30]. His plasma HDL-C and apoA-I concentrations were 5.4 and 13.3 mg/dL, respectively. Heterozygous E136X carriers ($n = 17$) had markedly low plasma HDL-C levels (males: 23.2 ± 7.7 mg/dL; females: 23.2 ± 3.7 mg/dL), and among nine carriers aged 35 years or older, five men developed premature CAD [31]. The proband of Q138fsX40$_{Tomioka}$ was a 64-year-old Japanese with marked plasma HDL-C (4 mg/dL) and apoA-I (5 mg/dL) deficiency, prior myocardial infarction, and moderate corneal opacities [32]. Coronary angiography revealed extensive atherosclerosis in all three major vessels. The proband of A154fsX24$_{Shinbashi}$ was a 51-year-old woman who was hospitalized for severe heart failure with CAD (Table 2.1) [33]. She exhibited corneal opacities and planar xanthomas on the eyelids and elbows. Her plasma HDL-C and apoA-I concentrations were 3.1 mg/dL (0.08 mmol/L) and 1 mg/dL, respectively. A 50-year-old Japanese woman was the proband of H162fsX46$_{Sasebo}$ (Table 2.1) [34]. She had corneal opacities, and bilateral upper palpebral, elbow, neck, knee, and Achilles tendon xanthomas. Her plasma HDL-C and apoA-I concentrations were 5.4 mg/dL (0.14 mmol/L) and 0.8 mg/dL, respectively. The patient with K184fsX16 was a 69-year-old Japanese woman (Table 2.1) [35]. Her HDL-C level was 5.0 mg/dL (0.13 mmol/L). She exhibited corneal opacities and corneal rings, but had no xanthomas, thickening of Achilles tendons, or CAD. The proband of T202fsX27 was a 42-year-old man [36].

He had massive corneal opacities, complete absence of HDL-C (0 mg/dL), and half-normal LCAT activity, but no CAD. His plasma apoA-I concentration was 3 mg/dL.

The ratio of plasma cholesterol esters to total cholesterol is used to measure the capacity of LCAT to generate cholesterol esters in plasma. Studies on plasma apoA-I deficiency have shown that the ratio of plasma cholesterol esters to total cholesterol decreases by ~30% compared with the ratios in controls [22,28,39]. It has been argued that nonsynthetic forms of apoA-I may be atherogenic. APOA1/APOC3 gene inversion and gene deletion of the entire APOA1/APOC3/APOA4 locus prevented the synthesis of apoA-I, and premature CAD was observed. CAD was not always present in other patients with single gene defects of APOA1 (Table 2.1). Almost all cases with homozygous defects of APOA1 were characterized by the absence of plasma HDL-C and apoA-I, but premature CAD was only present in six probands (Q-2X, Q5fsX6, R10X, Q84X, Q138fsX40$_{\text{Tomioka}}$, and A154fsX24$_{\text{Shinbashi}}$). However, four probands (Q5fsX30, W8X, Q32X, and T202fsX27) were younger than 50 years of age, and these patients may have been too young for clinical manifestations of atherosclerosis to occur.

5. ApoA-I MISSENSE MUTATIONS WITH LOW HDL-C

The naturally occurring APOA1 variants that produce pathological phenotypes are shown in Figure 2.3. It has been estimated that structural mutations in APOA1 occur in 0.3% of the Japanese population and may affect the plasma HDL levels [40]. Among a total of 50 reported natural missense or in-frame deletion mutations of APOA1, 18 variants are associated with low HDL levels (Figure 2.3) [41–59]. Moreover, 12 mutations between residues 26 and 107 and eight mutations between residues 130 and 178 are associated with low HDL levels and amyloidosis [60–75].

The apoA-I V156E$_{\text{Oita}}$ variant was associated with apoA-I deficiency. Missense or in-frame deletion mutations were found as heterozygotes excluding V156E$_{\text{Oita}}$. The carriers of the homozygous V156E$_{\text{Oita}}$ variant were Japanese brothers aged 67 and 71 years [49]. They had corneal opacities, but no xanthomas. The younger brother underwent aortocoronary bypass surgery, but the elder brother had no signs of CAD. Their plasma HDL-C and apoA-I concentrations were 7.3–9.3 mg/dL (0.19–0.24 mmol/L) and 11 mg/dL, respectively. The LCAT activity and cholesterol esterification rate were about 40% of the corresponding normal control values.

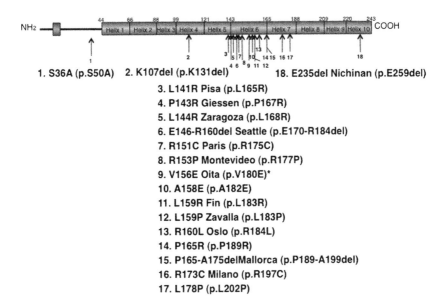

1. S36A (p.S50A)   2. K107del (p.K131del)   18. E235del Nichinan (p.E259del)
3. L141R Pisa (p.L165R)
4. P143R Giessen (p.P167R)
5. L144R Zaragoza (p.L168R)
6. E146-R160del Seattle (p.E170-R184del)
7. R151C Paris (p.R175C)
8. R153P Montevideo (p.R177P)
9. V156E Oita (p.V180E)*
10. A158E (p.A182E)
11. L159R Fin (p.L183R)
12. L159P Zavalla (p.L183P)
13. R160L Oslo (p.R184L)
14. P165R (p.P189R)
15. P165-A175delMallorca (p.P189-A199del)
16. R173C Milano (p.R197C)
17. L178P (p.L202P)

Figure 2.3 *Naturally Occurring ApoA-I Missense Mutations or In-frame Deletions that Produce Low Plasma HDL Cholesterol.* The C-terminal residues 44–243 of apoA-I contain 22- and 11-amino acid repeats, and are organized into 10 amphipathic α-helices (17). "p." indicates a protein sequence, and the translation initiator methionine is numbered as +1. *ApoA-I V156E$_{Oita}$ was found as a homozygous mutation and is associated with apoA-I deficiency.

The mean plasma HDL-C concentration (26.7 mg/dL) of 10 heterozygous carriers of apoA-I K107del was 36% lower than that of controls [42]. After screening for this apoA-I mutation in autopsy cases with aortic amyloidosis, one case with extensive aortic intimal amyloid deposits showed apoA-I K107del [69]. The proband and carriers of L159R$_{Fin}$ had diminished concentrations of plasma HDL-C (7.3–15.5 mg/dL) and apoA-I (21.9–47.8 mg/dL) [76]. There was no clinical evidence of CAD. Heterozygous carriers (n = 19) of apoA-I L159P$_{Zavalla}$ had reduced mean HDL-C (10.4 mg/dL) and apoA-I (38.7 mg/dL) levels [52]. LCAT activity was slightly lower in some carriers, but the cholesterol esterification rate did not differ from that of controls. Two sisters aged 68 and 72 years with apoA-I$_{Zavalla}$ had clinically manifested CAD. Four subjects with apoA-I E235del$_{Nichinan}$ were heterozygous carriers, and their mean plasma concentrations of apoA-I and HDL-C were 30% and 32% lower than those in control subjects, respectively [59]. The apoA-I E235del$_{Nichinan}$ variant was associated with reduced lipid-binding properties [77]. Some apoA-I mutations supposedly underlie CAD (e.g., L141R$_{Pisa}$ and L159P$_{Zavalla}$) [25,52]. whereas others show no effect (e.g., L144R$_{Zaragoza}$, P165-A175del$_{Mallorca}$, and E235del$_{Nichinan}$) [45,56,59] or even protection against CAD (e.g., R173C$_{Milano}$).

A catalogue of the naturally occurring human apoA-I mutations associated with low HDL concentrations suggests that disruption of the native structure of apoA-I dramatically alters the intravascular metabolism of HDL [78]. While most of the apoA-I helices participate in lipid binding, various helical regions also have other functions. For example, the helices spanning residues 143–187 are critical for LCAT activation [79–81], while the helices from residue 190 to the C-terminus are involved in interactions with cells and phospholipid bilayers [82,83]. The helical wheel diagram in Figure 2.4 shows that many apoA-I mutations are located on helix 6. Among the documented apoA-I mutants associated with low HDL-C levels that span residues 107–235 of apoA-I but are predominantly clustered on or in the vicinity of helix 6 (Figure 2.4), 81% of the mutations are amino acid substitutions/deletions within helices 6 and 7 [78].

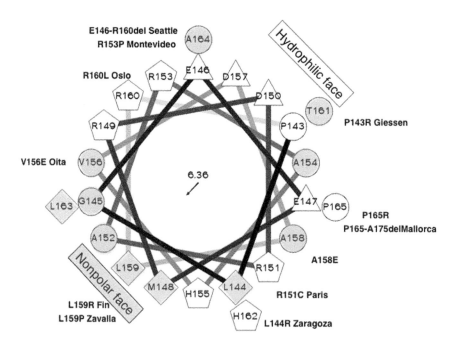

Figure 2.4 *Helical Wheel Representation of Helix 6 (Residues 143–165) in ApoA-I. Val156 and Ala158 are Located at Nearly Opposite Edges of the Amphiphatic Face.* Note that the substitution of a charged amino acid (V156K or A158E) decreases the hydrophobicity in the amphiphatic boundary, as indicated by the arrow.

The associations between mutant forms of helices 6 and 7 and aberrant HDL metabolism have been studied in mice using several different approaches. For example, transgenic mice expressing a mutant form of apoA-I that lacked the entire proline-punctuated 22-amino acid helix 6 (residues 143–165) were created [84,85]. In other studies, an adenoviral construct expressing a dominant-negative helix 6 point mutation, $L159R_{Fin}$ apoA-I, was expressed in apoA-I-knockout mice, and a targeted replacement for the dominant-negative mutant apoA-I $R173C_{Milano}$, a helix 7 point mutation, was examined [86,87]. Both of these mutants decreased the plasma HDL-C and apoA-I levels. Koukos and colleagues [88] performed adenovirus-mediated gene transfer of mutants into apoA-I-knockout mice, and revealed that apoA-I $L141R_{Pisa}$ and $L159R_{Fin}$ inhibit an early step in the biogenesis of HDL through inefficient esterification of the cholesterol in pre-β1-HDL particles by endogenous LCAT, and that both defects can be corrected by treatment with LCAT [88]. They also showed that the apoA-I levels in apoA-I -knockout mice expressing the apoA-I $R160L_{Oslo}$ and $R151C_{Paris}$ mutants were reduced by 68% and 55%, respectively, and that apoA-I $R151C_{Paris}$ generated subpopulations of different sizes that migrated between pre-β- and α-HDL particles and formed mostly spherical and a few discoidal particles [87]. An artificial point mutation (pR160V/H162A) and deletion (L144-P165del) in helix 6 were also associated with abolished LCAT activity and reduced plasma spherical HDL [89]. In each of these studies, the data suggested that mutations within apoA-I helices 6 and/or 7 cause decreased production of HDL and contribute to the dominant-negative phenotype observed in individuals heterozygous for these mutations. Recently, apoA-I S36A was identified in a patient with severe hypoalphalipoproteinemia. In addition, it was reported that the region around S36 in the N-terminus of human apoA-I is necessary for LCAT activation [75]. It has been shown that N-terminal deletion destabilizes apoA-I and leads to unfolding of the α-helices in the C-terminal domain, which is responsible for specific interactions with ABCA1 [90]. For example, the C-terminal domain of apoA-I is critical for effective lipid binding, and its deletion greatly inhibits cellular phospholipid/free cholesterol efflux and nascent HDL production [91].

6. ApoA-I MUTATIONS RELATED TO AMYLOIDOSIS

ApoA-I amyloidosis is an autosomal dominant amyloidosis caused by mutations in the *APOA1* gene. The first apoA-I variant associated with amyloidosis was apoA-I $G26R_{Iowa}$ found in an Iowa kindred with British ancestry

[92]. Since the identification of apoA-I G26R$_{Iowa}$, 21 apoA-I variants have been found to be associated with amyloidosis, of which 15 are single amino acid substitutions, two are deletions, one is a deletion/insertion, and three are frameshift mutations (Figure 2.5). Currently, hereditary apoA-I amyloidosis mutations in apoA-I that promote apoA-I processing by metalloproteinases and release of N-terminal 9–11-kDa fragments that become deposited as amyloid have been identified [93]. Normal apoA-I is itself weakly amyloidogenic, and acts as a precursor of the small amyloid deposits that occur quite frequently in aortic atherosclerotic plaques [69]. In hereditary apoA-I amyloidosis, which is an autosomal dominant disorder, the N-terminal fragments of variant apoA-I are deposited as extracellular fibrils in organs such as the kidney, liver, heart, spleen, skin, larynx, nerves, gastrointestinal tract, ovaries, and testes, leading to organ damage. The fibrillar deposits in the heart contain mainly the amino acid fragment 1–93, while those in the other organs contain slightly shorter or longer fragments ranging from approximately amino acids 1–80 to 1–100 [94]. Many amyloidogenic mutations in apoA-I are clustered in two hotspots encompassing residues 26–107 and 170–178. Analyses of individual mutation sites have suggested a molecular mechanism by which each amyloidogenic mutation

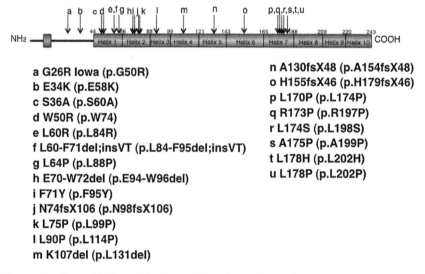

a G26R Iowa (p.G50R)
b E34K (p.E58K)
c S36A (p.S60A)
d W50R (p.W74)
e L60R (p.L84R)
f L60-F71del;insVT (p.L84-F95del;insVT)
g L64P (p.L88P)
h E70-W72del (p.E94-W96del)
i F71Y (p.F95Y)
j N74fsX106 (p.N98fsX106)
k L75P (p.L99P)
l L90P (p.L114P)
m K107del (p.L131del)

n A130fsX48 (p.A154fsX48)
o H155fsX46 (p.H179fsX46)
p L170P (p.L174P)
q R173P (p.R197P)
r L174S (p.L198S)
s A175P (p.A199P)
t L178H (p.L202H)
u L178P (p.L202P)

Figure 2.5 *Naturally Occurring ApoA-I Mutations that Produce Hereditary Amyloidosis.* N-terminal peptide fragments comprising residues 1 to 93–100 of apoA-I have been found in the amyloid deposits of patients, illustrating the propensity of this region for fibrillogenesis. "p." indicates a protein sequence, and the translation initiator methionine is numbered as +1.

destabilizes the molecular packing in lipid-free and HDL-bound apoA-I, thereby promoting dissociation of apoA-I from HDL followed by proteolysis and misfolding of the N-terminal fragment. Although an extensive visceral amyloid load is frequently observed as a postmortem finding or by means of serum amyloid P component scintigraphy in patients with progressive kidney or heart disease [62,65], severe deterioration of liver function is not frequently observed. The penetrance of hereditary apoA-I amyloidosis is high, although very substantial visceral amyloid deposits may be present for decades before any symptoms develop. Depending on the mutation, patients can present with massive abdominal visceral amyloid involvement, predominant cardiomyopathy, or a familial amyloidotic polyneuropathy-like syndrome.

The Iowa-type hereditary systemic amyloidosis was first described in 1969 [92]. In this large kindred, the onset of symptoms occurs in the third or fourth decade and the disease is characterized by lower extremity peripheral neuropathy, nephropathy, and peptic ulcer disease [95]. A 29-year-old Polish woman was the proband of E34K [93]. She presented with edema and was discovered to have subnephrotic-range proteinuria (1.68 g/24 h) and stage 3 chronic kidney disease (CKD). A kidney biopsy was performed in 2008 and revealed amyloidosis, which was thought on initial immunohistochemistry at her local hospital to be AA type. A 77-year-old English woman was the proband of S36A [93]. She presented with hypertension and was discovered to be nephrotic (7.9 g/24 h; serum albumin: 24 g/L) with CKD stage 3 on a background of Type 2 diabetes mellitus and ischemic heart disease. A renal biopsy specimen revealed amyloid deposits in the glomeruli and interstitium. A man with hereditary non-neuropathic systemic amyloidosis had amyloid fibril protein subunits consisting of the apoA-I (W50R) variant [61]. A mutation for L60R was identified in an English family with autosomal dominant non-neuropathic systemic amyloidosis. Major visceral amyloidosis has been confirmed in three generations [62]. The propositus was completely asymptomatic at 24 years of age, when his extensive splenic and hepatic amyloidosis was discovered by scintigraphy. He subsequently developed progressive hypertension, thrombocytopenia, and easy bruising. In view of his active lifestyle and the grave risk of splenic rupture, a splenectomy was undertaken. An amyloid-laden spleen (1.5 kg) was removed and the platelet count subsequently returned to the normal range. A Spanish family with autosomal dominant non-neuropathic hereditary amyloidosis with a unique hepatic presentation and death from liver failure, usually by the sixth decade, was reported [63]. The diagnosis of amyloidosis was first made in two asymptomatic female first cousins with abnormal liver function tests detected during routine screening at 60 and 55 years of age, and in whom liver biopsies were then undertaken. The male proband for L60-F71del;insVT, in whom chronic liver disease had been diagnosed at 40 years of age, presented with bleeding esophageal varices at 61 years of age and rapidly developed fatal hepatic encephalopathy. The proband of L64P was a 58-year-old man of Italian background [64]. He was found to have hypercholesterolemia in 1999, followed by marked protein-uria and edema 2 years later. A renal biopsy performed in January 2002 showed extensive deposition of amyloid within the glomeruli, interstitial tissue, and blood vessel walls. A family with autosomal dominant systemic amyloidosis involving E70-W72del in three generations, who presented with renal involvement, was reported [65]. Two members of the current generation received renal transplants for end-stage renal failure 16 and 18 years ago, and remain very well clinically, despite massive visceral amyloido-sis. Two other members of this generation, aged 32 and 47 years, have massive systemic amyloid deposition, but no clinical disability. Individuals known to be affected in previous generations died of renal failure in early adult life. The amyloid deposits in the proband, one of the transplanted individuals, were composed of apoA-I. A 51-year-old English woman was the proband of F71Y [93]. She presented with a 4-year history of a gradually enlarging mass on the palate within an area of scarring from a childhood injury, but no other features to suggest a peripheral neuropathy or carpal tunnel syndrome. Two of the patients carried an adenine insertion at nucle-otide position 294 (c.293-294insA), which also leads to a frameshift (Asn74fsX106) [96]. A proband of Asn74fsX106 was a 48-year-old woman, who underwent a hysteroadnexectomy because of a serous borderline tumor of the right ovary. Tissue samples were obtained from the hysteroadnexec-tomy specimens. She has been on hemodialysis twice a week since 2000. Another proband of Asn74fsX106 was a 67-year-old woman. A rectal biopsy from her was submitted for immunohistochemical classification of amyloid. The proband of L75P associated with amyloid deposits in the liver was found in an Italian population [66]. The propositus of L90P was a 54-year-old woman who presented in 1991 with recent onset of exertional dyspnea and cutaneous lesions for many years [68]. The skin lesions, which were yellow and maculopapular, first appeared on the forehead and extended rapidly to the face, neck, shoulders, and axillary and antecubital areas. The proband of A130fsX48 was a normotensive woman who was admitted with nephrotic syndrome at 58 years of age, and underwent a kidney biopsy [96].

A 56-year-old English man was the proband of H155MfsX46 [93]. He presented with edema and was discovered to have proteinuria, advanced CKD, and mildly obstructive derangement of liver function tests. A renal biopsy specimen revealed amyloid deposits. There was mild breathlessness on exertion, but no clinical evidence of peripheral or autonomic neuropathy. A 52-year-old man, the proband of L170P, had suffered from swallowing complaints for 6 years [96]. A biopsy from the vocal cord revealed amyloid deposits in the larynx. The patient showed no symptoms of systemic amyloidosis. All laboratory values were normal. Ultrasound of the abdomen as well as an electrocardiogram and echocardiography were also normal. There was no family history of amyloidosis. The proband of R173P was a 33-year-old Caucasian woman who was referred to a dermatologist for evaluation of a diffuse rash with the appearance of acanthosis nigricans in the axillae [71]. A skin biopsy stained with Congo red revealed amyloid deposits. The proband's father was diagnosed as having systemic amyloidosis with multi-organ involvement. He died at 63 years of age with cardiomyopathy, and liver and renal failure. The family, originating in central Italy, included two affected members, the proband of L174S and his maternal uncle, who died at 52 years of age [72]. Biopsies of the heart, skin, and gingiva were taken from the uncle and stained with Congo red. A 35-year-old English man was the proband of A175P [93]. He presented with a hoarse voice and no other significant history. He underwent a direct laryngoscopy and excision of multiple nodules on the vocal cords. The proband of L178H was found in a French kindred associated with cardiac and larynx amyloidosis and skin lesions with onset during the fourth decade [74]. The patient with L178P was a 67-year-old man who had been dysphonic for 10 months [97]. There was a clinical suggestion of autonomic and cardiac amyloid and histological corroboration of systemic amyloidosis in the abdominal fat.

The features of amyloidosis with apoA-I G26R$_{Iowa}$ included peripheral neuropathy, peptic ulcer disease, nephropathy, and death from renal failure [92]. The amyloid fibrils from an affected member of the original Iowa kindred showed a protein that was identical to the N-terminal portion of wild-type apoA-I, apart from a substitution at position 26 (G26R), with no normal apoA-I [60]. The three variants discovered in the N-terminal portion, apoA-I E34K, W50R, and L60R [61,62,93], had the following similarities with apoA-I G26R$_{Iowa}$: the affected subjects were heterozygotes; in each subject, arginine or lysine replaced a residue in the N-terminal, shifting the net charge by one or two positive units; the amyloid fibrils consisted of N-terminal fragments of the variant apoA-I. The apoA-I L60-F71del;INS,

V, T, and E70-W72del mutants also resulted in a net charge gain of +1. Although charge substitution does not preclude a conformational change, as has been predicted for many mutations of transthyretin as another amyloidogenic protein, it is obvious that a change in charge is not required for fibrillogenesis. Furthermore, the mutant residues are not structurally required for fibril formation, based on the absence of mutant fragments in the amyloid fibrils of skin deposits, but their presence in the cardiac amyloid deposits [68]. K107del has been associated with increased susceptibility to amyloid deposition in the aortic intima and ischemic heart disease [69]. Several variants, including proline substitutions at residues 90, 173, and 175 and histidine substitution at residue 178, are associated with hoarseness caused by laryngeal amyloid deposits, which may be misdiagnosed as localized primary amyloidosis (AL type) [68,71,73,74,97].

7. CYSTEINE MUTANTS OF apoA-I

ApoA-I$_{Milano}$ was the first described mutant of apoA-I. In the 1980s, a small Italian community was found to have a mutant version of apoA-I associated with a decreased risk of arteriosclerosis, heart attack, and stroke. Individuals with this mutation are characterized by very low HDL-C, but no increased risk of atherosclerosis [57]. ApoA-I is a lipid-binding protein that forms complexes with other proteins (like apoA-II) and lipids to form HDL particles. ApoA-I can self-associate, and normally forms modest amounts of dimers, trimers, and tetramers, as well as being in the monomeric form, to produce HDL particles in a range of sizes [98]. In apoA-I$_{Milano}$, the basic amino acid arginine at position 173 is mutated to the sulfur-containing amino acid cysteine (R173C). This results in a greater ability to form stable dimers compared with normal apoA-I, since the two cysteines can form a chemical bond together via their sulfur groups [98]. ApoA-I$_{Milano}$ can lead to the formation of homodimers and heterodimers with apoA-II. Since the affected individuals are heterozygous for this mutation, the full impact of apoA-I$_{Milano}$ on HDL-C metabolism is unknown. ApoA-I$_{Milano}$ shows faster catabolism than normal apoA-I, thereby explaining the accompanying low HDL-C [99]. Reduced plasma HDL-C and a decreased risk of vascular disease were observed in another cysteine variant, apoA-I$_{Paris}$, in which Arg151 is substituted with cysteine [48,100]. Both apoA-I R173C$_{Milano}$ and apoA-I R151C$_{Paris}$ are rare cysteine variants that produce low HDL-C in the absence of cardiovascular disease in humans. Although there are no differences between lipid-free apoA-I$_{Milano}$, apoA-I$_{Paris}$, and normal apoA-I in mediating the efflux of cholesterol from macrophages, apoA-I$_{Milano}$ is twice as effective as apoA-I$_{Paris}$ in preventing lipoxygenase-mediated oxidation of phospholipids [101].

8. ANTI-ATHEROGENIC apoA-I AND apoA-I$_{Milano}$

Liver-specific overexpression of a wild-type *ApoA-I* transgene was found to increase HDL-C and apoA-I levels, thereby reducing atherosclerosis in C57BL/6 mice [102] and hyperlipidemic mice [103,104]. A significant reduction in atherosclerosis was found after somatic gene transfer of human apoA-I into low-density lipoprotein (LDL) receptor-knockout mice fed a high-fat diet, and mice lacking the *ApoA-I* gene had significantly increased atherosclerosis. Over the last few decades, several sets of apoA-I variants with different functions and structural correlations to HDL have been generated and characterized, and their clinical potentials have been assessed in animal and human models. This essential field is generally referred to as HDL therapy [105,106]. Recombinant apoA-I$_{Milano}$ displays remarkable atheroprotective activities, and suggests the possibility of directly reducing the burden of atherosclerosis in experimental models. ApoA-I$_{Milano}$ has been studied extensively with regard to its effects on atherosclerosis, including infusion and genetic expression in animals [99,107]. It was previously shown that intravenous recombinant apoA-I$_{Milano}$ inhibits progression and induces rapid regression and remodeling of atherosclerosis in hypercholesterolemic rabbits and mice, while attenuating endothelial dysfunction [107–111]. Based on these promising preclinical results, a small phase II human trial of patients with acute coronary syndromes was conducted, in which once-weekly intravenous injections of recombinant apoA-I$_{Milano}$/phospholipid complexes administered to patients with CAD were shown to induce rapid coronary atheroma regression within 5 weeks [112]. Gene transfer with hepatic expression of wild-type apoA-I or apoA-I$_{Milano}$ was found to result in plasma apoA-I levels that were sufficient to significantly slow the development of atherosclerosis in mice, and both forms were similarly effective in their ability to retard atherosclerosis development in LDL receptor-knockout mice [113]. A study revealed a superior atheroprotective effect of apoA-I$_{Milano}$ compared with wild-type apoA-I using transduced macrophages after bone marrow transplantation into *ApoA-I/ApoE* double-knockout mice [113]. It has been suggested that apoA-I$_{Milano}$ may be a gain-of-function mutation and the prototype of apoA-I mimetics. However, apoA-I has 243 amino acid residues, not only making it difficult and expensive to synthesize, but also necessitating its intravenous administration. It is also likely that long periods of intravenous administration will be required, making this an unlikely therapy for the millions of patients with atherosclerosis.

Many of the apoA-I mimetic peptides have been shown to possess anti-inflammatory properties. Based on studies in a number of animal models and early human clinical trials, apoA-I mimetics appear to have promise as diagnostic and therapeutic agents. A potent apoA-I mimetic peptide, 4F, was found to exert significant activity in various inflammatory states in animals [114]. Another 24-amino acid apoA-I mimetic peptide is designated FAMP [68]. Ga-labeled FAMP was dramatically taken up by atherosclerotic tissues in the blood vessels and aorta of WHHL-MI rabbits, and may contribute to the development of a tool for the diagnosis of plaque with PET imaging [115]. ApoA-I mimetics have been constructed from a number of peptides and proteins with varying structures, all of which bind to lipids found in HDL.

9. CONCLUSION

ApoA-I is the principal protein component of all HDL subclasses. The lipid-free apoA-I molecule contains amphipathic α-helical segments. The lipid affinity of apoA-I confers detergent-like properties and it can solubilize vesicular phospholipids to create discoidal HDL particles. The flexible apoA-I molecule adapts to the surface of spherical HDL particles by bending and forming a stabilizing trefoil scaffold structure. The characteristics of apoA-I enable it to partner with ABCA1 in mediating the efflux of cellular phospholipids and cholesterol. The mutations that affect the complete translation of apoA-I mRNA (nonsense and frameshift mutations) are always associated with reduced levels of apoA-I and HDL-C. The missense mutations and some in-frame deletions are associated with amyloidosis or low levels of apoA-I and HDL-C. It has been reported that apoA-I_{Milano} and apoA-I mimetic peptides have anti-inflammatory or atheroprotective effects. Research toward practical use of apoA-I_{Milano} and apoA-I mimetic peptides is expected.

REFERENCES

[1] Gauthamadasa K, Rosales C, Pownall HJ, Macha S, Jerome WG, Huang R, et al. Speciated human high-density lipoprotein protein proximity profiles. Biochem 2010;49:10656–65.
[2] Favari E, Bernini F, Tarugi P, Franceschini G, Calabresi L. The C-terminal domain of apolipoprotein A-I is involved in ABCA1-driven phospholipid and cholesterol efflux. Biochem Biophys Res Commun 2002;299:801–5.

[3] Liu L, Bortnick AE, Nickel M, Dhanasekaran P, Subbaiah PV, Lund-Katz S, et al. Effects of apolipoprotein A-I on ATP-binding cassette transporter A1-mediated efflux of macrophage phospholipid and cholesterol: formation of nascent high density lipoprotein particles. J Biol Chem 2003;278:42976–84.

[4] Hassan HH, Denis M, Lee DY, Iatan I, Nyholt D, Ruel I, et al. Identification of an ABCA1-dependent phospholipid-rich plasma membrane apolipoprotein A-I binding site for nascent HDL formation: implications for current models of HDL biogenesis. J Lipid Res 2007;48:2428–42.

[5] Remaley AT, Thomas F, Stonik JA, Demosky SJ, Bark SE, Neufeld EB, et al. Synthetic amphipathic helical peptides promote lipid efflux from cells by an ABCA1-dependent and an ABCA1-independent pathway. J Lipid Res 2003;44:828–36.

[6] Gonzalez MC, Toledo JD, Tricerri MA, Garda HA. The central type Y amphipathic alpha-helices of apolipoprotein AI are involved in the mobilization of intracellular cholesterol depots. Arc Biochem Biophys 2008;473:34–41.

[7] Fitzgerald ML, Morris AL, Rhee JS, Andersson LP, Mendez AJ, Freeman MW. Naturally occurring mutations in the largest extracellular loops of ABCA1 can disrupt its direct interaction with apolipoprotein A-I. J Biol Chem 2002;277:33178–87.

[8] Vedhachalam C, Ghering AB, Davidson WS, Lund-Katz S, Rothblat GH, Phillips MC. ABCA1-induced cell surface binding sites for ApoA-I. Arterioscler Thromb Vasc Biol 2007;27:1603–9.

[9] Zannis VI, Chroni A, Krieger M. Role of apoA-I, ABCA1, LCAT, and SR-BI in the biogenesis of HDL. J Mol Med 2006;84:276–94.

[10] Gelissen IC, Harris M, Rye KA, Quinn C, Brown AJ, Kockx M, et al. ABCA1 and ABCG1 synergize to mediate cholesterol export to apoA-I. Arterioscler Thromb Vasc Biol 2006;26:534–40.

[11] Williamson R, Lee D, Hagaman J, Maeda N. Marked reduction of high density lipoprotein cholesterol in mice genetically modified to lack apolipoprotein A-I. Proc Natl Acad Sci USA 1992;89:7134–8.

[12] Karathanasis SK. Apolipoprotein multigene family: tandem organization of human apolipoprotein AI, CIII, and AIV genes. Proc Natl Acad Sci USA 1985;82: 6374–8.

[13] Pennacchio LA, Olivier M, Hubacek JA, Cohen JC, Cox DR, Fruchart JC, et al. An apolipoprotein influencing triglycerides in humans and mice revealed by comparative sequencing. Science 2001;294:169–73.

[14] Fielding CJ, Fielding PE. Molecular physiology of reverse cholesterol transport. J Lipid Res 1995;36:211–28.

[15] Thomas MJ, Bhat S, Sorci-Thomas MG. Three-dimensional models of HDL apoA-I: implications for its assembly and function. J Lipid Res 2008;49:1875–83.

[16] Tanaka M, Koyama M, Dhanasekaran P, Nguyen D, Nickel M, Lund-Katz S, et al. Influence of tertiary structure domain properties on the functionality of apolipoprotein A-I. Biochemistry 2008;47:2172–80.

[17] Segrest JP, Jones MK, De Loof H, Brouillette CG, Venkatachalapathi YV, Anantharamaiah GM. The amphipathic helix in the exchangeable apolipoproteins: a review of secondary structure and function. J Lipid Res 1992;33:141–66.

[18] Silva RA, Hilliard GM, Fang J, Macha S, Davidson WS. A three-dimensional molecular model of lipid-free apolipoprotein A-I determined by cross-linking/mass spectrometry and sequence threading. Biochemistry 2005;44:2759–69.

[19] Norum RA, Lakier JB, Goldstein S, Angel A, Goldberg RB, Block WD, et al. Familial deficiency of apolipoproteins A-I and C-III and precocious coronary-artery disease. N Engl J Med 1982;306:1513–9.

[20] Schaefer EJ, Ordovas JM, Law SW, Ghiselli G, Kashyap ML, Srivastava LS, et al. Familial apolipoprotein A-I and C-III deficiency, variant II. J Lipid Res 1985;26:1089–101.

[21] Pisciotta L, Fasano T, Calabresi L, Bellocchio A, Fresa R, Borrini C, et al. A novel mutation of the apolipoprotein A-I gene in a family with familial combined hyperlipidemia. Atherosclerosis 2008;198:145–51.

[22] Ng DS, Leiter LA, Vezina C, Connelly PW, Hegele RA. Apolipoprotein A-I Q[-2]X causing isolated apolipoprotein A-I deficiency in a family with analphalipoproteinemia. J Clin Invest 1994;93:223–9.

[23] Lackner KJ, Dieplinger H, Nowicka G, Schmitz G. High density lipoprotein deficiency with xanthomas. A defect in reverse cholesterol transport caused by a point mutation in the apolipoprotein A-I gene. J Clin Invest 1993;92:2262–73.

[24] Nakata K, Kobayashi K, Yanagi H, Shimakura Y, Tsuchiya S, Arinami T, et al. Autosomal dominant hypoalphalipoproteinemia due to a completely defective apolipoprotein A-I gene. Biochem Biophys Res Commun 1993;196:950–5.

[25] Miccoli R, Bertolotto A, Navalesi R, Odoguardi L, Boni A, Wessling J, et al. Compound heterozygosity for a structural apolipoprotein A-I variant, apo A-I(L141R)Pisa, and an apolipoprotein A-I null allele in patients with absence of HDL cholesterol, corneal opacifications, and coronary heart disease. Circulation 1996;94:1622–8.

[26] Takata K, Saku K, Ohta T, Takata M, Bai H, Jimi S, et al. A new case of apoA-I deficiency showing codon 8 nonsense mutation of the apoA-I gene without evidence of coronary heart disease. Arterioscler Thromb Vasc Biol 1995;15:1866–74.

[27] Al-Sarraf A, Al-Ghofaili K, Sullivan DR, Wasan KM, Hegele R, Frohlich J. Complete Apo AI deficiency in an Iraqi Mandaean family: case studies and review of the literature. J Clin Lipidol 2010;4:420–6.

[28] Romling R, von Eckardstein A, Funke H, Motti C, Fragiacomo GC, Noseda G, et al. A nonsense mutation in the apolipoprotein A-I gene is associated with high-density lipoprotein deficiency and periorbital xanthelasmas. Arterioscler Thromb 1994;14:1915–22.

[29] Matsunaga T, Hiasa Y, Yanagi H, Maeda T, Hattori N, Yamakawa K, et al. Apolipoprotein A-I deficiency due to a codon 84 nonsense mutation of the apolipoprotein A-I gene. Proc Natl Acad Sci USA 1991;88:2793–7.

[30] Matsunaga A, Sasaki J, Han H, Huang W, Kugi M, Koga T, et al. Compound heterozygosity for an apolipoprotein A1 gene promoter mutation and a structural nonsense mutation with apolipoprotein A1 deficiency. Arterioscler Thromb Vasc Biol 1999;19:348–55.

[31] Dastani Z, Dangoisse C, Boucher B, Desbiens K, Krimbou L, Dufour R, et al. Novel nonsense apolipoprotein A-I mutation (apoA-I(E136X)) causes low HDL cholesterol in French Canadians Atherosclerosis 2006;185:127–36.

[32] Wada M, Iso T, Asztalos BF, Takama N, Nakajima T, Seta Y, et al. Marked high density lipoprotein deficiency due to apolipoprotein A-I Tomioka (codon 138 deletion). Atherosclerosis 2009;207:157–61.

[33] Ikewaki K, Matsunaga A, Han H, Watanabe H, Endo A, Tohyama J, et al. A novel two nucleotide deletion in the apolipoprotein A-I gene, apoA-I Shinbashi, associated with high density lipoprotein deficiency, corneal opacities, planar xanthomas, and premature coronary artery disease. Atherosclerosis 2004;172:39–45.

[34] Moriyama K, Sasaki J, Takada Y, Matsunaga A, Fukui J, Albers JJ, et al. A cysteine-containing truncated apo A-I variant associated with HDL deficiency. Arterioscler Thromb Vasc Biol 1996;16:1416–23.

[35] Yokota H, Hashimoto Y, Okubo S, Yumoto M, Mashige F, Kawamura M, et al. Apolipoprotein A-I deficiency with accumulated risk for CHD but no symptoms of CHD. Atherosclerosis 2002;162:399–407.

[36] Funke H, von Eckardstein A, Pritchard PH, Karas M, Albers JJ, Assmann G. A frameshift mutation in the human apolipoprotein A-I gene causes high density lipoprotein deficiency, partial lecithin:cholesterol-acyltransferase deficiency, and corneal opacities. J Clin Invest 1991;87:371–6.

[37] Karathanasis SK, Ferris E, Haddad IA. DNA inversion within the apolipoproteins AI/CIII/AIV-encoding gene cluster of certain patients with premature atherosclerosis. Proc Natl Acad Sci USA 1987;84:7198–202.

[38] Ordovas JM, Cassidy DK, Civeira F, Bisgaier CL, Schaefer EJ. Familial apolipoprotein A-I, C-III, and A-IV deficiency and premature atherosclerosis due to deletion of a gene complex on chromosome 11. J Biol Chem 1989;264:16339–42.

[39] Parks JS, Li H, Gebre AK, Smith TL, Maeda N. Effect of apolipoprotein A-I deficiency on lecithin:cholesterol acyltransferase activation in mouse plasma. J Lipid Res 1995;36:349–55.

[40] Yamakawa-Kobayashi K, Yanagi H, Fukayama H, Hirano C, Shimakura Y, Yamamoto N, et al. Frequent occurrence of hypoalphalipoproteinemia due to mutant apolipoprotein A-I gene in the population: a population-based survey. Hum Mol Genet 1999;8:331–6.

[41] Rall Jr SC, Weisgraber KH, Mahley RW, Ogawa Y, Fielding CJ, Utermann G, et al. Abnormal lecithin:cholesterol acyltransferase activation by a human apolipoprotein A-I variant in which a single lysine residue is deleted. J Biol Chem 1984;259: 10063–70.

[42] Tilly-Kiesi M, Zhang Q, Ehnholm S, Kahri J, Lahdenpera S, Ehnholm C, et al. ApoA-IHelsinki (Lys107-->0) associated with reduced HDL cholesterol and LpA-I:A-II deficiency. Arterioscler Thromb Vasc Biol 1995;15:1294–306.

[43] Miccoli R, Zhu Y, Daum U, Wessling J, Huang Y, Navalesi R, et al. A natural apolipoprotein A-I variant, apoA-I (L141R)Pisa, interferes with the formation of alpha-high density lipoproteins (HDL) but not with the formation of pre beta 1-HDL and influences efflux of cholesterol into plasma. J Lipid Res 1997;38:1242–53.

[44] Utermann G, Haas J, Steinmetz A, Paetzold R, Rall Jr SC, Weisgraber KH, et al. Apolipoprotein A-IGiessen (Pro143----Arg). A mutant that is defective in activating lecithin:cholesterol acyltransferase. Eur J Biochem 1984;144:325–31.

[45] Recalde D, Velez-Carrasco W, Civeira F, Cenarro A, Gomez-Coronado D, Ordovas JM, et al. Enhanced fractional catabolic rate of apo A-I and apo A-II in heterozygous subjects for apo A-I(Zaragoza) (L144R). Atherosclerosis 2001;154:613–23.

[46] Deeb SS, Cheung MC, Peng RL, Wolf AC, Stern R, Albers JJ, et al. A mutation in the human apolipoprotein A-I gene. Dominant effect on the level and characteristics of plasma high density lipoproteins. J Biol Chem 1991;266:13654–60.

[47] Lindholm EM, Bielicki JK, Curtiss LK, Rubin EM, Forte TM. Deletion of amino acids Glu146-->Arg160 in human apolipoprotein A-I (ApoA-ISeattle) alters lecithin:cholesterol acyltransferase activity and recruitment of cell phospholipid. Biochemistry 1998;37:4863–8.

[48] Daum U, Langer C, Duverger N, Emmanuel F, Benoit P, Denefle P, et al. (R151C) Paris is defective in activation of lecithin: cholesterol acyltransferase but not in initial lipid binding, formation of reconstituted lipoproteins, or promotion of cholesterol efflux. J Mol Med 1999;77:614–22.

[49] Huang W, Sasaki J, Matsunaga A, Nanimatsu H, Moriyama K, Han H, et al. A novel homozygous missense mutation in the apo A-I gene with apo A-I deficiency. Arterioscler Thromb Vasc Biol 1998;18:389–96.

[50] Mahley RW, Innerarity TL, Rall Jr SC, Weisgraber KH. Plasma lipoproteins: apolipoprotein structure and function. J Lipid Res 1984;25:1277–94.

[51] Miettinen HE, Gylling H, Miettinen TA, Viikari J, Paulin L, Kontula K. Apolipoprotein A-IFin. Dominantly inherited hypoalphalipoproteinemia due to a single base substitution in the apolipoprotein A-I gene. Arterioscler Thromb Vasc Biol 1997;17:83–90.

[52] Miller M, Aiello D, Pritchard H, Friel G, Zeller K. Apolipoprotein A-I(Zavalla) (Leu159-->Pro): HDL cholesterol deficiency in a kindred associated with premature coronary artery disease. Arterioscler Thromb Vasc Biol 1998;18:1242–7.

[53] Leren TP, Bakken KS, Daum U, Ose L, Berg K, Assmann G, et al. Heterozygosity for apolipoprotein A-I(R160L)Oslo is associated with low levels of high density lipoprotein cholesterol and HDL-subclass LpA-I/A-II but normal levels of HDL-subclass LpA-I. J Lipid Res 1997;38:121–31.

[54] von Eckardstein A, Funke H, Henke A, Altland K, Benninghoven A, Assmann G. Apolipoprotein A-I variants. Naturally occurring substitutions of proline residues affect plasma concentration of apolipoprotein A-I. J Clin Invest 1989;84:1722–30.

[55] Daum U, Leren TP, Langer C, Chirazi A, Cullen P, Pritchard PH, et al. Multiple dysfunctions of two apolipoprotein A-I variants, apoA-I(R160L)Oslo and apoA-I(P165R), that are associated with hypoalphalipoproteinemia in heterozygous carriers. J Lipid Res 1999;40:486–94.

[56] Martin-Campos JM, Julve J, Escola JC, Ordonez-Llanos J, Gomez J, Binimelis J, et al. ApoA-I(MALLORCA) impairs LCAT activation and induces dominant familial hypoalphalipoproteinemia. J Lipid Res 2002;43:115–23.

[57] Weisgraber KH, Bersot TP, Mahley RW, Franceschini G, Sirtori CR. A-IMilano apoprotein. Isolation and characterization of a cysteine-containing variant of the A-I apoprotein from human high density lipoproteins. J Clin Invest 1980;66:901–7.

[58] Hovingh GK, Brownlie A, Bisoendial RJ, Dube MP, Levels JH, Petersen W, et al. A novel apoA-I mutation (L178P) leads to endothelial dysfunction, increased arterial wall thickness, and premature coronary artery disease. J Am Coll Cardiol 2004;44:1429–35.

[59] Han H, Sasaki J, Matsunaga A, Hakamata H, Huang W, Ageta M, et al. A novel mutant, ApoA-I nichinan (Glu235-->0), is associated with low HDL cholesterol levels and decreased cholesterol efflux from cells. Arterioscler Thromb Vasc Biol 1999;19: 1447–55.

[60] Nichols WC, Dwulet FE, Liepnieks J, Benson MD. Variant apolipoprotein AI as a major constituent of a human hereditary amyloid. Biochem Biophys Res Commun 1988;156:762–8.

[61] Booth DR, Tan SY, Booth SE, Hsuan JJ, Totty NF, Nguyen O, et al. A new apolipoprotein AI variant, Trp50Arg, causes hereditary amyloidosis. QJM 1995;88:695–702.

[62] Soutar AK, Hawkins PN, Vigushin DM, Tennent GA, Booth SE, Hutton T, et al. Apolipoprotein AI mutation Arg-60 causes autosomal dominant amyloidosis. Proc Natl Acad Sci USA 1992;89:7389–93.

[63] Booth DR, Tan SY, Booth SE, Tennent GA, Hutchinson WL, Hsuan JJ, et al. Hereditary hepatic and systemic amyloidosis caused by a new deletion/insertion mutation in the apolipoprotein AI gene. J Clin Invest 1996;97:2714–21.

[64] Murphy CL, Wang S, Weaver K, Gertz MA, Weiss DT, Solomon A. Renal apolipoprotein A-I amyloidosis associated with a novel mutant Leu64Pro. Am J Kidney Dis 2004;44:1103–9.

[65] Persey MR, Booth DR, Booth SE, van Zyl-Smit R, Adams BK, Fattaar AB, et al. Hereditary nephropathic systemic amyloidosis caused by a novel variant apolipoprotein A-I. Kidney Int 1998;53:276–81.

[66] Coriu D, Dispenzieri A, Stevens FJ, Murphy CL, Wang S, Weiss DT, et al. Hepatic amyloidosis resulting from deposition of the apolipoprotein A-I variant Leu75Pro. Amyloid 2003;10:215–23.

[67] Obici L, Palladini G, Giorgetti S, Bellotti V, Gregorini G, Arbustini E, et al. Liver biopsy discloses a new apolipoprotein A-I hereditary amyloidosis in several unrelated Italian families. Gastroenterology 2004;126:1416–22.

[68] Hamidi Asl L, Liepnieks JJ, Hamidi Asl K, Uemichi T, Moulin G, Desjoyaux E, et al. Hereditary amyloid cardiomyopathy caused by a variant apolipoprotein A1. Am J Pathol 1999;154:221–7.

[69] Amarzguioui M, Mucchiano G, Haggqvist B, Westermark P, Kavlie A, Sletten K, et al. Extensive intimal apolipoprotein A1-derived amyloid deposits in a patient with an apolipoprotein A1 mutation. Biochem Biophys Res Commun 1998;242:534–9.

[70] Mucchiano GI, Haggqvist B, Sletten K, Westermark P. Apolipoprotein A-1-derived amyloid in atherosclerotic plaques of the human aorta. J Pathol 2001;193:270–5.

[71] Hamidi Asl K, Liepnieks JJ, Nakamura M, Parker F, Benson MD. A novel apolipoprotein A-1 variant, Arg173Pro, associated with cardiac and cutaneous amyloidosis. Biochem Biophys Res Commun 1999;257:584–8.

[72] Obici L, Bellotti V, Mangione P, Stoppini M, Arbustini E, Verga L, et al. The new apolipoprotein A-I variant leu(174) --> Ser causes hereditary cardiac amyloidosis, and the amyloid fibrils are constituted by the 93-residue N-terminal polypeptide. Am J Pathol 1999;155:695–702.

[73] Lachmann HJ, Booth DR, Booth SE, Bybee A, Gilbertson JA, Gillmore JD, et al. Misdiagnosis of hereditary amyloidosis as AL (primary) amyloidosis. N Engl J Med 2002;346:1786–91.

[74] de Sousa MM, Vital C, Ostler D, Fernandes R, Pouget-Abadie J, Carles D, et al. Apolipoprotein AI and transthyretin as components of amyloid fibrils in a kindred with apoAI Leu178His amyloidosis. Am J Pathol 2000;156:1911–7.

[75] Weers PM, Patel AB, Wan LC, Guigard E, Kay CM, Hafiane A, et al. Novel N-terminal mutation of human apolipoprotein A-I reduces self-association and impairs LCAT activation. J Lipid Res 2011;52:35–44.

[76] Miettinen HE, Jauhiainen M, Gylling H, Ehnholm S, Palomaki A, Miettinen TA, et al. (Leu159-->Arg) mutation affects lecithin cholesterol acyltransferase activation and subclass distribution of HDL but not cholesterol efflux from fibroblasts. Arterioscler Thromb Vasc Biol 1997;17:3021–32.

[77] Huang W, Sasaki J, Matsunaga A, Han H, Li W, Koga T, et al. A single amino acid deletion in the carboxy terminal of apolipoprotein A-I impairs lipid binding and cellular interaction. Arterioscler Thromb Vasc Biol 2000;20:210–6.

[78] Sorci-Thomas MG, Thomas MJ. The effects of altered apolipoprotein A-I structure on plasma HDL concentration. Trends Cardiovasc Med 2002;12:121–8.

[79] Minnich A, Collet X, Roghani A, Cladaras C, Hamilton RL, Fielding CJ, et al. Site-directed mutagenesis and structure-function analysis of the human apolipoprotein A-I. Relation between lecithin-cholesterol acyltransferase activation and lipid binding. J Biol Chem 1992;267:16553–60.

[80] Sorci-Thomas M, Kearns MW, Lee JP. Apolipoprotein A-I domains involved in lecithin-cholesterol acyltransferase activation. Structure: function relationships. J Biol Chem 1993;268:21403–9.

[81] Holvoet P, Zhao Z, Vanloo B, Vos R, Deridder E, Dhoest A, et al. Phospholipid binding and lecithin-cholesterol acyltransferase activation properties of apolipoprotein A-I mutants. Biochemistry 1995;34:13334–42.

[82] Sviridov D, Pyle LE, Fidge N. Efflux of cellular cholesterol and phospholipid to apolipoprotein A-I mutants. J Biol Chem 1996;271:33277–83.

[83] Ji Y, Jonas A. Properties of an N-terminal proteolytic fragment of apolipoprotein AI in solution and in reconstituted high density lipoproteins. J Biol Chem 1995;270: 11290–7.

[84] Sorci-Thomas MG, Thomas M, Curtiss L, Landrum M. Single repeat deletion in ApoA-I blocks cholesterol esterification and results in rapid catabolism of delta6 and wild-type ApoA-I in transgenic mice. J Biol Chem 2000;275:12156–63.

[85] Baralle M, Baralle FE. Genetics and molecular biology. Curr Opin Lipidol 2000;11:653–6.

[86] McManus DC, Scott BR, Franklin V, Sparks DL, Marcel YL. Proteolytic degradation and impaired secretion of an apolipoprotein A-I mutant associated with dominantly inherited hypoalphalipoproteinemia. J Biol Chem 2001;276:21292–302.

[87] Parolini C, Chiesa G, Zhu Y, Forte T, Caligari S, Gianazza E, et al. Targeted replacement of mouse apolipoprotein A-I with human ApoA-I or the mutant ApoA-IMilano. Evidence of APOA-IM impaired hepatic secretion. J Biol Chem 2003;278:4740–6.

[88] Koukos G, Chroni A, Duka A, Kardassis D, Zannis VI. LCAT can rescue the abnormal phenotype produced by the natural apoA-I Mutations (Leu141Arg)(Pisa) and (Leu159Arg)(FIN). Biochemistry 2007;8(46):10713–21.

[89] Chroni A, Duka A, Kan HY, Liu T, Zannis VI. Point mutations in apolipoprotein A-I mimic the phenotype observed in patients with classical lecithin:cholesterol acyltransferase deficiency. Biochemistry 2005;44:14353–66.

[90] Tanaka M, Dhanasekaran P, Nguyen D, Ohta S, Lund-Katz S, Phillips MC, et al. Contributions of the N- and C-terminal helical segments to the lipid-free structure and lipid interaction of apolipoprotein A-I. Biochemistry 2006;45:10351–8.

[91] Vedhachalam C, Liu L, Nickel M, Dhanasekaran P, Anantharamaiah GM, Lund-Katz S, et al. Influence of ApoA-I structure on the ABCA1-mediated efflux of cellular lipids. J Biol Chem 2004;279:49931–9.

[92] Van Allen MW, Frohlich JA, Davis JR. Inherited predisposition to generalized amyloidosis. Clinical and pathological study of a family with neuropathy, nephropathy, and peptic ulcer. Neurology 1969;19:10–25.

[93] Rowczenio D, Dogan A, Theis JD, Vrana JA, Lachmann HJ, Wechalekar AD, et al. Amyloidogenicity and clinical phenotype associated with five novel mutations in apolipoprotein A-I. Am J Pathol 2011;179:1978–87.

[94] Obici L, Franceschini G, Calabresi L, Giorgetti S, Stoppini M, Merlini G, et al. Structure, function and amyloidogenic propensity of apolipoprotein A-I. Amyloid 2006;13:191–205.

[95] Nichols WC, Gregg RE, Brewer Jr HB, Benson MD. A mutation in apolipoprotein A-I in the Iowa type of familial amyloidotic polyneuropathy. Genomics 1990;8: 318–23.

[96] Eriksson M, Schonland S, Yumlu S, Hegenbart U, von Hutten H, Gioeva Z, et al. Hereditary apolipoprotein AI-associated amyloidosis in surgical pathology specimens: identification of three novel mutations in the APOA1 gene. J Mol Diagn 2009;11: 257–62.

[97] Hazenberg AJ, Dikkers FG, Hawkins PN, Bijzet J, Rowczenio D, Gilbertson J, et al. Laryngeal presentation of systemic apolipoprotein A-I-derived amyloidosis. Laryngoscope 2009;119:608–15.

[98] Calabresi L, Vecchio G, Frigerio F, Vavassori L, Sirtori CR, Franceschini G. Reconstituted high-density lipoproteins with a disulfide-linked apolipoprotein A-I dimer: evidence for restricted particle size heterogeneity. Biochemistry 1997;36:12428–33.

[99] Chiesa G, Sirtori CR. Apolipoprotein A-I(Milano): current perspectives. Curr Opin Lipidol 2003;14:159–63.

[100] Bruckert E, von Eckardstein A, Funke H, Beucler I, Wiebusch H, Turpin G, et al. The replacement of arginine by cysteine at residue 151 in apolipoprotein A-I produces a phenotype similar to that of apolipoprotein A-IMilano. Atherosclerosis 1997;128:121–8.

[101] Bielicki JK, Oda MN. Apolipoprotein A-I(Milano) and apolipoprotein A-I(Paris) exhibit an antioxidant activity distinct from that of wild-type apolipoprotein A-I. Biochemistry 2002;41:2089–96.

[102] Rubin EM, Krauss RM, Spangler EA, Verstuyft JG, Clift SM. Inhibition of early atherogenesis in transgenic mice by human apolipoprotein AI. Nature 1991;353:265–7.

[103] Paszty C, Maeda N, Verstuyft J, Rubin EM. Apolipoprotein AI transgene corrects apolipoprotein E deficiency-induced atherosclerosis in mice. J Clin Invest 1994;94: 899–903.

[104] Plump AS, Scott CJ, Breslow JL. Human apolipoprotein A-I gene expression increases high density lipoprotein and suppresses atherosclerosis in the apolipoprotein E-deficient mouse. Proc Natl Acad Sci USA 1994;91:9607–11.

[105] Newton RS, Krause BR. HDL therapy for the acute treatment of atherosclerosis. Atheroscler Suppl 2002;3:31–8.

[106] Brewer Jr HB. Focus on high-density lipoproteins in reducing cardiovascular risk. Am Heart J 2004;148:S14–8.

[107] Shah PK, Nilsson J, Kaul S, Fishbein MC, Ageland H, Hamsten A, et al. Effects of recombinant apolipoprotein A-I(Milano) on aortic atherosclerosis in apolipoprotein E-deficient mice. Circulation 1998;97:780–5.

[108] Ameli S, Hultgardh-Nilsson A, Cercek B, Shah PK, Forrester JS, Ageland H, et al. Recombinant apolipoprotein A-I Milano reduces intimal thickening after balloon injury in hypercholesterolemic rabbits. Circulation 1994;90:1935–41.

[109] Shah PK, Yano J, Reyes O, Chyu KY, Kaul S, Bisgaier CL, et al. High-dose recombinant apolipoprotein A-I(milano) mobilizes tissue cholesterol and rapidly reduces plaque lipid and macrophage content in apolipoprotein e-deficient mice. Potential implications for acute plaque stabilization. Circulation 2001;103:3047–50.

[110] Chiesa G, Monteggia E, Marchesi M, Lorenzon P, Laucello M, Lorusso V, et al. Recombinant apolipoprotein A-I(Milano) infusion into rabbit carotid artery rapidly removes lipid from fatty streaks. Circ Res 2002;90:974–80.

[111] Kaul S, Coin B, Hedayiti A, Yano J, Cercek B, Chyu KY, et al. Rapid reversal of endothelial dysfunction in hypercholesterolemic apolipoprotein E-null mice by recombinant apolipoprotein A-I(Milano)-phospholipid complex. J Am Coll Cardiol 2004;44:1311–9.

[112] Nissen SE, Tsunoda T, Tuzcu EM, Schoenhagen P, Cooper CJ, Yasin M, et al. Effect of recombinant ApoA-I Milano on coronary atherosclerosis in patients with acute coronary syndromes: a randomized controlled trial. JAMA 2003;290:2292–300.

[113] Lebherz C, Sanmiguel J, Wilson JM, Rader DJ. Gene transfer of wild-type apoA-I and apoA-I Milano reduce atherosclerosis to a similar extent. Cardiovasc Diabetol 2007;6:15.

[114] Navab M, Shechter I, Anantharamaiah GM, Reddy ST, Van Lenten BJ, Fogelman AM. Structure and function of HDL mimetics. Arterioscler Thromb Vasc Biol 2010;30:164–8.

[115] Kawachi E, Uehara Y, Hasegawa K, Yahiro E, Ando S, Wada Y, et al. Novel molecular imaging of atherosclerosis with gallium-68-labeled apolipoprotein A-I mimetic peptide and positron emission tomography. Circulation J 2013;77:1482–9.

CHAPTER 3

The Complexity of High-Density Lipoproteins

Bela F. Asztalos, Mariko Tani, Brian Ishida
Lipid Metabolism Laboratory, Tufts University, Boston, MA, USA

Contents

Abstract

High-density lipoprotein (HDL) is the most complex class of lipoproteins. HDL is comprised of several subclasses that are different in size, protein and lipid composition, physiological functions, and pathophysiological significance. Although HDL has been studied for at least half a century, its roles in diseases are poorly understood. Recently, with the rapid development of analytical techniques, much has been learned about HDL composition, especially about HDL proteomics. However, HDL lipidomics and metabolomics are still in their infancy. At least two dozen biologically important functions of HDL have been identified. This chapter summarizes what HDL is, what we know about its composition, how it can be separated into subclasses, and what the roles of HDL particles are in HDL structure and function.

The HDL Handbook

1. HDL AND CARDIOVASCULAR DISEASE

There is a significant correlation between low-density lipoprotein choles-terol (LDL-C) levels and cardiovascular disease (CVD) risk. However, a significant residual CVD risk remains in patients even after lowering LDL-C below 70 mg/dL. Large cohort and prospective studies have consistently shown that high-density lipoprotein cholesterol (HDL-C) is inversely cor-related with CVD risk [1–3]. Therefore, strategies for increasing HDL-C were considered promising in reducing the residual CVD risk in statin-treated patients [4–6].

In the Veterans Affairs HDL Intervention Trial (VA-HIT), a multivari-able analysis, adjusting for traditional coronary heart disease (CHD) risk factors, suggested an independent association of increased HDL-C with reduction in CHD death and nonfatal myocardial infarction. However, the VA-HIT investigators also showed that the 6% increase in HDL-C explained only a small portion of the beneficial effect of gemfibrozil [7]. Recently, Briel *and colleagues* reported no correlation between increases in HDL-C and CVD risk in a large meta-analysis comprised of data from approxi-mately 300,000 subjects who received lipid-modulating therapies [8]. How-ever, it is worth noting that 1) raising HDL-C was not the primary objective of any of these lipid-modulating therapies, and 2) the absolute increases in circulating HDL-C levels in the studies were modest, raising the possibility that the analysis lacked statistical power.

The lack of CVD-risk reduction by cholesteryl ester transfer protein (CETP) inhibition and nicotinic acid (AimHigh, DalOutcome, HPS2-THRIVE trials) has increased concerns regarding HDL-C as a suitable tar-get for the prevention of CVD [5,9,10]. However, raising HDL-C levels showed benefit for certain subgroups of these trials. Further clinical trials with new compounds for CETP-inhibition might conclude unequivocally whether CETP inhibition significantly decreases CVD risk or not.

We do not think that the failure of these recent clinical trials to show clinical evidence of raising HDL-C proves a lack of therapeutic potential for HDL manipulation. Rather, it underscores a lack of fundamental under-standing of HDL composition and functions.

2. WHAT IS HDL?

HDL is the most complex lipoprotein class in terms of composition and function. HDL is comprised of a heterogeneous group of lipoprotein par-ticles. All of these particles float in the d = 1.063 − 1.21 g/mL range after ultracentrifugation with a 100,000 G-force. By traditional nomenclature, HDL contains apolipoproteins (apo), charged lipids (phospholipids [PL] and free cholesterol [FC]) on the surface, and neutral lipids (triglycerides [TG] and cholesteryl ester [CE]) in the core. On average, HDL contains about 50% protein and 50% lipids. However, this ratio varies widely because larger, more lipidated HDL particles contain approximately 35% proteins and 65% lipids. In contrast, some smaller HDL particles, such as preβ-1, are poorly lipidated and contain a higher proportion of proteins.

The major apolipoproteins of HDL are apoA-I and apoA-II. The concentration of ApoA-II is about 30% of that of apoA-I in HDL. ApoA-I is essential for HDL formation; therefore, apoA-I-deficient patients have no measurable HDL-C [11]. Minor apolipoproteins of HDL include apoA-IV, A-V, C-I, C-II, C-III, C-IV, D, E, F, H, J, L, M, O, and P. Some of them (e.g., apoE and apoA-IV) are not incorporated into apoA-I-containing HDL.

HDL can be separated into several subclasses based on its apolipoprotein contents. At first, HDL was separated into fractions containing only apoA-I but no apoA-II (LpA-I) and fractions containing both apoA-I- and apoA-II (LpA-I:A-II) [12]. Kostner and Alaupovic, using immune-absorption chromatography, separated more than 10 distinct HDL subclasses, each of which contains a variety of different apolipoproteins [13]. Nowadays, there are better separation techniques available. Therefore, in our opinion, it would be more appropriate to rename HDL subclasses by their apolipoprotein content—for example, apoA-I HDL, apoA-I:A-II HDL, apoE HDL, apoA-IV HDL, and so on. All of these HDL subclasses differ not only in composition and metabolism, but also in physiological function and pathophysiological significance. With the introduction of liquid chromatography and tandem mass spectroscopy (LC-MS/MS), significantly more proteins have been found on HDL, suggesting that there are significantly more (10-fold) HDL particles with different protein composition than previously thought. Moreover, HDL composition can vary markedly from person to person and also under various pathological conditions.

3. HDL METABOLISM

ApoA-I has a molecular weight of about 28 KD and is secreted by the liver and the small intestine as a monomer in a lipid-free form. ApoA-I self-associates to form a homodimer when its circulating concentration reaches > 10 µg/mL. Two apoA-I molecules form a belt-shaped disc—secured with salt bridges—in which the two apoA-I molecules are positioned in an anti-parallel fashion [14,15]. This apoA-I belt migrates as preβ-1 on nondenaturing 2-dimensional (2d) (agarose/polyacrylamide) gel electrophoresis with an estimated molecular size of 5.6 nm. By electron microscopy, preβ-1 particles appear as flattened discs that arrange in stacks ranging from a couple of discs to a rope-like structure containing tens of stacked discs. ApoA-I is the only protein present in preβ-1 and the apoA-I belt harbors some PL and trace amounts of FC. There are several different hypotheses about the lipid to protein ratio of preβ-1 HDL. Estimates range from less than 10 PL per apoA-I to more than 100 PL per apoA-I. However, none of these speculations are supported by direct evidence.

We have investigated HDL particles in subjects with various mutations and polymorphisms in genes encoding key players in HDL metabolism. The data led to a step-by-step working model of HDL metabolism in humans as shown in Figure 3.1 [16,17].

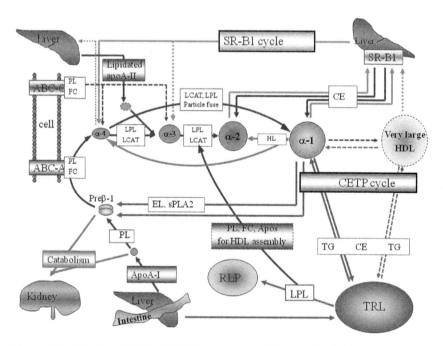

Figure 3.1 *A Working Model of HDL Metabolism and Remodeling in Humans Based on Information Learned from Various Genetically Determined HDL Deficiency States.* Green color represents pathways in HDL maturation. Tangerine and brown colors represent HDL remodeling by the SR-BI and CETP cycle, respectively. Blue color represents HDL catabolism. Major pathways are indicated by solid lines; minor pathways are represented by dotted lines. For color version of this figure, the reader is referred to the online version of this book.

3.1. Step 1: ApoA-I Synthesis

ApoA-I, a key molecule in the formation and functionality of HDL, is synthesized in the liver and intestine. Marked HDL–deficient states (HDL-C < 5 mg/dL) and undetectable plasma apoA-I levels have been reported in humans as a result of mutations at the APOA1/C3/A4 gene locus (Figure 3.2) [11,18–25]. These mutations caused amino acid alterations of the carboxyl terminal region containing helices 6 (amino acid residues 145–164) and/or 7 (167–183) of apoA-I [26–28]. This region is considered to be important for the protein–protein interaction between apoA-I and ATP-binding cassette transporter (ABC) A1. It is also critical for lecithin:cholesterol acyltransferase (LCAT) activation, which is required for lipid efflux from peripheral tissues, and is responsible for the conversion of discoidal HDL into spherical HDL particles [28].

When apoA-I enters the circulation, it associates with some PL to form a discoidal-shaped particle containing two apoA-I molecules (dimer). Both the lipid-free monomer and the apoA-I-PL dimer have preβ-mobility but differ in size, as determined by gel-filtration and nondenaturing polyacrylamide gel electrophoresis (PAGE). Preβ-1 particles are apoA-I-PL-containing dimer particles with a modal diameter of 5.6 nm.

3.2. Step 2: Cellular Cholesterol Efflux via ABCA1

In the absence of functional ABCA1, only preβ-1 HDL particles are present in plasma. ABCA1 affects cellular lipid efflux in the presence of apoA-I. Cellular lipid transport from the Golgi to the plasma membrane is defective in patients with ABCA1 deficiency (Tangier disease) and in Abca1$^{-/-}$ mice,

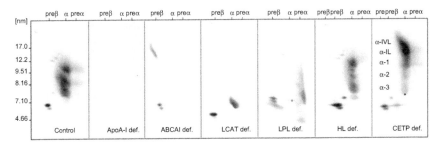

Figure 3.2 *Apo-I-Containing HDL Subpopulation Profiles of Subjects with Genetic Disorders Influencing HDL Metabolism.* HDL was separated by nondenaturing 2d gel electrophoresis. Gels were electro transferred to nitrocellulose membranes. The distribution of apolipoproteins was detected by immuno-blotting with monospecific first antibodies and ^{125}I-labeled second antibodies.

resulting in retention of caveolin-1, the main structural protein of caveolae, in the Golgi complex [29]. Overexpression of caveolin-1 increases the number of caveolae and enhances HDL-mediated cholesterol efflux [30].

HDL-mediated cholesterol efflux is impaired in fibroblasts in patients with Tangier disease. Consequently, only native preβ-1 HDL can be detected as preβ-1 particles are rapidly catabolized rather than matured to larger particles in these patients (Figure 3.2) [31,32]. It has been shown that preβ-1 is the only HDL particle that is able to promote cellular cholesterol efflux via the ABCA1 pathway [33]. As preβ-1 pulls PLs and cholesterol from the cells via ABCA1, it is transformed into α-4 (modal diameter = 7.3 nm), a discoidal LpA-I particle containing two apoA-I molecules, PL and FC.

The role of ABCA1 in the development of atherosclerosis is controversial. About 99% of cholesterol transported to the bile via the liver is carried by LDL. Therefore, direct contribution of HDL to reverse cholesterol transport (RCT) is negligible. Many argue that HDL removes cholesterol from macrophages in the vessel wall, which results in decreased cholesterol deposition. However, it has also been indicated that decreased expression of ABCA1 in the liver increases susceptibility to atherosclerosis, while selective removal of macrophage ABCA1 does not [34]. These data indicate that ABCA1 is essential for the maturation of HDL. It is worth noting that HDL-mediated efflux of excess cholesterol from macrophages can decrease the inflammatory response of those cells. This reduces the attraction of other macrophages into the vessel wall and, ultimately, results in decreased plaque formation.

3.3. Step 3: Cholesterol Esterification by LCAT

LCAT is synthesized in the liver. After secretion, LCAT either binds to lipoproteins or is present in a lipid-free form in plasma [35]. LCAT synthesizes the majority of cholesteryl esters in plasma by transferring a fatty acid from lecithin (phosphatidyl choline) to the 3-hydroxyl group of cholesterol. LCAT maintains the unesterified-cholesterol gradient between peripheral cells and HDL. Efflux of FC from cells occurs by passive diffusion of FC between cellular membranes and acceptors and by mechanisms facilitated by ABCs and scavenger receptor (SR) BI. In the presence of LCAT, the bidirectional movement of cholesterol between cells and HDL results in net cholesterol efflux; therefore, LCAT plays a central role in the initial steps of RCT [36,37]. Although LCAT activity is not essential for the transformation of preβ-1 HDL into α-mobility HDL, it is necessary for the maturation of HDL from the small, lipid-poor α-4 to larger HDL particles [16]. This concept is supported by data obtained from LCAT-deficient subjects who have only preβ-1 and α-4 HDL particles (Figure 3.2).

3.4. Step 4: Lipoprotein Lipase (LPL)-Mediated Lipolysis

LPL hydrolyses triglycerides (TG) in apoB–containing lipoproteins, resulting in the release of surface PL, FC, and apolipoproteins. These molecules can be used for HDL particle biosynthesis where the concerted actions of LCAT and LPL increase HDL particle size.

LPL also facilitates changes in lipid composition and fusion of α-4 HDL with apoA-II to form α-3 HDL. ApoA-II is secreted from the liver in a lipidated form containing two apoA-II molecules (apoA-II dimer). The newly formed LpA-I:A-II particles are α-3 particles with a modal diameter of 7.9 nm. Patients with homozygous LPL deficiency have a unique HDL subpopulation profile with a low level of small α-mobility particles (Figure 3.2).

3.5. Step 5: ABCG1-Mediated Cholesterol Efflux

It is indicated that the FC pool available for efflux is increased in cells expressing ABCG1. Cholesterol efflux is linked to the ability of ABCG1 to relocate cholesterol from inside the cell to the cell surface (membrane) [38,39]. Further evidence that ABCG1 increases the availability of cholesterol in the plasma membrane comes from the observation that the cholesterol oxidase-sensitive pool of membrane cholesterol is expanded upon ABCG1 expression.

In addition to efflux of cholesterol, ABCG1 expression stimulates the release of cellular PLs to HDL_3 and to human serum. Kobayashi and colleagues have shown that ABCG1 in Human Embryonic Kidney (HEK) 293 cells mediates the efflux of cholesterol as well as choline PLs to HDL_3 and that ABCG1 differs from ABCA1 in the type of PL transported [40].

It is hypothesized that the binding of extracellular acceptor particles is necessary for cellular-cholesterol efflux mediated by ABCA1 and SR-BI, but not by ABCG1. Moreover, while the small, LpA-I preβ-1 is a good acceptor of cholesterol via the ABCA1 pathway, larger, more lipidated, LpA-I:A-II α-2 or α-3 HDL particles more effectively promote cellular cholesterol efflux via the ABCG1 pathway [41].

3.6. Step 6: CETP Cycle

CETP mediates the exchange of cholesteryl ester (CE) from HDL for TGs from apoB-containing lipoproteins, although not stoichiometrically. Although HDL carries cholesterol directly to the bile via the liver SR-BI receptor (Figure 3.1), this is only a minor part of RCT: the majority of cholesterol is transported to the bile via LDL as an action of CETP. As a result of CETP activity, HDL becomes smaller and TG-enriched. Moreover, the CETP-mediated TG-enrichment of α-1 makes this HDL particle a good substrate for hepatic lipase (HL). The concerted actions of CETP and HL transform large α-1 particles into preβ-1 particles, decreasing the ability of HDL to transport CE into the bile via the SR-BI pathway.

CETP plays a significant role in modulating HDL metabolism, affecting HDL-C level and the distribution, size, and apolipoprotein composition of HDL particles. CETP deficiency causes the appearance of very large HDL particles. The apoA-I in these particles associates with apoA-II, apoCs, and apoE (Figure 3.2) [42]. Therefore, CETP is essential for the formation of heterogeneous HDL both in size and in lipoprotein composition.

3.7. Step 7: Hepatic Lipase-Mediated Lipolysis

Very large HDL particles (TG-enriched α-1) are good substrates for HL. HL has both TG and phospholipase activity. HDL particles in HL-deficient subjects are enriched in TG and PL [43]. Through the concerted actions of HL and CETP, the core of α-1 shrinks and the particles are transformed into small preβ-1 HDL. Compared to unaffected subjects, homozygous HL-deficient subjects have increased preβ-1, slightly decreased α-4 and α-3, substantially decreased α-2, and normal to slightly elevated α-1 HDL particles, indicating that a partial digestion of α-1 is needed for the formation of α-2. This clearly indicates that HDL maturation is not a linear process from the smallest to the largest particles (Figure 3.2).

3.8. Step 8: SR-BI Cycle

The last step in HDL-mediated RCT is the selective transfer of CE to the bile via the SR-BI pathway in the liver (SR-BI cycle). Selective cholesterol removal from HDL via the SR-BI pathway decreases the concentrations of the large, lipid-enriched α-1 and increases the concentrations of the smaller (α-3 and α-4) particles. However, preβ-1 particles are not produced via this pathway (Figure 3.2) [44].

Despite the fact that HDL mediates only about 1% of direct total RCT, this step is very significant because HDL carries cholesterol from macrophages in the vessel wall directly inhibiting plaque formation. The indirect action of HDL in RCT is mediated through transfer of CE to LDL which then delivers CE to the liver. It has been shown that there are a number of pathways for the flux of cholesterol between cells and serum lipoproteins. At least three protein-mediated pathways have been identified and these proteins demonstrate both similarities and differences. ABCG1 and SR-BI require cholesterol acceptors that contain PL. The efficiency of the acceptor is determined by the composition and size of the extracellular particle. Therefore, the best substrates for cell–cholesterol efflux are the large, lipid-enriched α-1 and α-2 HDL particles. In contrast to ABC transporters, SR-BI mediates both cellular efflux and influx of free and esterified cholesterol.

3.9. Step 9: HDL Catabolism

In addition to LPL and HL, other lipases—endothelial lipase (EL) and secretory lipoprotein-associated phospholipase (LpPL) A_2—and transfer proteins participate in HDL remodeling. EL has high phospholipase and very low TG lipase activity and does not dissociate lipid-free/lipid-poor apoA-I from HDL [45,46]. Conversely, HL has mainly TG lipase activity. Through the concerted actions of HL and CETP, the core of large HDL particles shrinks as apoA-I is shed from these particles.

Phospholipid transfer protein (PLTP) transfers phospholipids and unesterified cholesterol from very low-density lipoprotein (VLDL) to HDL as well as among HDL particles. PLTP remodels HDL simultaneously into larger and smaller particles by triggering particle fusion and the dissociation of lipid-free/lipid-poor apoA-I [47]. The role of PLTP in atherogenesis is controversial. It has been reported that its expression in macrophages both enhances and inhibits atherosclerosis in mice [48,49]. So far subjects with EL-, sPLA2-, or PLTP-deficiency have not been identified; therefore, direct in vivo evidence for the above-mentioned steps in HDL remodeling in humans is not available.

The kidney and liver are major sites of HDL catabolism. The structural integrity of the apolipoprotein components of HDL are major factors in determining the half-life of HDL [50–53]. SR-BI has been shown to function as an HDL receptor that mediates selective cholesterol uptake [54]. However, unlike the catabolism of LDL by LDL receptors, the catabolism of HDL by SR-BI does not involve holoparticle uptake and lysosomal degradation as transgenic mice, deficient in SR-BI, were shown to have elevated levels of plasma HDL-C yet not to exhibit any change in the level of plasma apoA-I [55]. However, larger HDL particles can be taken up by the liver and both the protein and lipid components are catabolized [56].

4. MODERATORS OF HDL PARTICLE DISTRIBUTION AND COMPOSITION

HDL particle distribution and composition are influenced in a number of different ways. It is well-known, although not well-documented, that HDL particle distribution (and therefore overall HDL particle size) has a stronger correlation with TG levels than with HDL-C or apoA-I levels [33]. HDL particle distribution is also significantly influenced by several inflammatory conditions. For example, serum amyloid A (SAA) can replace apoA-I in HDL particles, resulting in size and functional changes in HDL [57,58]. How other inflammatory proteins, such as C-reactive protein (CRP), LpPLA$_2$, secretory PLA$_2$, EL, or the cell master inflammatory regulator, NFκB, influence HDL metabolism is poorly understood. It has also been indicated that diabetes, insulin resistance, and uncontrolled high blood-glucose levels significantly influence HDL metabolism. Hypertriglyceridemia, which accompanies these conditions, only partially explains the substantial differences in HDL size between normal and affected subjects. It is worth noting that it is not the level of TGs but the metabolism of TG-rich lipoproteins (TRL) that influences HDL particle distribution and overall HDL size.

This phenomenon can be demonstrated by comparing the different effects of statins and fibrates on HDL particle distribution and overall HDL size. Both drugs decrease plasma TG levels by about the same rate (30%), but statins increase overall HDL size by selectively increasing the concentrations of the largest (α-1) HDL particles; while fenofibrate and gemfibrozil increase the concentrations of smaller HDL particles (α-2 and α-3) and (slightly) decrease the concentration of the large α-1 particles [59,60]. Statins increase HDL by decreasing TRL concentration [61]. As a result, CETP activity decreases and HDL becomes less enriched in TG, which in turn decreases HL activity on HDL. Consequently, HDL size increases and the fractional catabolic rate (FCR) of apoA-I decreases. Statins also decrease inflammation substantially as evidenced by decreased CRP levels. Decreased inflammation is associated with increased HDL size and decreased apoA-I FCR. On the contrary, fibrates work through the peroxisome proliferator-activated receptor (PPAR)-α pathway, increasing apoA-I and apoA-II production (about 2%), but not influencing HDL catabolism. In contrast to statins, fibrates do not decrease LDL-C; however, they do influence lipase activity more than statins. Also, PPARα is a master regulator of hundreds of cell functions, some of which might interfere with lipoprotein metabolism.

5. HDL STRUCTURE

The double-belt model of lipid-poor HDL (preβ-1 and small α-mobility particles), as proposed by Segrest and colleagues [62], is supported by other investigators [63–65]. Recently, Segrest and colleagues published a detailed validation of previous computer models for the tertiary and quaternary structure of the apoA-I belt (Figure 3.3) [66]. This model describes two apoA-I belts positioned in an anti-parallel way. However, determining the molecular structure of larger, highly lipidated HDL particles is more diffi-cult. The most widely accepted model is the Trefoil Model [67]. In this model, three apoA-I molecules are shaped in a way that all three share an anti-parallel part with the other two apoA-I molecules (Figure 3.4). There are, however, HDL particles containing at least four apoA-I molecules. In order to accommodate four apoA-I molecules without increasing particle size, the authors proposed the opposite poles of the particle in opposing

Figure 3.3 *The Anti-parallel Double Belt Model of Two apoA-I Molecules. [The figure is a modification of Davidson et al. model published in JBC 2007 [63]].*

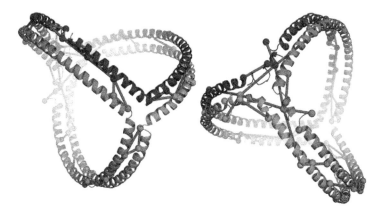

Figure 3.4 *The Threfoil Model of Three apoA-I Molecules on a Spherical HDL Particle. [The figure is a modification of Davidson et al. Nat Struct Mol Biol 2011 [67]].*

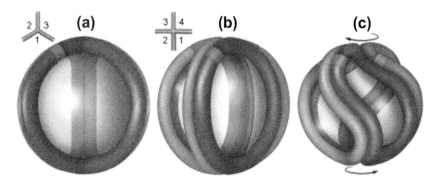

Figure 3.5 *The Twisted Model of Four apoA-I Molecules on a Spherical HDL Particle.*
(a) Schematic representation of the three molecule Trefoil Model as originally proposed with each molecule of apoA-I. The inset shows a schematic top view showing the bend angles of each apoA-I. (b) An idealized, fully extended tetrameric complex. (c) Twisted tetrameric complex with a reduced particle diameter as proposed for LpA-I$_{2a}$. The figures were generated using Adobe Photoshop. *[The figure is a modification of Davidson et al.* Nat Struct Mol Biol *2011 [67]].*

directions (Figure 3.5). Despite the tremendous efforts, this model describes lipoprotein particles containing only apoA-I.

In human plasma, most of the lipidated HDL particles contain apoA-I and apoA-II as structural proteins. It is speculated that apoA-II interacts with apoA-I, alters its conformation and, very likely, its functions as well. It is also hypothesized that apoA-II, the most hydrophobic of the HDL proteins, is deeply embedded into the PL monolayer and alters the curvature of the surface which might have an effect on apoA-I conformation [65]. Independent of these different molecular mechanisms, it is widely accepted that apoA-II stabilizes HDL and makes LpA-I:A-II HDL particles less prone to remodeling.

However, the situation is even more complex because many of the HDL particles carry a plethora of other apolipoproteins and plasma proteins.

6. HDL PROTEOMICS

Figure 3.6 shows the distribution of the six most abundant apolipoproteins in HDL in a healthy subject, as determined by 2d nondenaturing gradient PAGE and immune-detection. This composite picture shows that some HDL proteins—apoA-IV and the majority of apoE—are not incorporated into the apoA-I-containing HDL particles in humans. The distribution of apolipoproteins and other associated proteins in HDL particles are unique and vary widely between subjects [16,32,42]. The presence and/or concentration of specific apolipoproteins on an HDL particle determine the function and metabolism

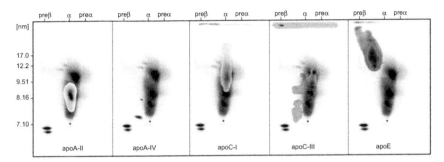

Figure 3.6 *Distribution of the Major Apolipoproteins in HDL.* HDL particles were separated by nondenaturing 2d gel electrophoresis followed by the transfer of gels to nitrocellulose membrane and immune detection of each apolipoprotein presented on the figure. Grey represents the distribution of apoA-I-containing particles. Different colors represent the distribution of different apolipoproteins superimposed on the A-I image. *[from Asztalos et al. JLR 2007 [16]].* For color version of this figure, the reader is referred to the online version of this book.

of the particle. It is worth noting that in many cases, these apolipoproteins have opposing effects on the same enzymes or lipid transfer proteins.

HDL composition and function are extraordinarily complex due to the presence of proteins other than the generally accepted apolipoproteins of HDL and the other known HDL–associated enzymes, transfer proteins, and acute phase proteins. Drs. Heinecke, Davidson, Chapman, and their colleagues have recently published data on HDL proteomics [68,69]. The Heinecke group identified 48 different proteins in the HDL_2 and HDL_3 fractions, while the Chapman group found 28 different proteins in the HDL fraction. The large discrepancy in HDL–associated protein numbers is presumably due to the different methods the two groups used for HDL isolation. The Heinecke group separated whole HDL and HDL_3 using horizontal rotor ultracentrifugation [68]. The Chapman group separated five HDL subfractions (light HDL_2b and HDL_2a, and dense HDL_3a, HDL_3b, and HDL_3c) by isopycnic density gradient ultracentrifugation [69]. Both groups measured the tryptic digested peptides by LC–ESI–MS/MS, but the Chapman group used a somewhat different and less sensitive detection method. Heinecke's laboratory published 13 novel HDL–associated proteins, eight of which were not present in HDL_3. Moreover, they illustrated that the HDL–associated proteins were participants in four physiological functions: 1) lipid metabolism, 2) proteinase regulation, 3) complement regulation, and 4) acute-phase response. Chapman's group emphasized the abundance patterns of representative proteins across HDL subpopulations: class A, limited to dense HDL_3; class B, preferably in dense HDL_3 but present in all subfractions; class C, evenly distributed; class D, preferably in light HDL_2 but present in all subfractions; and class E, limited to light HDL_2.

Our laboratory has also investigated proteomics on HDL particles separated by preparative 2d nondenaturing PAGE [70]. This method has the added benefit of avoiding ultracentrifugation. Three of the most abundant apoA-I-containing HDL subclasses (large α-1, medium α-2, and small α-4-α-3) were separated and electro-eluted from the gel under nondenaturing conditions. To remove comigrating proteins, each fraction was absorbed on a mono-specific anti-apoA-I column and eluted with low pH buffer after washing off proteins nonspecifically bound. After tryptic digestion, peptides were analyzed by LC-ESI-MS/MS. We identified 92 different proteins associated with the three HDL subfractions. A total of 67 different proteins were associated with the large α-1 HDL subfraction, 51 of which were associated exclusively with α-1. Twenty-eight different proteins were associated with the medium LpA-I:A-II α-2 HDL subfraction, 16 of which were associated exclusively with this fraction. Twenty-one different proteins were associated with the small combined α-4-α-3 HDL subfractions, nine of which were specific to this HDL subfraction.

Besides lipoprotein metabolism, several of these HDL-associated proteins are involved in: 1) complement regulation ($n = 21$; complement-3C, C4-A, C4-B, H, B, clusterin, clusterin precursors, C1, C4-binding protein, and C5 in decreasing order of abundance); 2) plasma protease activity regulation ($n = 16$; serine protease C1-inhibitor; inter-α-trypsin inhibitor heavy chain H1, H2, and H4; plasmin; serine peptidase inhibitor; vitronectin; antithrombin III; heparin cofactor II; N-acethylmuranoyl-L alanine amidase; and kallikrein were the 10 most abundant); 3) blood coagulation ($n = 7$; fibrinogen β-chain, prothrombin, apoH or β2-glycoprotein 1, C-1 inhibitor vitamin K-dependent protein S, protein S100A-9, coagulation factor XII and XIII); 4) angiogenesis ($n = 3$); 5) cell adhesion ($n = 5$); 6) hypertension (angiotensinogen); 7) cell oxidation ($n = 3$); and 8) hemoglobin metabolism (hemopexin). The number and abundance of the HDL-associated proteins suggest that many of these proteins are on separate, although similar HDL particles, resulting in a high variability in protein composition within HDL subclasses. The binding of a protein, by weak or stronger forces, is presumably influenced by HDL size and by other proteins as well as the lipid composition of the particle. Some proteins attach to HDL by hydrophobic forces while others might use anchor proteins to attach themselves to the particles.

Recently, there has been much interest in determining HDL proteomics in patients with various diseases, including CVD, diabetes, kidney disease, and autoimmune diseases. It became important to identify the proteins and other factors that alter HDL function and that are selectively enriched or depleted in subjects who are at risk for CHD. Several groups have identified nearly 100 proteins associated with HDL using MS-based proteomics [69–72].

Proteomic studies indicate that HDL carries a unique cargo of proteins in patients with CHD and that these proteins might contribute to the altered functional properties of HDL. Oxidation of apolipoproteins and compositional changes could affect the ability of HDL particles to remove cholesterol from macrophages. Vaisar and colleagues carried out shotgun proteomics, which indicated that HDL_3 from CHD patients was enriched in apoC-IV, paraoxonase (PON) 1, complement C3, apoA-IV, and apoE [72]. They also showed that the HDL_2 of CHD patients contained oxidized methionine residues of apoA-I and elevated levels of apoC-III.

The levels of specific HDL proteins, primarily apoL-1, PON1, and PON3, correlated with the potent capacity of HDL_3 to protect LDL from oxidation [73]. HDL is also enriched in SAA1 (an inflammatory protein) in patients with inflammatory disease, correlating with a reduced anti-inflammatory capacity of these HDLs [73].

Recently, a substantial difference in the HDL proteome by mass between coronary artery disease (CAD) patients and unaffected subjects was demonstrated [74]. It was shown that HDL_3 from CHD patients contained a low level of clusterin, which is known to contribute to the cardioprotective properties of HDL. It was also demonstrated that HDL from CAD patients stimulated potential endothelial pro-apoptotic pathways, while HDL from unaffected subjects activated endothelial anti-apoptotic pathways. The remodeled HDL proteome was associated with the pro-apoptotic function of HDL.

A significant problem with proteomic analyses remains that almost all MS-based proteomic studies on HDL to date have utilized density gradient ultracentrifugation techniques for HDL isolation prior to analysis. These involve high share forces (100,000 G) and high salt concentrations that can disrupt HDL composition and protein–protein and lipid–protein interactions and subsequently alter particle functions.

7. HDL FUNCTIONS

The functions of HDL are based on the chemical composition of the particles. The major obstacle for studying HDL and its various subclasses is separating the particles, especially preparatively, for further analyses without altering physical-chemical properties and biological functions. We summarize currently used technologies for HDL subclass separation [75–78] in Table 3.1.

Removing excess cholesterol from lipid-laden macrophages and peripheral tissue is still considered to be a main atheroprotective function of HDL. However, several other proposed functions of HDL might contribute to potential atheroprotective activities.

RCT, which refers to the overall flux of cholesterol from the periphery to the liver and its fecal excretion, was first proposed in the 1960s [79]. HDL can efflux cholesterol from any cell that expresses ABCA-1, ABCG-1, or SR-BI receptors. The most well known and accepted anti-atherogenic function of HDL is its ability to mediate cholesterol efflux from macrophages in the vessel wall, thereby playing an important role in the maintenance of the net cholesterol balance in the arterial wall. Cholesterol efflux from lipid-laden macrophages contributes a relatively minor amount to the cholesterol content of HDL particles and does not have significant influence on HDL remodeling. Cholesterol, carried to the bile by HDL, is no more than 1% of the excreted cholesterol since not all of the cholesterol effluxed by HDL is carried directly to the bile via the SR-BI mechanism: a significant part of HDL-C ends up in LDL as a result of CETP activity.

It has been proposed that the anti-inflammatory function of HDL depends on its capacity to decrease cell cholesterol content. Successful signal transduction relies on proper lipid rafts, containing high concentrations of cholesterol and sphingolipids; therefore it is speculated that perturbation of lipid rafts by HDL decreases cell inflammatory response. Moreover, HDL-mediated efflux of potentially oxidized lipids, many of them inflammatory, from cells and LDL also decreases inflammation. It has been demonstrated in murines that HDL potently blocks experimental endotoxinemia, indicating that HDL binds lipopolysaccharide (LPS), which leads to lower systemic proinflammatory cytokine levels [80]. ApoA-I was identified as the LPS-binding molecule in HDL [81]. Furthermore, HDL was shown to suppress cytokine and chemokine production on monocytes, macrophages, and monocyte-derived dendritic cells, and to downregulate costimulatory molecules and to inhibit antigen presentation [82–84].

One of the initial steps in atherogenesis is the activation of the vascular endothelium, upregulation of adhesion molecules (e.g., E-selectin, intercellular adhesion molecule [ICAM] 1, and vascular cell adhesion molecule [VCAM] 1), and secretion of pro-inflammatory mediators (e.g., monocyte chemoattractant protein [MCP] 1). HDL can inhibit cytokine-induced

Table 3.1 Separating Methods for Lipoprotein Subclasses: Principals, Advantages and Disadvantages of Each Method

Method	Particles	Advantage	Disadvantage
Flotation rate by ultracentrifugation: 1) sequential, 2) isopycnic density-gradient, 3) vertical rotor (VAP).	Vary depending on the applied method and salt gradient.	Preparative, separates all lipoprotein subclasses.	High G-force and high salt (KBr) concentration alter the composition, size, and functionality of HDL.
Net surface charge (agarose, homogeneous native PAGE, isotachophoresis).	Preβ-, α-, and preα-mobility subclasses.	Lipoproteins are not significantly altered.	Low resolution, not preparative, requires specific detection or pre-purification of HDL.
Particle charge and mass ratio (1d or 2d PAGE using native gradient gels (detection by staining or immunoblot).	Preβ, α, and preα-migrating particles with different sizes.	High resolution, little alteration in HDL, quantitative using image analysis.	Not preparative, needs specific detection method.
Particle size by high performance liquid chromatography (FPLC).	Vary.	Fast, preparative.	Low resolution, HDL particles are co-eluted with other lipoproteins and plasma proteins.
Protein composition by immunoaffinity chromatography.	Vary depending on applied antibody.	Fast, preparative.	Highly purified antibodies against the protein of interest are required, specificity is questionable.
Particle size by nuclear magnetic resonance (NMR) measuring the terminal methyl groups on fatty acids.	Large, medium, and small HDL; LDL1, LDL2, VLDL1, VLDL2.	Fast, estimates all lipoprotein subclasses in one step.	Low resolution. Low specificity.
Ion mobility using an ion separation/particle detector system that sorts ions by size and which can count lipoprotein particles over a wide range of sizes.	Large, medium, and small HDL.	Fast.	Method was originally developed for protein separation and has been used for lipoprotein separation only recently; therefore, specificity is not well established.

expression of VCAM-1 and ICAM-1 on human umbilical vein endothelial cells, reducing adhesion and transmigration of monocytes [85,86].

Moreover, HDL is the major carrier of lipid hydroperoxides in human plasma [87]. Lipid hydroperoxides are taken up faster by the liver than non-oxidized lipids [88], suggesting an important role of HDL in reducing oxidative stress. The anti-oxidative properties of HDL involve HDL-associated enzymes, such as PON1, LpPLA$_2$, and LCAT, which have been reported to hydrolyze oxidized phospholipids into lysophosphatidylcholine. In addition, sulfur-containing amino acids (methionine) of HDL-associated apoproteins are able to detoxify redox-active phospholipid hydroperoxides (PLOOH) into redox-inactive phospholipid hydroxides (PLOH). Therefore, it is speculated that a significant portion of HDL's anti-oxidative capacity comes from its main protein: apoA-I.

HDL has been shown to induce endothelial repair by increasing endothelial nitric oxide (NO) production. Several different mechanisms may account for the HDL-promoted increase in NO production. It has been suggested that HDL acts by preventing the detrimental effects of oxidized LDL on endothelial NO-synthase and that HDL binding to endothelial SR-BI directly stimulates NO production [89,90]. Moreover, HDL was shown to remove oxysterols from endothelial cells in an ABCG1-dependent manner, thereby improving endothelial nitric oxide synthase (eNOS) activity [91].

HDL can also improve vascular health by reducing endothelial cell apoptosis, while promoting endothelial cell proliferation and migration. These steps are crucial for neovascularization after a vascular injury. The major pathway for improving vascular health is the removal of oxidized LDL from the vessel wall. In addition, HDL influences vascular smooth muscle cell migration by its sphingosin-1 phosphate cargo [92,93]. An inverse association between HDL-C levels and platelet aggregation has been documented, suggesting an anti-platelet function of HDL. The mechanism involves the downregulation and release of platelet-activating factor from endothelial cells or by activating eNOS [89].

In addition to its direct actions on cells that govern vascular health and disease, HDL may favorably influence glucose homeostasis and thereby has potentially novel functions that indirectly promote vascular health. Both lipid-free apoA-I and HDL stimulate glucose uptake by skeletal muscle myocytes via increasing adenosine monophosphate (AMP)-activated protein kinase activity. HDL also promotes glycogen synthesis by skeletal muscle myocytes, and, via SR-BI, it stimulates glucose uptake by adipocytes

[94,95]. HDL also promotes insulin secretion by pancreatic β cells. This process requires ABCA1-mediated cholesterol efflux, as well as SR-BI expression in pancreatic β-cell [96]. Mice with pancreatic β-cell-specific deletion of ABCA1 display glucose intolerance [97]. In contrast to HDL, VLDL and LDL particles attenuate insulin mRNA levels and β-cell proliferation and cause β-cell apoptosis. HDL antagonizes the pro-apoptotic signaling by VLDL and LDL in an Akt-dependent manner [98,99].

The impact of HDL on glucose homeostasis has also been demonstrated in individuals with Type 2 diabetes. Patients' plasma glucose decreased while insulin level and AMP-activated protein kinase activity in skeletal muscle increased after receiving reconstituted HDL intravenously [100].

It is worth mentioning that if HDL "fails its mission", it is called dysfunctional HDL. Although this terminology is widely used, there is no clear determination of what a dysfunctional HDL is. First, a dysfunctional HDL is the result of a malfunctioning system. The potent atheroprotective functions of HDL particles originate from their unique composition and structure. Therefore, if HDL composition or structure is altered, HDL may lose its normal functions and can even acquire deleterious functions.

The most studied HDL function is its capability of removing cholesterol from cells via at least four mechanisms (ABCA1, ABCG1, SRB1, and augmenting passive cholesterol diffusion from cells by removing extracellular cholesterol and maintaining cholesterol gradient between the interstitial space and cells). The first three mechanisms are HDL-subclass specific. Only preβ-1 is able to efflux cholesterol via the ABCA1 pathway [33]. Cholesterol efflux via ABCG1 is mediated by the larger α-3 and α-2 HDL subclasses. SR-BI mediates bidirectional cholesterol flux and equally works with α-1 and α-2 HDL subclasses. It is clear that changes in the proportion of HDL particles (HDL subpopulation profile) alter the efflux capacity of HDL. However, it is not clear how qualitative (compositional) change, without quantitative (proportional) change, influences HDL's efflux capacity. The compositional changes might involve the protein or the lipid components of HDL or both. Moreover, changes in HDL functions can be the result of chemical modifications of HDL molecules without change in particle composition. The most common alterations are glycation and oxidation of its proteins or lipids.

It is not clear whether glycation of HDL, as measured by glycated apoA-I, changes the cholesterol efflux capacity of HDL. However, in patients with Type 2 diabetes, the content of oxidized fatty acids is increased and the anti-inflammatory and antioxidant activities of HDL are impaired [101]. Pan and colleagues showed that HDL in diabetic patients (diabetic HDL) was dysfunctional in promoting endothelial cell proliferation, migration, and adhesion to matrix, which was associated with the downregulation of SR-BI [102]. It was also found that oxHDL was independently, significantly, and positively correlated with fasting plasma glucose level, suggesting that high plasma glucose levels may contribute to HDL oxidation [103]. Interestingly, it has been reported that diabetic HDL carries higher levels of sphingosine-1 phosphate compared to normal HDL, which has the potential of contributing to protective effects on endothelial cells by inducing cyclooxygenase-2 (COX-2) expression and prostacyclin-2 (PGI-2) release.

Inflammation alters the composition and function of circulating HDL as shown in studies investigating the acute phase response [104,105]. Inflammation can significantly change HDL composition by incorporating SAA1, an acute phase protein carried in the circulation mainly on HDL. SAA1 is a very lipophilic protein and replaces apoA-I on HDL particles. Acute-phase HDL shows a reduced ability to: 1) promote cholesterol efflux from macrophages [106], 2) protect LDL against oxidation, and 3) inhibit cytokine-induced adhesion molecule expression [104].

The majority of the anti-oxidative capacity of HDL is derived from its PON1 activity. However, it is hypothesized that apoA-I and HDL-associated LCAT also significantly contribute to HDL's anti-oxidative capacity. Both the protein and lipid components of HDL can be oxidatively modified. The most deleterious is when apoA-I is oxidized by MPO. MPO produces chlorinated and nitrated apoA-I, which in turn alters normal HDL composition and functions. A fairly recent study demonstrates that MPO binds to apoA-I on HDL and oxidatively modifies it, resulting in loss in cholesterol efflux and LCAT activating capacity [107]. Excess cholesterol in cells triggers cell stress and an inflammatory response. When HDL effluxes excess cell-cholesterol it reduces both. When HDL loses its cholesterol efflux capacity it also loses some of its anti-inflammatory capacity. Moreover, it has recently been demonstrated that HDL nitration and chlorination catalyzed by MPO impairs its effects in promoting endothelial repair [108].

Lipoproteins, including HDL, bind LPS and diminish its ability to stimulate cytokine production in macrophages [109]. HDL-bound LPS does not interact with the cellular receptors on macrophages that induce cytokine production and secretion, thereby decreasing the uptake of LPS by macrophages, which results in a diminished stimulation of cytokine production and reduced toxicity. It is widely accepted that HDL's LPS-binding capacity is positively associated with apoA-I concentration. However, it is also suggested that the second most abundant protein on HDL, apoA-II, may help maintain sensitive host responses to LPS by suppressing LPS binding protein (LBP)–mediated inhibition [110]. In patients with inflammatory conditions, when SAA1 replaces apoA-I, but not apoA-II, HDL becomes dysfunctional or inflammatory. Inflammatory HDL has diminished cholesterol efflux capacity [111]. In addition, SAA1- and apoA-II-enriched HDL might even be proinflammatory. Since the large HDL subclass (HDL$_2$/α-1) carries a large number of proteins involved in the regulation of inflammatory response (complement factors, proteases, and protease inhibitors) [69–71], it is probable that the decrease of this HDL subclass significantly alters HDL's anti-inflammatory capacity.

The hypothesis that HDL is rendered dysfunctional in disease was further supported by the observation that, despite high levels of HDL, CAD patients had less anti-oxidative activity as measured by inhibition of phospholipid oxidation [112], In a series of studies, cohorts of systemic lupus erythematosus, end-stage renal disease, and metabolic syndrome patients showed HDL with decreased anti-oxidant activity [113].

Although these results are very intriguing, more precise characterization of dysfunctional HDL and for novel assays to assess the multiple functional properties of HDL is necesssary.

8. THE FUTURE OF HDL RESEARCH

Efforts for characterizing the HDL proteome confirmed that the role of HDL is much more complex than previously thought. Besides identifying proteins in HDL, it may be important to identify the distribution of a given protein across HDL subspecies or their associations with other proteins. Particular HDL complexes could turn out to be better biomarkers than individual proteins.

The recent failures of niacin and some CETP inhibitors—all capable of significantly raising HDL-C—have cast some doubt on whether HDL plays a truly protective role in CVD. We believe, as many others do, that measurement of one component of HDL (i.e., cholesterol) is not sufficient to capture the cardioprotective potential of HDL. Specific functions, such as cholesterol efflux capacity or anti-inflammatory, anti-oxidatory, and host-defense indices, may be better indicators than the HDL-C value for predicting CAD risk in an individual. It is important to identify those HDL subspecies that are most relevant to CVD protection and focus on therapeutics that specifically target those beneficial fractions.

REFERENCES

[1] Miller NE, Thelle DS, Forde OH, Mjos OD. The Tromsø heart-study. High-density lipoprotein and coronary heart-disease: a prospective case-control study. Lancet 1977;1(8019):965–8.

[2] Gordon DJ, Probstfield JL, Garrison RJ, Neaton JD, Castelli WP, Knoke JD, et al. High-density lipoprotein cholesterol and cardiovascular disease. Four prospective American studies. Circulation 1989;79(1):8–15.

[3] Assmann G, Schulte H. Relation of high-density lipoprotein cholesterol and triglycerides to incidence of atherosclerotic coronary artery disease (the PROCAM experience). Prospective Cardiovascular Münster study. Am J Cardiol 1992;70(7): 733–7.

[4] Wierzbicki AS, Mikhailidis DP, Wray R, Schacter M, Cramb R, Simpson WG, et al. Statin-fibrate combination: therapy for hyperlipidemia: a review. Curr Med Res Opin 2003;19(3):155–68.

[5] AIM-HIGH InvestigatorsBoden WE, Probstfield JL, Anderson T, Chaitman BR, Desvignes-Nickens P, Koprowicz K, et al. Niacin in patients with low HDL cholesterol levels receiving intensive statin therapy. N Engl J Med 2011;365(24):2255–67; Erratum in: N Engl J Med 2012;367(2):189.

[6] Barter PJ, Caulfield M, Eriksson M, Grundy SM, Kastelein JJ, Komajda M ILLUMINATE Investigators, et al. Effects of torcetrapib in patients at high risk for coronary events. N Engl J Med 2007;357(21):2109–22.

[7] Robins SJ, Collins D, Wittes JT, Papademetriou V, Deedwania PC, Schaefer EJVA-HIT Study Group, et al. Veterans Affairs High-Density Lipoprotein Intervention Trial. Relation of gemfibrozil treat and lipid levels with major coronary events: VA-HIT: a randomized controlled trial. JAMA 2001;285(12):1585–91.

[8] Briel M, Ferreira-Gonzalez I, You JJ, Karanicolas PJ, Akl EA, Wu P, et al. Association between change in high density lipoprotein cholesterol and cardiovascular disease morbidity and mortality: systematic review and meta-regression analysis. BMJ 2009;338; b92.

[9] Ballantyne CM, Miller M, Niesor EJ, Burgess T, Kallend D, Stein EA. Effect of dalcetrapib plus pravastatin on lipoprotein metabolism and high-density lipoprotein composition and function in dyslipidemic patients: results of a phase IIb dose-ranging study. Am Heart J 2012;163(3):515–21.

[10] HPS2-THRIVE Collaborative Group. HPS2-THRIVE randomized placebo-controlled trial in 25 673 high-risk patients of ER niacin/laropiprant: trial design, pre-specified muscle and liver outcomes, and reasons for stopping study treatment. Eur Heart J 2013;34(17):1279–91.

[11] Santos RD, Schaefer EJ, Asztalos BF, Polisecki E, Wang J, Hegele RA, et al. Characterization of high density lipoprotein particles in familial apolipoprotein A-I deficiency. J Lipid Res 2008;49(2):349–57.

[12] Puchois P, Kandoussi A, Fievet P, Fourrier JL, Bertrand M, Koren E, et al. Apolipoprotein A-I containing lipoproteins in coronary artery disease. Atherosclerosis 1987;68(1–2):35–40.

[13] Kostner G, Alaupovic P. Studies of the composition and structure of plasma lipoproteins. Separation and quantification of the lipoprotein families occurring in the high density lipoproteins of human plasma. Biochemistry 1972;11(18):3419–28.

[14] Phillips JC, Wriggers W, Li Z, Jonas A, Schulten K. Predicting the structure of apolipoprotein A-1 in reconstituted high-density lipoprotein disks. Biophys J 1997;73: 2337–46.

[15] Segrest JP, Jones MK, Klon AE, Sheldahl CJ, Hellinger M, De Loof HM, et al. A detailed molecular belt model for apolipoprotein A-I in discoidal high density lipoprotein. J Biol Chem 1999;274(45):31755–8.

[16] Asztalos BF, Schaefer EJ, Horvath KV, Yamashita S, Miller M, Franceschini G, et al. Role of LCAT in HDL remodeling: investigation of LCAT deficiency states. J Lipid Res 2007;48(3):592–9.

[17] Schaefer EJ, Santos RD, Asztalos BF. Marked HDL deficiency and premature coronary heart disease. Curr Opin Lipidol 2010;21(4):289–97; Review.

[18] Schaefer EJ, Heaton WH, Wetzel MG, Brewer Jr HB. Plasma apolipoprotein A-I absence associated with marked reduction of high density lipoproteins and premature coronary artery disease. Arteriosclerosis 1982;2:16–26.

[19] Schaefer EJ. The clinical, biochemical, and genetic features in familial disorders of high density lipoprotein deficiency. Arteriosclerosis 1984;4:303–22.

[20] Schaefer EJ, Ordovas JM, Law S, Ghiselli G, Kashyap ML, Srivastava LS, et al. Familial apolipoprotein A-I and C-III deficiency, variant II. J Lipid Res 1985;26:1089–101.

[21] Ordovas JM, Cassidy DK, Civeira F, Bisgaier CL, Schaefer EJ. Familial apolipoprotein A-I, C-III, and A-IV deficiency with marked high density lipoprotein deficiency and premature atherosclerosis due to a deletion of the apolipoprotein A-I, C-III, and A-IV gene complex. J Biol Chem 1989;264:16339–42.

[22] Norum RA, Lakier JB, Goldstein S, Angel A, Goldberg AB, Block WD, et al. Familial deficiency of apolipoproteins A-I and C-III and precocious coronary artery disease. N Engl J Med 1982;306:1513–9.

[23] Karathanasis SK, Norum RA, Zannis VI, Breslow JL. An inherited polymorphism in the human apolipoprotein A-I gene locus related to the development of atherosclerosis. Nature 1983;301:718–20.

[24] Karathanasis SK, Ferris E, Haddad EA. DNA inversion within the apolipoprotein AI/CIII/AIV-encoding gene cluster of certain patients with premature atherosclerosis. Proc Natl Acad Sci USA 1987;84:7198–202.

[25] Forte TM, Nichols AV, Krauss RM, Norum RA. Familial apolipoprotein AI and apolipoprotein CIII deficiency. Subclass distribution, composition, and morphology of lipoproteins in a disorder associated with premature atherosclerosis. J Clin Invest 1984;74:1601–13.

[26] Sorci-Thomas MG, Thomas MJ. The effects of altered apolipoprotein A-I structure on plasma HDL concentration. Trends Cardiovasc Med 2002;12:121–8.

[27] Zannis VI, Chroni A, Krieger M. Role of apoA-I, ABCA1, LCAT, and SR-BI in the biogenesis of HDL. J Mol Med 2006;84:276–94.

[28] Frank PG, Marcel YL, Apolipoprotein A-I: structure-function relationships. J Lipid Res 2000;41:853–72.

[29] Orso E, Broccardo C, Kaminske WE, Bottcher A, Liebisch G, Drobnik W, et al. Transport of lipids from golgi to plasma membrane is defective in Tangier disease patients and ABCA-1-deficient mice. Nat Genet 2000;24:192–6.

[30] Lin YC, Ma C, Hsu WC, Lo HF, Yang VC. Molecular interaction between caveolin-1 and ABCA1 on high-density lipoprotein-mediated cholesterol efflux in aortic endothelial cells. Cardiovasc Res 2007;75:575–83.

[31] Rust S, Rosier M, Funke H, Real J, Amoura Z, Piette JC, et al. Tangier disease is caused by mutations in the gene encoding ATP-binding cassette transporter 1. Nat Genet 1999;22:352–5.

[32] Asztalos BF, Brousseau ME, McNamara JR, Horvath KV, Roheim PS, Schaefer EJ. Subpopulations of high density lipoproteins in homozygous and heterozygous Tangier disease. Atherosclerosis 2001;156(1):217–25.

[33] Asztalos BF, de la Llera-Moya M, Dallal GE, Horvath KV, Schaefer EJ, Rothblat GH. Differential effects of HDL subpopulations on cellular ABCA1- and SR-BI-mediated cholesterol efflux. J Lipid Res 2005;46(10):2246–53.

[34] Brunham LR, Singaraja RR, Duong M, Timmins JM, Fievet C, Bissada N, et al. Tissue-specific roles of ABCA1 influence susceptibility to atherosclerosis. Arterioscler Thromb Vasc Biol 2009;29(4):548–54.

[35] McLean J, Fielding C, Drayna D, Dieplinger H, Baer B, Kohr W, et al. Cloning and expression of human lecithin-cholesterol acyltransferase cDNA. Proc Natl Acad Sci USA 1986;83:2335–9.

[36] Fielding CJ, Fielding PE. Molecular physiology of reverse cholesterol transport. J Lipid Res 1995;36:211–28.

[37] Czarnecka H, Yokoyama S. Regulation of cellular cholesterol efflux by lecithin:cholesterol acyltransferase reaction through nonspecific lipid exchange. J Biol Chem 1996;266:2023–8.

[38] Kennedy MA, Venkateswaran A, Tarr PT, Xenarios I, Kudoh J, Shimizu N, et al. Characterization of the human ABCG1 gene: liver X receptor activates an internal promoter that produces a novel transcript encoding an alternative form of the protein. J Biol Chem 2001;276:39438–47.

[39] Klucken J, Buchler C, Orso E, Kaminski WE, Porsch-Ozcurumez M, Liebisch G, et al. ABCG1 (ABC8), the human homolog of the Drosophila white gene, is a regulator of macrophage cholesterol and phospholipid transport. Proc Natl Acad Sci USA 2000;97(2):817–22.

[40] Kobayashi A, Takanezawa Y, Hirata T, Shimizu Y, Misasa K, Kioka N, et al. Efflux of sphingomyelin, cholesterol, and phosphatidylcholine by ABCG1. J Lipid Res 2006;47(8):1791–802.

[41] Sankaranarayanan S, Oram JF, Asztalos BF, Vaughan AM, Lund-Katz S, Adorni MP, et al. Effects of acceptor composition and mechanism of ABCG1-mediated cellular free cholesterol efflux. J Lipid Res 2009;50(2):275–84.

[42] Asztalos BF, Horvath KV, Kajinami K, Nartsupha C, Cox CE, Batista M, et al. Apolipoprotein composition of HDL in cholesteryl ester transfer protein deficiency. J Lipid Res 2004;45(3):448–55.

[43] Zambon A, Deeb SS, Bensadoun A, Foster KE, Brunzell JD. In vivo evidence of a role for hepatic lipase in human apoB-containing lipoprotein metabolism, independent of its lipolytic activity. J Lipid Res 2000;41(12):2094–9.

[44] Webb NR, de Beer MC, Asztalos BF, Whitaker N, van der Westhuyzen DR, de Beer FC. Remodeling of HDL remnants generated by scavenger receptor class B type I. J Lipid Res 2004;45(9):1666–73.

[45] Jaye M, Lynch KJ, Krawiec J, Marchadier D, Maugeais C, Doan K, et al. A novel endothelial-derived lipase that modulates HDL metabolism. Nat Genet 1999;21:424–8.

[46] Jahangiri A, Rader DJ, Marchadier D, Curtiss LK, Bonnet DJ, Rye KA. Evidence that endothelial lipase remodels high density lipoproteins without mediating the dissociation of apolipoprotein A-I. J Lipid Res 2005;46:896–903.

[47] Settasatian N, Duong M, Curtiss LK, Ehnholm C, Jauhiainen M, Huuskonen J, et al. The mechanism of the remodeling of high density lipoproteins by phospholipid transfer protein. J Biol Chem 2001;276:26898–905.

[48] van Haperen R, Samyn H, Moerland M, van Gent T, Peeters M, Grosveld F, et al. Elevated expression of phospholipid transfer protein in bone marrow derived cells causes atherosclerosis. PLoS One 2008;3:e2255.

[49] Valenta DT, Ogier N, Bradshaw G, Black AS, Bonnet DJ, Lagrost L, et al. Atheroprotective potential of macrophage-derived phospholipid transfer protein in low-density lipoprotein receptor-deficient mice is overcome by apolipoprotein AI overexpression. Arterioscler Thromb Vasc Biol 2006;26:1572–8.

[50] Caiazza D, Jahangiri A, Rader DJ, Marchadier D, Rye KA. Apolipoproteins regulate the kinetics of endothelial lipase-mediated hydrolysis of phospholipids in reconstituted high-density lipoproteins. Biochemistry 2004;43:11898–905.

[51] Glass C, Pittman RC, Civen M, Steinberg D. Uptake of high-density lipoprotein-associated apoprotein A-I and cholesterol esters by 16 tissues of the rat in vivo and by adrenal cells and hepatocytes in vitro. J Biol Chem 1985;260(2):744–50.

[52] Le NA, Ginsberg HN. Heterogeneity of apolipoprotein A-I turnover in subjects with reduced concentrations of plasma high density lipoprotein cholesterol. Metabolism 1988;37(7):614–7.

[53] Jäckle S, Rinninger F, Lorenzen T, Greten H, Windler E. Dissection of compartments in rat hepatocytes involved in the intracellular trafficking of high-density lipoprotein particles or their selectively internalized cholesteryl esters. Hepatology 1993;17(3): 455–65.

[54] Acton S, Rigotti A, Landschulz KT, Xu S, Hobbs HH, Krieger M. Identification of scavenger receptor SR-BI as a high density lipoprotein receptor. Science 1996;271(5248):518–20.

[55] Rigotti A, Trigatti BL, Penman M, Rayburn H, Herz J, Krieger M. A targeted mutation in the murine gene encoding the high density lipoprotein (HDL) receptor scavenger receptor class B type I reveals its key role in HDL metabolism. Proc Natl Acad Sci USA 1997;94(23):12610–5.

[56] Hammad SM, Barth JL, Knaak C, Argraves WS. Megalin acts in concert with cubilin to mediate endocytosis of high density lipoproteins. J Biol Chem 2000;275(16): 12003–8.

[57] de Beer MC, Yuan T, Kindy MS, Asztalos BF, Roheim PS, de Beer FC. Characterization of constitutive human serum amyloid A protein (SAA4) as an apolipoprotein. J Lipid Res 1995;36(3):526–34.

[58] Jahangiri A, de Beer MC, Noffsinger V, Tannock LR, Ramaiah C, Webb NR, et al. HDL remodeling during the acute phase response. Arterioscler Thromb Vasc Biol 2009;29(2):261–7.

[59] Asztalos BF, Horvath KV, McNamara JR, Roheim PS, Rubinstein JJ, Schaefer EJ. Effects of atorvastatin on the HDL subpopulation profile of coronary heart disease patients. J Lipid Res 2002;43(10):1701–7.

[60] Asztalos BF, Collins D, Horvath KV, Bloomfield HE, Robins SJ, Schaefer EJ. Relation of gemfibrozil treatment and high-density lipoprotein subpopulation profile with cardiovascular events in the Veterans Affairs High-Density Lipoprotein Intervention Trial. Metabolism 2008;57(1):77–83.

[61] Lamon-Fava S, Diffenderfer MR, Barrett PH, Buchsbaum A, Matthan NR, Lichtenstein AH, et al. Effects of different doses of atorvastatin on human apolipoprotein B-100, B-48, and A-I metabolism. J Lipid Res 2007;48(8):1746–53.

[62] Segrest JP, Jones MK, Klon AE, Sheldahl CJ, Hellinger M, De Loof H, et al. A detailed molecular belt model for apolipoprotein A-I in discoidal high density lipoprotein. J Biol Chem 1999;274:31755–8.

[63] Davidson WS, Thompson TB. The structure of apolipoprotein A-I in high density lipoproteins. J Biol Chem 2007;282:22249–53.

[64] Thomas MJ, Bhat S, Sorci-Thomas MG. Three-dimensional models of HDL apoA-I: implications for its assembly and function. J Lipid Res 2008;49(9):1875–83.

[65] Gao X, Yuan S, Jayaraman S, Gursky O. Role of apolipoprotein A-II in the structure and remodeling of human high-density lipoprotein (HDL): protein conformational ensemble on HDL. Biochemistry 2012;51(23):4633–41.

[66] Segrest JP, Jones MK, Catte A, Thirumuruganandham SP. Validation of previous computer models and MD simulations of discoidal HDL by a recent crystal structure of apoA-I. J Lipid Res 2012;53(9):1851–63.

[67] Huang R, Silva RA, Jerome WG, Kontush A, Chapman MJ, Curtiss LK, et al. Apolipoprotein A-I structural organization in high-density lipoproteins isolated from human plasma. Nat Struct Mol Biol 2011;18(4):416–22.

[68] Vaisar T, Pennathur S, Green PS, Gharib SA, Hoofnagle AN, Cheung MC, et al. Shotgun proteomics implicates protease inhibition and complement activation in the anti-inflammatory properties of HDL. J Clin Invest 2007;117(3):746–56.

[69] Davidson WS, Silva RA, Chantepie S, Lagor WR, Chapman MJ, Kontush A. Proteomic analysis of defined HDL subpopulations reveals particle-specific protein clusters: relevance to antioxidative function. Arterioscler Thromb Vasc Biol 2009;29(6):870–6.

[70] Asztalos BF. High density lipoprotein particles. In: Schaefer EJ, editor. High Density Lipoproteins, Dyslipidemia, and Coronary Heart Disease. New York: Springer; 2010. p. 25–32.

[71] Gordon S, Durairaj A, Lu JL, Davidson WS. High-density lipoprotein proteomics: identifying new drug targets and biomarkers by understanding functionality. Curr Cardiovasc Risk Rep 2010;4(1):1–8.

[72] Vaisar T, Mayer P, Nilsson E, Zhao XQ, Knopp R, Prazen BJ. HDL in humans with cardiovascular disease exhibits a proteomic signature. Clin Chim Acta 2010; 411(13-14):972–9.

[73] Tölle M, Huang T, Schuchardt M, Jankowski V, Prüfer N, Jankowski J, et al. High-density lipoprotein loses its anti-inflammatory capacity by accumulation of pro-inflammatory-serum amyloid A. Cardiovasc Res 2012;94(1):154–62.

[74] Riwanto M, Rohrer L, Roschitzki B, Besler C, Mocharla P, Mueller M, et al. Altered activation of endothelial anti- and proapoptotic pathways by high-density lipoprotein from patients with coronary artery disease: role of high-density lipoprotein-proteome remodeling. Circulation 2013;127(8):891–904.

[75] Chapman MJ, Goldstein S, Lagrange D, Laplaud PM. A density gradient ultracentrifugal procedure for the isolation of the major lipoprotein classes from human serum. J Lipid Res 1981;22(2):339–58.

[76] Asztalos BF, Roheim PS, Milani RL, Lefevre M, McNamara JR, Horvath KV, et al. Distribution of ApoA-I-containing HDL subpopulations in patients with coronary heart disease. Arterioscler Thromb Vasc Biol 2000;20(12):2670–6.

[77] Ala-Korpela M, Korhonen A, Keisala J, Hörkkö S, Korpi P, Ingman LP, et al. 1H NMR-based absolute quantitation of human lipoproteins and their lipid contents directly from plasma. J Lipid Res 1994;35(12):2292–304.

[78] Caulfield MP, Li S, Lee G, Blanche PJ, Salameh WA, Benner WH, et al. Direct determination of lipoprotein particle sizes and concentrations by ion mobility analysis. Clin Chem 2008;54:1307–16.

[79] Glomset JA, Wright JL. Some properties of a cholesterol esterifying enzyme in human plasma. Biochim Biophys Acta 1964;89:266–76.

[80] Levine DM, Parker TS, Donnelly TM, Walsh A, Rubin AL. In vivo protection against endotoxin by plasma high density lipoprotein. Proc Natl Acad Sci USA 1993; 90(24):12040–4.

[81] Grunfeld C, Feingold KR. HDL and innate immunity: a tale of two apolipoproteins. J Lipid Res 2008;49(8):1605–6.

[82] Barter PJ, Nicholls S, Rye KA, Anantharamaiah GM, Navab M, Fogelman AM. Anti-inflammatory properties of HDL. Circ Res 2004;95(8):764–72.

[83] Säemann MD, Poglitsch M, Kopecky C, Haidinger M, Hörl WH, Weichhart T. The versatility of HDL: a crucial anti-inflammatory regulator. Eur J Clin Invest 2010;40(12):1131–43.

[84] Norata GD, Catapano AL. HDL and adaptive immunity: a tale of lipid rafts. Atherosclerosis 2012;225(1):34–5.

[85] Cockerill GW, Rye KA, Gamble JR, Vadas MA, Barter PJ. High-density lipoproteins inhibit cytokine-induced expression of endothelial cell adhesion molecules. Arterioscler Thromb Vasc Biol 1995;15(11):1987–94.

[86] Calabresi L, Franceschini G, Sirtori CR, De Palma A, Saresella M, Ferrante P, et al. Inhibition of VCAM-1 expression in endothelial cells by reconstituted high density lipoproteins. Biochem Biophys Res Commun 1997;238(1):61–5.

[87] Bowry VW, Stanley KK, Stocker R. High density lipoprotein is the major carrier of lipid hydroperoxides in human blood plasma from fasting donors. Proc Natl Acad Sci USA 1992;89(21):10316–20.

[88] Sattler W, Stocker R. Greater selective uptake by Hep G2 cells of high-density lipoprotein cholesteryl ester hydroperoxides than of unoxidized cholesteryl esters. Biochem J 1993;294:771–8.

[89] Uittenbogaard A, Shaul PW, Yuhanna IS, Blair A, Smart EJ. High density lipoprotein prevents oxidized low density lipoprotein-induced inhibition of endothelial nitric-oxide synthase localization and activation in caveolae. J Biol Chem 2000;275(15): 11278–83.

[90] Yuhanna IS, Zhu Y, Cox BE, Hahner LD, Osborne-Lawrence S, Lu P, et al. High-density lipoprotein binding to scavenger receptor-BI activates endothelial nitric oxide synthase. Nat Med 2001;7(7):853–7.

[91] Terasaka N, Yu S, Yvan-Charvet L, Wang N, Mzhavia N, Langlois R, et al. ABCG1 and HDL protect against endothelial dysfunction in mice fed a high-cholesterol diet. J Clin Invest 2008;118(11):3701–13.

[92] Tamama K, Tomura H, Sato K, Malchinkhuu E, Damirin A, Kimura T, et al. High-density lipoprotein inhibits migration of vascular smooth muscle cells through its sphingosine 1-phosphate component. Atherosclerosis 2005;178(1):19–23.

[93] Damirin A, Tomura H, Komachi M, Liu JP, Mogi C, Tobo M, et al. Role of lipoprotein-associated lysophospholipids in migratory activity of coronary artery smooth muscle cells. Am J Physiol Heart Circ Physiol 2007;292(5):H2513–22.

[94] Han R, Lai R, Ding Q, Wang Z, Luo X, Zhang Y, et al. Apolipoprotein A-I stimulates AMP-activated protein kinase and improves glucose metabolism. Diabetologia 2007;50(9):1960–8.

[95] Zhang Q, Zhang Y, Feng H, Guo R, Lin L, Wan R, et al. High density lipoprotein (HDL) promotes glucose uptake in adipocytes and glycogen synthesis in muscle cells. PLoS One 2011;6:e23556.

[96] Fryirs MA, Barter PJ, Appavoo M, Tuch BE, Tabet F, Heather AK, et al. Effects of high-density lipoproteins on pancreatic beta-cell insulin secretion. Arterioscler Thromb Vasc Biol 2010;30(8):1642–8.

[97] Brunham LR, Kruit JK, Pape TD, Timmins JM, Reuwer AQ, Vasanji Z, et al. Beta-cell ABCA1 influences insulin secretion, glucose homeostasis and response to thiazolidinedione treatment. Nat Med 2007;13(3):340–7.

[98] Roehrich ME, Mooser V, Lenain V, Herz J, Nimpf J, Azhar S, et al. Insulin-secreting beta-cell dysfunction induced by human lipoproteins. J Biol Chem 2003;278: 18368–75.

[99] Rütti S, Ehses JA, Sibler RA, Prazak R, Rohrer L, Georgopoulos S, et al. Low- and high-density lipoproteins modulate function, apoptosis, and proliferation of primary human and murine pancreatic beta-cells. Endocrinology 2009;150:4521–30.

[100] Drew BG, Duffy SJ, Formosa MF, Natoli AK, Henstridge DC, Penfold SA, et al. High-density lipoprotein modulates glucose metabolism in patients with type 2 diabetes mellitus. Circulation 2009;119:2103–11.

[101] Navab M, Reddy ST, Van Lenten BJ, Fogelman AM. HDL and cardiovascular disease: atherogenic and atheroprotective mechanisms. Nat Rev Cardiol 2011;8(4): 222–32.

[102] Pan B, Ma Y, Ren H, He Y, Wang Y, Lv X, et al. Diabetic HDL is dysfunctional in stimulating endothelial cell migration and proliferation due to down regulation of SR-BI expression. PLoS One 2012(11);7:e48530.

[103] Kotani K, Sakane N, Ueda M, Mashiba S, Hayase Y, Tsuzaki K, et al. Oxidized high-density lipoprotein is associated with increased plasma glucose in non-diabetic dyslipidemic subjects. Clin Chim Acta 2012;414:125–9.

[104] Van Lenten BJ, Hama SY, de Beer FC, Stafforini DM, McIntyre TM, Prescott SM, et al. Anti-inflammatory HDL becomes pro-inflammatory during the acute phase response. Loss of protective effect of HDL against LDL oxidation in aortic wall cell cocultures. J Clin Invest 1995;96(6):2758–67.

[105] Van Lenten BJ, Navab M, Shih D, Fogelman AM, Lusis AJ. The role of high-density lipoproteins in oxidation and inflammation. Trends Cardiovasc Med 2001;11(3-4): 155–61.

[106] Artl A, Marsche G, Lestavel S, Sattler W, Malle E. Role of serum amyloid A during metabolism of acute-phase HDL by macrophages. Arterioscler Thromb Vasc Biol 2000;20(3):763–72.

[107] Zheng L, Nukuna B, Brennan ML, Sun M, Goormastic M, Settle M, et al. Apolipoprotein A-I is a selective target for myeloperoxidase-catalyzed oxidation and functional impairment in subjects with cardiovascular disease. J Clin Invest 2004;114(4):529–41.

[108] Pan B, Yu B, Ren H, Willard B, Pan L, Zu L, et al. Eugene Chen Y, Pennathur S, Zheng L. High density lipoprotein (HDL) nitration and chlorination catalyzed by myeloperoxidase impairs its effects in promoting endothelial repair. Free Radic Biol Med 2013;60:272–81.

[109] Cavaillon JM, Fittin C, Cavaillon NH, Kirsch SJ, Warren HS. Cytokine response by monocytes and macrophages to free and lipoprotein-bound lipopolysaccharide. Infect Immun 1990;58:2375–82.

[110] Thompson PA, Berbée JF, Rensen PC, Kitchens RL. Apolipoprotein A-II augments monocyte responses to LPS by suppressing the inhibitory activity of LPS-binding protein. Innate Immun 2008;14(6):365–74.

[111] de la Llera Moya M, McGillicuddy FC, Hinkle CC, Byrne M, Joshi MR, Nguyen V, et al. Inflammation modulates human HDL composition and function in vivo. Atherosclerosis 2012;222(2):390–4.

[112] Ansell BJ, Navab M, Watson KE, Fonarow GC, Fogelman AM. Anti-inflammatory properties of HDL. Rev Endocr Metab Disord 2004;5(4):351–8.

[113] Navab M, Reddy ST, Van Lenten BJ, Anantharamaiah GM, Fogelman AM. The role of dysfunctional HDL in atherosclerosis. J Lipid Res 2009;50(Suppl):S145–9.

CHAPTER 4

Reverse Cholesterol Transport in HDL Metabolism: Modulation of Structural and Functional Features of HDL Particles

Elise F. Villard, Maryse Guerin
UPMC Université Pierre et Marie Curie, Hôpital de la Pitié, Paris, France

Contents

Abstract

On the basis of epidemiological studies, a low high-density lipoprotein cholesterol (HDL-C) level is considered as an independent risk factor for the development of cardiovascular disease (CVD). Inversely, increased HDL-C levels correlate with decreased CVD risk. In consequence, emphasis has been given to HDL particles and their role in protection from atherosclerosis development. The reverse cholesterol transport (RCT) pathway represents the primary mechanism by which HDL may induce plaque regression. This pathway regulates the recycling and elimination of excess cholesterol from peripheral tissues and thus represents a rescue mechanism from deposition of cholesterol in the intima of vessel wall. Targeting HDL to decrease CVD risk and cardiovascular events has become a major focus for CVD research. However, identification of the real HDL-correlated parameter allowing the decrease of CV risk is not an easy task. HDL particles represent in fact a cluster of possible candidate targets. Level, functionality, structure, composition, and size of individual HDL subfractions are not independent one from the other. Physiological or pharmacological induced changes in one of these parameters directly impact the others, suggesting that therapeutic strategies based on HDL particles need a multilevel approach. The purpose of this chapter is to analyze the role of HDL particles in the RCT process, going through the relative importance of HDL structure and function.

1. OVERALL MECHANISM OF THE REVERSE CHOLESTEROL TRANSPORT

Since an excess of cholesterol in cells is toxic and because peripheral cells are not able to catabolize cholesterol, with the exception of steroidogenic tissues, each cell must balance its internal and external sources in order to avoid lack of or overaccumulation of cholesterol. In humans, an excess of sterol production or absorption can be deleterious by contributing to the development of atherosclerosis following accumulation of lipid-rich macrophages at the level of the vessel wall. In this context, there is a physiological need for a centripetal movement of cholesterol from peripheral cells via the plasma compartment to the liver; this is called reverse cholesterol transport (RCT).

RCT constitutes an efficient pathway by which excess cholesterol can be removed from the body after degradation as bile acids or recycled in newly secreted plasma lipoproteins. The general sequence of events occurring in RCT is shown in Figure 4.1. Apolipoprotein A-I (apoA-I) secreted by the liver and the intestine takes out cholesterol from cells. This represents the first step of RCT, called cellular cholesterol efflux, which is mediated by a transporter belonging to the ATP binding cassette family, ABCA1. Following acquisition of lipids and subsequent esterification of free cholesterol

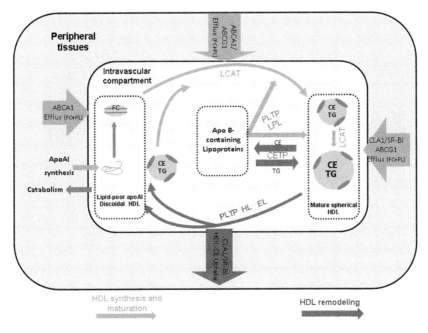

Figure 4.1 *Schematic Representation of the Reverse Cholesterol Transport Pathway.* ApoA-I secreted by the liver and the intestine is lipidated by ABCA1. Subsequent esterification of FC by LCAT allows maturation of nascent HDL into spherical HDL particles. FC efflux from plasma membrane equally occurs through ABCG1 and CLA-1/SR-BI. Cholesterol associated with mature HDL can deliver cholesterol to the liver by a selective uptake process and direct interaction with the CLA-1/SR-BI receptor. HDL-CE can be transferred to apoB-containing lipoproteins (TGRL, LDL) through the action of the CETP, with ultimate uptake by LDL receptor in the liver. This latter process represents the major route for the return of cholesterol to the liver in humans.

by lecithin:cholesterol acyltransferase (LCAT), partially lipidated apoA-I matures into spherical particles, thus preventing re-entry of free cholesterol into the cells. Cholesterol efflux from plasma membrane also occurs toward mature HDL through another member of the ABC family (ABCG1) and the CLA-1 (CD36 and LIMPII analogous-1; human ortholog of the scavenger receptor class B type I [SR-BI]). Passive diffusion equally contributes to free cholesterol desorption from plasma membrane. Cholesterol associated with mature HDL can deliver cholesterol to the liver by a selective uptake process and direct interaction with the CLA-1/SR-BI receptor, or, to a lesser degree, via the LDL receptor (LDLR), when HDL particles contain apoE. It has been demonstrated that the membrane-associated ecto-F_1-ATPase, which binds apolipoprotein A-I, contributes to the uptake of both protein and lipid moieties of the HDL particle, a process called holo-HDL endocytosis. Cholesteryl esters (CE) of HDL particles can be transferred to apoB-containing lipoproteins through the action of the cholesteryl ester transfer protein (CETP), with ultimate uptake by LDLR in the liver. The latter process, which accounts for up to 50% of RCT in humans, represents the major route for the return of cholesterol to the liver.

2. RELATIONSHIP BETWEEN HDL STRUCTURE AND FUNCTION IN REVERSE CHOLESTEROL TRANSPORT

As a result of differences in their relative content in apolipoproteins and lipids, circulating HDL particles are highly heterogeneous in their physico-chemical properties, metabolism, and biological activity. Several enzymes and proteins are associated with HDL particles and are responsible for various functions of HDL particles. Lipids of HDL represent phospholipids and free cholesterol in the amphipathic outer layer and esterified cholesterol and triglycerides within the neutral lipid core of the particles. Distinct content in lipids and proteins result in numerous HDL subclasses, each characterized by differences in shape, density, size, charge, and anti-atherogenicity.

2.1. HDL Lipid Component Influencing RCT

2.1.1. Phospholipids

Phosphatidylcholine (PC) and sphingomyelin (SM) are the two major phospholipids of HDL. These lipids are located in the surface monolayer of the particle together with the free cholesterol and apoA-I. A significant portion of the surface constituents of HDL is derived from the hydrolysis of fasting and postprandial triglyceride-rich lipoproteins (TGRL), and transfer of redundant constituents is enhanced by phospholipid transfer protein (PLTP). PC represents the main phospholipid subclass present in HDL. Phospholipids on the lipoprotein surface serve to solubilize cholesterol. ABCG1 has been reported to mediate cellular efflux of both cholesterol and phospholipids (PC and SM) to preβ and mature HDL. It is likely that ABCG1 preferentially effluxes SM as compared to PC [1]. There are several studies supporting the concept that HDL-PLs play a key role in HDL function, that is, FC efflux and CE uptake in RCT. Indeed, the ability of a given particle to accept FC is related to the amount of PL and to the types of PL present in the HDL. In addition, HDL-PLs strongly correlate with FC efflux to serum. Moreover, total serum from hypertriglyceridemic subjects displayed a similar FC efflux capacity as compared to control normolipidemic subjects despite low HDL-C level but normal HDL-PL level [2]. The capacity of total serum to promote FC efflux via CLA-1/SR-BI can be markedly enhanced by the enrichment with PL. The phospholipid's fatty acyl composition of lipoproteins is known to have subtle but measurable effects on the fluidity of the lipoprotein phospholipidic layer as a result of its impact on apoA-I conformation [3]. These changes may affect the ability of HDL particles to accommodate FC molecules that have desorbed from peripheral cells. Sola and colleagues [4] demonstrated that increasing the percentage of saturated acyl chains in HDL reduces its ability to act as an FC acceptor as a consequence of a decrease in fluidity of the particle. Enrichment of HDL with sphingomyelin causes a decreased rate of FC desorption from the surface of the HDL, resulting from the high affinity of SM for cholesterol. Finally, it has been shown that degradation of SM in HDL favors LCAT activity, suggesting that SM may represent an important factor in influencing RCT at different levels [5].

2.1.2. Triglycerides

Plasma HDL-C levels are inversely correlated with plasma TG concentrations. Equally, apoA-I residence time is inversely correlated to plasma TG levels in normolipidemic subjects, and fractional catabolic rate of apoA-I is increased in Type 2 diabetes associated with hypertriglyceridemia. A low HDL-C phenotype in hypertriglyceridemic states may arise from insufficient availability of surface components necessary for HDL formation due to dysfunctional lipolysis of TGRL, or from increased CETP activity, which reduces the cholesterol content of HDL with a concomitant increase in TG. Thus, enrichment of the TG of HDL particles may alter their metabolism and function. In particular, HDL-CE selective uptake is strongly influenced by the

CE/TG ratio of HDL core with an impaired CLA-1/SR-BI dependent HDL-CE delivery as a function of the increase in HDL-TG content [6].

2.2. HDL-Associated Apolipoproteins Influencing RCT

2.2.1. Apolipoprotein A-I

Apolipoprotein A-I fractional catabolic rate mostly determines circulating HDL levels. ApoA-I and apoA-II are the major protein moieties of HDL, and approximately 90% of total apoA-I and apoA-II is found within HDL density range. ApoA-I is synthesized in the liver and small intestine, has a molecular weight of 28.000 kDa, and, in association with lipids, antiparallel dimers of apoA-I form an extended "belt" around the periphery of both spherical lipoproteins and bilayer disc complexes with hydrophobic regions of protein in contact with a lipid surface [7]. There is a strong link between plasma levels of HDL-C and apoA-I residence time, which shows a significant inverse correlation with the apoA-I fractional catabolic rate (FCR). ApoA-I residence time positively correlates with major lipids of HDL (phospholipids and cholesterol) and, as indicated earlier, inversely with TG. Horowitz and colleagues [8] demonstrated that subjects with low plasma HDL-C levels displayed an increased apoA-I FCR as a result of a more loosely bound and more easily exchangeable apoA-I which is more rapidly cleared by the kidney. Small HDL particles have a decreased core lipid content that could alter the particle's stability and apoA-I conformation, which has been shown to be very sensitive to core lipid composition [9]. Subjects with elevated concentrations of TGRL are known to have decreased concentrations of plasma apoA-I. In particular, such patients display increased plasma concentrations of lipoprotein-unassociated apoA-I. In this context, apoA-I may be dissociated from HDL during lipolysis or CETP-mediated neutral lipid transfer between HDL and TGRL. Subsequently, lipoprotein-unassociated apoA-I might be rapidly cleared by the kidney resulting in low plasma apoA-I concentration in hypertriglyceridemic states. From studies conducted in transgenic mice, it has been proposed that HDL carrying only apolipoprotein A-I has an anti-atherogenic effect, in contrast to HDL transporting both apoA-I and apoA-II. Moreover, in humans preβ-HDL, which is considered as the preferential acceptor of cholesterol upon efflux from peripheral tissues, has until now been shown to contain only apoA-I.

2.2.2. Apolipoprotein A-II

Apolipoprotein A-II is the second major protein constituent of HDL, accounting for about 20% of total HDL protein content. The mean apoA-II plasma concentration in normolipidemic subjects is about 30–35 mg/dL although more than 20% of patients with coronary artery disease display concentrations of apoA-II between 40 and 60 mg/dL. ApoA-II is mainly associated with HDL, and only a small fraction is associated with chylomicrons and very low-density lipoprotein (VLDL). Intravascular remodeling of HDL particles appears to be determined by the impact of apoA-II on the activity of lipid transfer proteins, enzymes, or receptors involved in HDL metabolism and, equally, by its ability to displace apoA-I from the HDL surface. However, human apoA-II deficiency is associated with normal lipid and lipoprotein levels [10]. ApoA-II decreases the rates of CETP-mediated CE reciprocal exchange between apoA-II-enriched HDL3 and LDL, although no effect is observed when reconstituted HDL are used. PLTP-mediated interconversion of HDL is inhibited when HDL particles display a reduced apoA-I/apoA-II ratio, however the phospholipid transfer activity is maintained, suggesting that apoA-I is required for particle fusion. LCAT activity is higher in HDL reconstituted containing apoA-I than in those containing apoA-II.

Lipoprotein A-I/A-II (LpA-I/A-II) can be significantly less efficient than LpA-I in acting as a lipoprotein substrate at several stages of the reverse cholesterol transport pathway, including plasma cholesterol esterification, plasma cholesteryl ester transfer, and bile acid synthesis. LpA-I is recognized to be more active than LpA-I/A-II in the RCT pathway. CLA-1/SR-BI-dependent efflux as evaluated using rat hepatoma cell line (Fu5AH) is significantly lower with apoA-II-enriched HDL than with nonenriched homologous particles. Analysis of plasma efflux capacity revealed a greater association with LpA-I than with LpA-I/A-II, suggesting that LpA-I represent better cholesterol acceptors than LpA-I/A-II particles. In addition to its reduced ability to induce cholesterol efflux, apoA-II has been specifically found to induce decreased phospholipid efflux compared with that induced by apoA-I.

2.3. HDL-Associated Lipid Transfer Proteins

2.3.1. CETP

By transferring neutral lipids between plasma lipoproteins, CETP plays an important role in the RCT pathway and continuously remodels HDL particles into large and small particles in a process that generates lipid-poor, preß-migrating apoA-I. By the action of LCAT, small HDL3c are progressively transformed in CE-enriched HDL2. CETP mediates mainly the heterotransfer of TG and CE between HDL2a and triglyceride-rich lipoproteins; such exchange results in the formation of HDL2b subspecies. These latter particles are then transformed back to HDL3c by hydrolysis of TG and PL as a result of the combined action of PLTP, hepatic lipase (HL), and CETP. A variation in plasma CETP activity is mostly associated to variations in plasma HDL2 levels.

In vivo studies measuring the removal of cholesterol from macrophages and its transport to the liver until its further elimination into feces revealed the determinant role of CETP in the RCT pathway. Evaluation of RCT in apobec-1 knockout (KO) mice, displaying apoB-containing lipoprotein metabolism similar to that observed in the human situation and co-expressing the hCETP in the liver, showed reduced levels of HDL and accelerated cholesterol flux from macrophages to feces, consistent with increased RCT process [11]. Overexpression of CETP in mice enhances RCT as it allows an alternative pathway for cholesterol delivery to the liver via apoB-containing lipoproteins and LDLR [12].

2.3.2. HDL from CETP-Deficient Subjects

In CETP-deficient subjects most of the plasma cholesterol is associated to HDL particles, mostly large HDL2. In these subjects β-migrating lipoproteins are TG rich and CE depleted as compared to normal subjects, while the α-migrating lipoproteins are CE rich and TG depleted. The presence of apoE on large α-HDL has been reported only when CETP is totally absent. Such apoE-containing HDL displays a strong affinity for the LDLR, however apoE-rich HDL are slowly catabolized as LDLR is rapidly saturated in CETP-deficient subjects. Stable isotope studies showed that fractional catabolic rates of apoA-I and apoA-II are significantly reduced in CETP-deficient subjects, while normal production rates of both apoA-I and apoA-II are observed.

2.3.3. Functionality of HDL Particles in CETP-Deficient Subjects

Preβ-HDL particles are not increased in CETP-deficient subjects, suggesting that the ABCA1-mediated cholesterol efflux pathway might not represent the predominant cellular cholesterol efflux pathway, thus raising the question of whether the lipoprotein profile of CETP-deficient subjects would be pro- or anti-atherogenic. Earlier studies have demonstrated that HDL from CETP-deficient subjects was defective to mediate cholesterol efflux from cholesterol-loaded macrophages, leading to the hypothesis that enrichment in CE and the increased size of HDL in homozygous subjects might not be favorable for the anti-atherogenic activity of these particles

[13]. More recently, HDL from CETP-deficient subjects has been demonstrated to possess an increased capacity to mediate cholesterol efflux through the ABCG1 pathway [14]. In this context, it has been equally reported that HDL particles isolated from patients with partial CETP-deficiency are not dysfunctional in their capacity to mediate either cellular cholesterol efflux or CE hepatic uptake [15].

2.3.4. PLTP

PLTP impacts HDL particles by transferring phospholipids from TGRL during lipolysis and equally through remodeling of HDL size and composition, which generates preβ-HDL in a process called HDL conversion. PLTP mediates fusion of large HDL particles with a concomitant release of lipid-poor apoA-I, such process being enhanced in TG-rich HDL. Overexpression of hPLTP in mice favors preβ-HDL formation with concomitant reduction in HDL-C levels [16]. It is relevant to consider that PLTP rather than CETP is responsible for preβ-HDL formation in mouse models. Indeed, it has been shown that CETP did not quantitatively affect the process of preβ-HDL formation in double transgenic mice for hPLTP and hCETP [17]. The determinant role of PLTP in the transfer of phospholipids from TGRL to HDL has been revealed in PLTP KO mice fed with a high-fat diet in which an accumulation of TGRL surface remnant is observed. Mice with a targeted disruption of the *PLTP* gene display reduced HDL and apoA-I levels as a consequence of hypercatabolism of PL-poor HDL [18]. By contrast, hPLTP transgenic mice have increased plasma levels of preβ1-HDL, apoA-I, and PL and decreased plasma HDL levels [16]. In LDLR KO mice, overexpression of hPLTP results in increased atherosclerosis due to the formation of dysfunctional enlarged HDL particles that are less efficient to mediate cholesterol efflux from macrophages [19].

2.4. Plasma Enzymes Influencing RCT

2.4.1. LCAT

LCAT is responsible for the esterification of free cholesterol and thus for the maturation of small preβ-HDL into larger CE-enriched HDL particles. Once esterified, cholesterol moves from the surface to the hydrophobic core of the HDL particles. As a result of cholesterol esterification, LCAT maintains a gradient of free cholesterol between cell membranes and lipoproteins. The activity of LCAT is essential to maintain a normal HDL metabolism and optimal functional properties of HDL particles. In humans, LCAT deficiency is responsible for low HDL-C levels and important changes in HDL particle distribution and composition [20]. HDL particles from transgenic mice overexpressing hLCAT are better acceptors for free cholesterol from fibroblast as compared to control HDL. It is likely that LCAT-mediated changes in HDL lipid composition favors cholesterol accommodation within the particle. The flux of CE to the liver is increased in hLCAT transgenic mice as a result of increased CE content in HDL and not because of an increased catabolic rate of these particles [21]. While it is clear that LCAT deficiency in humans and mice is associated with reduced HDL-C levels, it is not defined whether LCAT overexpression or deficiency is pro- or anti-atherogenic in mouse models for atherosclerosis. In studies where hLCAT overexpression is 10- to 20-fold as compared to control mice, diet-induced atherosclerosis is minimal, however, when hLCAT is about 50- to 100-fold versus control, large HDL-CE rich particles that are dysfunctional in CE delivery to the liver appear, resulting in increased atherosclerosis.

2.4.2. LPL

LPL is the primary enzyme required for chylomicron (CM) catabolism and thus indirectly acts on HDL composition. LPL has predominantly a triglyceride lipase activity on TRL, allowing the release of fatty acids. Hydrolysis of TG transforms TRL particles in smaller TG-depleted remnants. Redundant lipids surface (PL and FC) and apolipoproteins are then transferred to HDL, contributing to the formation of mature HDL particles. HDL-C levels correlate with post-heparin plasma LPL activity and HDL levels are considered as an index for LPL activity in vivo [22].

2.4.3. HL

HL is a lipolytic enzyme with greater activity toward HDL rather than VLDL or CMs. In contrast to LPL, post-heparin HL activity correlates inversely with HDL-C in humans and particularly with HDL2 levels, thus HL represents an important determinant of plasma HDL concentrations. In patients with total HL deficiency, apoA-I and HDL catabolism is reduced, while in partial HL deficiency, HDL composition and metabolism are normal, thus suggesting that partial HL activity ensures adequate processing of HDL particles. However, both total and partial HL deficiencies are associated with normal HDL-mediated cholesterol efflux from human fibroblasts with unaltered preβ-HDL levels as compared to controls [23]. Introduction of the HL gene in rabbits, which naturally lack HL, reduces HDL particle size and increases the rate of apoA-I catabolism [24].

2.4.4. EL

Endothelial lipase (EL) is synthesized by endothelial cells and is active at the vascular endothelial surface. EL has primarily phospholipase A1 activity. The impact of EL on HDL metabolism has been demonstrated from in vivo studies. Inactivation of the *EL* gene in mice results in an elevation of plasma HDL-C levels, increased particle size, and impaired HDL clearance as compared to wild-type (WT) mice [25]. Injection of anti-EL antibody in WT mice, HL$^{-/-}$, or in apoA-I transgenic mice increased HDL-C and PL levels [26]. By contrast, recombinant EL adenovirus administration to WT mice decreased HDL-C levels, plasma PL, and apoA-I, suggesting that EL activity can generate smaller lipid-depleted HDL with poorly bound apoA-I [27].

3. METABOLIC DISORDERS ASSOCIATED WITH A LOW HDL-C PHENOTYPE AND DYSFUNCTIONAL HDL PARTICLES

3.1. Familial Hypercholesterolemia

Familial hypercholesterolemia (FH) is a common inherited dominant autosomal disease caused by defective hepatic clearance of circulating LDL particles. Mutations in genes that encode for the LDLR, the apolipoprotein B-100 (ApoB), or the proprotein convertase subtilisin/kexin 9 (PCSK9) explained 80% of FH cases [28]. FH subjects are susceptible to the development of atherosclerosis due to elevated LDL-C levels and cholesterol deposition in the arterial wall. In addition, a low HDL-C phenotype is frequently observed in FH subjects, which equally contributes to the severity of atherosclerosis [29]. The reduction of HDL-C levels in FH might result from both increased catabolic and decreased production rates of apoA-I. In addition, plasma CETP activity is elevated in FH patients, primarily as a result of an important CE transfer from HDL to small dense LDL [30]. The altered lipoprotein metabolism in FH patients leads to modifications in HDL subspecies distribution as compared to normolipidemic subjects. Plasma from FH patients displayed increased levels of small preβ1-HDL and reduced concentration of large HDL2. Lipid content of HDL particles is equally affected in FH. HDL are enriched in CE and depleted in phospholipids with an increase in the SM/PC ratio. Plasma from heterozygote FH patients displays defective capacity to mediate cellular efflux through CLA-1/SR-BI pathway. Such altered capacity primarily result from a marked reduction in the ability of HDL2, and to a lesser extent those of HDL3, to mediate the SR-BI-dependent efflux. Interestingly, the capacity of HDL2 to promote cholesterol efflux from human THP-1 macrophages is reduced and negatively correlated with carotid intima-media thickness [30]. Equally, TG-enriched HDL3 isolated from heterozygous FH patients has been shown to display a reduced capacity to mediate cellular cholesterol efflux from human macrophages, such dysfunction being associated with reduced ability of HDL3 to inhibit the release of pro-inflammatory cytokine from endothelial cells [31]. Both in vitro and in vivo studies have demonstrated that HDL particles isolated from FH patients possess a reduced capacity to deliver CE to the liver for its excretion [30].

Monocytes isolated from FH patients display an increased lipid content as compared to monocytes isolated from normolipidemic subjects which result from a downregulation of *ABCA1* gene expression as well as from an increase oxidized LDL uptake [32]. However, human monocyte-derived macrophages, isolated from FH patients possess a similar efflux capacity toward apoA-I, isolated mature HDL, or whole plasma, suggesting that defective efflux from macrophage in FH primarily results from dysfunctional circulating HDL particles rather than from an altered cellular efflux capacity [30].

3.2. Familial Hypoalphalipoproteinemias

Inherited low HDL-C phenotype is characterized by a quantitative reduction of all individual HDL subspecies and a qualitative modification in the HDL profile as shown by a shift of HDL distribution toward smaller HDL particles. Although familial hypoalphalipoproteinemias are commonly associated with mutations within the *APOAI*, *ABCA1*, and *LCAT* genes, activities of key proteins involved in HDL metabolism are equally affected. Both CETP and HL activities are elevated while LPL activity is reduced in familial low HDL-C subjects, thus resulting in a significant reduction of free cholesterol, esterified cholesterol, and phospholipid content in HDL particles, while the triglyceride content increased [33]. On a quantitative basis, familial hypoalphalipoproteinemias are characterized by a major reduction in circulating levels of large HDL2 particles while plasma levels of preβ-HDL are moderately reduced [34,35]. As a consequence, the mean HDL particle size is significantly reduced in the plasma of subjects with familial low HDL-C. HDL particle size, preβHDL, and HDL2b levels have been reported to be negatively associated with intima media thickness in familial low HDL-C subjects, while HDL-C is not [35]. In addition, the overall efflux capacity of sera to mediate cholesterol efflux from THP-1 human macrophage cholesterol efflux is impaired [34].

Although contrasting data have been reported on the capacity of macrophages from patients displaying hypoalphalipoproteinamias to mediate ABCA1-dependent efflux, a defective cellular efflux capacity may equally contribute to atherosclerosis progression. Some studies described an impaired ABCA1-dependent efflux to apoA-I despite an elevated ABCA1 expression, suggesting defective ABCA1 function in macrophages of low HDL-C subjects [33]. Other studies reported that human monocytes derived macrophages isolated from subjects with very low HDL-C levels present a similar efflux capacity toward apoA-I, HDL, or serum than those isolated from controls [34,36]. In fact, approximately one-third of low-HDL subjects display a defective macrophage cholesterol efflux as compared to controls [36]. The majority of these latter subjects do not harbor functional mutation in the *ABCA1* gene, thus suggesting that low cellular efflux phenotype is heritable and that another gene or genes may determine macrophages' efflux capacity. Macrophage efflux capacity to apoA-I and HDL is correlated with the distribution of plasma HDL subspecies, the strongest correlation coefficient being observed with the smaller HDL subclass. Cellular efflux mechanism might therefore be genetically determined by genetic variants that impact HDL metabolism and structure [37].

3.3. Type 2 Diabetes

Type 2 diabetes mellitus is accompanied by low HDL-C phenotype and elevated plasma triglyceride levels, responsible for increased cardiovascular risk. In this metabolic context, HDL-associated lipid transfer protein activities are accelerated, leading to active intravascular HDL remodeling and improvement of plasma capacity to promote macrophage cholesterol efflux. The impact of Type 2 diabetes on HDL efflux function and the underlying mechanisms may depend on disease grade, associated complications, and therapeutic management of patients. On the other hand, HDL undergoes diverse structural modifications responsible for impaired HDL functions (i.e., efflux, anti-oxidative and anti-inflammatory capacities) that may account for elevated cardiovascular disease outcomes in diabetic subjects [38].

3.3.1. Type 2 Diabetes and HDL Metabolism

In noninsulin-dependent diabetes, or Type 2 diabetes, dyslipidemia results from insulin resistance, hyperglycemia, and hyperinsulinemia. Such metabolic disorders induce triglyceride synthesis in the liver through the activation of ChREBP and SREBP1c transcription factors, which results in an elevation of CETP concentration and activity leading to the formation of

HDL particles enriched in TG and depleted in CE and decreased in HDL-C levels [39]. Hyperglycemia and insulin resistance are associated with an increase in HL activity, thus hydrolysis of TG by HL is enhanced in Type 2 diabetes and contributes significantly to the formation of smaller HDL particles [40]. CETP and PLTP activities are equally elevated in diabetic patients as compared to nondiabetic subjects. Increases in lipid transfer protein activities favor HDL remodeling, formation of small HDL particles and acceleration of HDL clearance and catabolism. Lipoprotein lipase, which contributes to the formation of mature HDL particles, is equally reduced in Type 2 diabetes, leading to the establishment of a low HDL-C phenotype in which small HDL particles predominate [41].

3.3.2. Type 2 Diabetes and HDL Efflux Capacity

Plasma from diabetic subjects displayed a reduced capacity to mediate CLA-1/SR-BI-dependent efflux as compared to nondiabetic subjects [42]. This is entirely consistent with the strong correlation described between CLA-1/SR-BI-dependent efflux and plasma HDL-C level or large HDL particles. By contrast, using fibroblasts as a cellular model for efflux assay plasmas from diabetic patients has been shown to display an equivalent efflux capacity to those from nondiabetic subjects [43]. It has been proposed that higher LCAT and PLTP activities represent main determinant factors responsible for the maintenance of plasma efflux capacity [43]. The degree of hypertriglyceridemia associated with Type 2 diabetes equally represents a modulator of plasma efflux ability [44]. Indeed, in diabetic subjects hypertriglyceridemia is responsible for a significant improvement of cholesterol efflux from fibroblast, together with an elevation in PLTP, CETP, and LCAT activities and an increase in preß-HDL formation. Although small HDL particles are predominant in the plasma from diabetic subjects, cholesterol efflux from macrophages strongly correlates with HDL particles of medium size and to a lesser extent with small HDL particles [45]. In addition, the capacity of plasma to mediate cholesterol efflux from human THP-1 macrophages is significantly improved in diabetic subjects as compared to nondiabetic control subjects despite a significant reduction in plasma levels of preß-HDL particles in diabetic subjects [46]. Hence, despite the fact that Type 2 diabetes and hypertriglyceridemia are associated with the appearance of a pro-atherogenic plasma lipid profile (low HDL-C phenotype with increase in plasma TG and small dense LDL), the combined actions of CETP and PLTP allow active HDL remodeling and thus preserve an efficient ability of plasma to stimulate cellular cholesterol efflux from human macrophages.

3.3.3. Type 2 Diabetes and HDL Dysfunctions

Glycation of HDL-associated apolipoproteins occurs in diabetes and results in partial inactivation of HDL-associated enzymes such as paraoxonase 1 (PON1), CETP, and LCAT activities [41,47–51]. In vitro glycation of HDL apoA-I significantly reduces the capacity of HDL particles to mediate cholesterol efflux from human THP-1 macrophages while incubation with glycation inhibitors (metformin and aminoguanidine) has been shown to restore HDL efflux function [52]. HDL particles isolated from diabetic subjects possess a reduced capacity to mediate cholesterol efflux from THP-1 human macrophages, which results in part from the reduction in HDL-associated PON1 activity [53]. In addition, hyperglycemia and advanced glycation end product (AGE) formation may also downregulate gene expression of cellular cholesterol receptors or transporters such as CLA-1/SR-BI and ABCA1. However, the deleterious effect of glycated-HDL has recently been challenged. Indeed, despite elevated AGE in plasma, the capacity of whole plasma or isolated HDL to mediate cholesterol efflux from THP-1 human macrophages has been shown to be greater for diabetic patients as compared to control [46]. In vivo studies conducted in mice models for diabetic dyslipidemia (db/db mouse) demonstrate that despite elevated plasma efflux capacity, the overall RCT mechanism is impaired as shown by a reduced fecal sterol excretion in db/db mice as compared to wild-type mice. Glycated HDL are more susceptible to in vitro oxidation as shown by an increase in lipid peroxidation products and thiobarbituric acid-reactive substances (TBARS) following incubation of HDL particles in the presence of glucose. Hyperglycemia and glycation equally contribute to HDL dysfunction by decreasing their anti-oxidative and anti-inflammatory properties [49,54]. Indeed, glycated HDL exhibits an impaired capacity to protect LDL against oxidative modifications and does not inhibit in vitro monocytes adhesion to aortic cells. Equally, injection of glycated apoA-I has no effect on VCAM-1 or ICAM-1 expression in endothelial cells from collared carotid arteries [49,55]. Finally, HDL particles isolated from diabetic subjects are less effective to stimulate nitric oxide production by endothelial cells, improve endothelium-dependent vasodilation and endothelial repair, and reduce endothelial oxidant stress [56].

3.4. Obesity

Current observations lead to the conclusion that the initial step of reverse cholesterol transport (i.e., human macrophage cholesterol efflux) is not dysfunctional in obese subjects. The elevated activity of CETP in obese subjects maintains an efficient production of preβ1-HDL particles that account for an elevation of plasma capacity to mediate ABCA1-dependent efflux. By contrast, anti-oxidative and anti-inflammatory atheroprotective properties of HDL particles are altered in obesity. Such dysfunction of HDL has been shown to be associated with clinical complications of obesity, such as infection, inflammation, diabetes, and CVD [57].

3.4.1. Obesity and HDL Metabolism

The pathophysiology of dyslipidemia (i.e., low HDL-C phenotype and elevated plasma TG levels) in obesity is close to that previously described in diabetic subjects. Briefly, obesity is accompanied by an overproduction of VLDL by the liver due to the high flux of nonesterified fatty acids (NEFA) from adipose tissue to the liver. Elevated circulating levels of TRL and increased CETP secretion by adipose tissue lead to an elevated plasma CETP activity that is responsible for the formation of CE-depleted and TG-enriched HDL particles. Equally, HL is increased in obese subjects, favoring the formation of small HDL by the dissociation of apoA-I. Such HDL remnants are rapidly cleared by the kidney, thus accounting for reduction in circulating HDL-C levels in obese subjects [58]. Plasma levels of adipocyte-derived protein called adiponectin are reduced in obesity and inversely correlated with fractional catabolic rate of apoA-I [59] and positively with plasma HDL-C levels. Finally, adipose tissue represents an important site for the storage of cholesterol in the body, and ABCA1-mediated cholesterol efflux from adipose tissue, which significantly contributes to the formation of nascent small HDL particles, might be impaired in obese subjects [60]. As a consequence of altered HDL metabolism, obesity is associated with low HDL-C plasma levels and a shift of HDL profile toward smaller HDL particles. The correlation between low HDL-C levels and obesity seems to be strongest with central obesity, evaluated by abdominal circumference, which is characterized by intra-abdominal or visceral fat deposition [61].

3.4.2. Obesity and HDL Efflux Capacity

Obesity not only affects HDL-C plasma levels, but the modulation of many key proteins involved in HDL metabolism equally contributes to change in HDL structure and function. Body mass index (BMI) is positively correlated with the smallest HDL particles, such as preβ1-HDL, and inversely with the large α1-HDL particles. Weight loss consecutive to energy restriction leads to redistribution of apoA-I-containing HDL subspecies in a pattern similar to that observed in lean subjects. In obese subjects, weight loss is associated with an increase in α-HDL at the expense of preβ1-HDL. Distinct HDL particle profiles between lean and obese subjects are associated with a modulation of HDL efflux capacity. Indeed, the ability of whole plasma to mediate cholesterol efflux from fibroblast is reduced in obese subjects as compared to lean subjects. In addition, in lean subjects, preβ2-HDL and α1-HDL subspecies appear to be the most active particles for removal of cholesterol from cells, while in obese subjects preβ1-HDL display a greater capacity to accept cellular cholesterol with larger α-HDL, displaying a reduced efflux capacity as compared to those from lean subjects. The cellular efflux mechanism has been deciphered using rat hepatoma (Fu5AH) and mouse macrophage (J774) cell lines, thus providing further explanation of the relative contribution of each efflux pathways between lean and obese subjects [62]. Sera from obese subjects exhibit a decreased SR-BI mediated cholesterol efflux capacity, consistent with the reduced HDL-C levels. More importantly, plasma efflux capacity via the ABCA1-dependent cholesterol efflux is increased in obese subjects as compared to lean subjects. Such improved ABCA1-dependent plasma efflux capacity in obese subjects is related to increased levels of preβ-HDL in these subjects. As obesity modifies both HDL structure and function, weight loss is expected to influence quantitative and qualitative features of HDL particles. Diet-induced weight loss in obese women is responsible for a decreased ABCA1-dependent plasma efflux capacity [63]. Bariatric surgery-induced weight loss in morbidly obese subjects increases the SR-BI dependent efflux to large HDL particles, while the ABCA1-dependent plasma efflux capacity is dramatically reduced [64]. Modifications of HDL subspecies profile and efflux capacity following weight loss induced by Roux-en-Y bypass results at least in part from a reduction in plasma CETP activity [64,65]. Recent studies have afforded new insights in the comprehension of the mechanism underlying variation in plasma efflux capacity from human macrophages following weight loss in obese subjects. Plasma efflux capacity of overweight subjects with insulin resistance is better than those from lean subjects or insulin-sensitive overweight patients [66]. In addition, plasma efflux capacity is positively correlated with circulating CETP concentrations and is inversely linked with insulin sensitivity. Finally, weight loss induces a significant reduction of plasma CETP concentration and reduces plasma or apoB-depleted plasma efflux from human macrophages in obese patients with Type 2 diabetes [67].

4. INFLAMMATORY STATES ASSOCIATED WITH ALTERED HDL FUNCTION IN REVERSE CHOLESTEROL TRANSPORT

Several diseases characterized by chronic inflammation are associated with reduced HDL-C plasma levels [68,69]. Metabolic diseases, such as metabolic syndrome, obesity, and Type 2 diabetes, which are characterized by a low HDL-C phenotype, are equally associated with low-grade inflammation [70,71]. Pro-inflammatory cytokines, such as TNFα and IL1-β, reduce apoA-I production and increase its catabolism, resulting in reduced plasma levels of both apoA-I and HDL-C [72,73]. In addition, many markers of inflammation are significantly correlated with the reduction in plasma HDL-C levels [74]. Hence, systemic inflammation is closely linked to decreased HDL-C levels. Acute phase inflammation is responsible for structural modifications of HDL, which result in the formation of dysfunctional HDL displaying defective anti-inflammatory and anti-oxidative properties as well as impaired efflux capacity [75,76]. As described earlier for obesity and diabetes, during acute phase HDL particles display several structural alterations which reduce their anti-inflammatory and anti-oxidative capacities as a result in part of an increase in serum amyloid A (SAA) content and defective activities of several HDL-associated enzymes such as PON1, PAF-AH, and LCAT [75]. However, in the context of metabolic diseases, the specific contribution of inflammation on structural and functional changes of HDL that alter RCT pathways in humans is difficult to decipher since dyslipidemia, oxidative stress, and inflammation all contribute in the formation of defective HDL particles.

During acute phase inflammation, SAA, an amphipathic, α-helical apolipoprotein transported in the circulation in association with HDL, is mostly produced by the liver though extra-hepatic tissue also contributes. In vitro, SAA has been shown to displace apoA-I on HDL particles. During the acute phase, SAA becomes the predominant apolipoprotein in HDL. While it remodels the lipoprotein particle by displacing primarily apoA-I, SAA may also significantly alter the atheroprotective properties of its physiological carrier. The capacity of SAA-HDL to mediate cellular cholesterol efflux is controversial. Many studies observed an increased capacity of SAA-HDL to promote macrophage cholesterol efflux as compared to normal HDL [77,78], while others reported dysfunctional SAA-HDL to promote efflux [79]. An important portion of SAA exists in a lipid-free/lipid-poor form which is able to promote ABCA1-dependent cellular cholesterol efflux. Equally, SAA has been shown to promote CLA-1/SR-BI-dependent FC efflux [80]. However, SAA (as commonly known for apoA-I) needs to be lipidated through a series of intermediate steps to form spherical SAA-enriched HDL particles before being able to mobilize cholesterol through SR-BI [80]. In HDL, as a result of apoA-I displacement, the activity of LCAT, which is known to be activated by apoA-I, could be impaired. In addition, altered binding properties of acute-phase HDL to hepatocytes might be responsible for altered cholesterol homeostasis. It has been demonstrated that HDL efflux capacity from human macrophages becomes impaired when SAA content in HDL exceeds 50% of the total HDL-protein content, however SAA can constitute up to 80% of HDL proteins only in extremely rare circumstances. SAA is known to be a marker for obesity, as its expression is well correlated with obesity [81].

Accumulating data suggest that HDL can be modified in vivo by myelo-peroxidase (MPO) during the inflammatory state [76]. Activated monocytes and neutrophiles produce MPO as a self-defense system catalyzing the formation of reactive oxygen species (ROS). HDL-apoA-I represents the preferential target for oxidation by ROS as compared with other plasma proteins [76]. MPO-dependent modification of HDL-apoA-I impairs the interaction with the ABCA1 transporter thus reducing ABCA1-dependent cellular cholesterol efflux [82].

Recent human studies on the acute phase have afforded a new integrated explanation for the impact of inflammation on HDL metabolism and RCT pathway. Acute phase consecutive to cardiac operation or endotoxin results not only in an elevation of circulating SAA levels but equally in increased plasma activity of sPLA2 and hepatic lipase while those of CETP and LCAT activity are decreased [83]. In vitro incubation of HDL in the presence of CETP and/or sPLA2 has been shown to transform the majority of apoA-I from an HDL-bound form to a lipid-poor form in both normal and acute phases. While CETP alone can generate lipid-poor HDL, sPLA2 acts in combination with CETP [84]. Modifications of plasma proteins during acute phase response may account for active HDL remodeling in vivo. Although the total HDL particle number does not significantly change during acute phase, plasma HDL subspecies profile is modified with a reduction in medium-sized HDL, small HDL, and preβ1-HDL particles from 12 to 24 hours after lipopolysaccharide (LPS) challenge [83]. Inflammation-induced HDL remodeling primarily reduces the ABCA1-dependent cholesterol efflux and secondarily the CLA1/SR-BI-dependent HDL cholesterol efflux in a time-dependent manner that results in a significant reduction in the overall efflux from human macrophages [79,83].

5. GENETIC VARIANTS ASSOCIATED WITH ALTERED HDL FUNCTION IN REVERSE CHOLESTEROL TRANSPORT

To date, 50 different genes have been reported to be associated with circulating HDL-C levels, its heritability being estimated to approximately 50% [85]. A plethora of studies demonstrating the genetic determination of HDL-C levels is currently available. However, only few studies have evaluated the genetic determination of HDL function (Table 4.1).

5.1. CETP

Plasma CETP activity is closely linked to HDL-C levels and HDL subclass distribution. Human CETP deficiency is associated with elevated HDL-C levels, a predominance of large HDL particles displaying an increased capacity to mediate cellular cholesterol efflux [14,15,86]. The association between CETP polymorphisms and HDL-C is robust and has been confirmed in many studies [87–89]. Several genetic variants have been identified on the human *CETP* gene. The *CETP* TaqIB SNP (rs708272), the most extensively studied CETP polymorphism which is associated with plasma CETP concentration, HDL-C levels, and HDL particle size, does not, by itself, determine CETP and HDL-C levels but represents a marker for at least three functional sites—*CETP* c. -629 C>A polymorphism (rs1800775), *CETP* c.-971 G>A (rs4783961), and *CETP* c.-1337 C>T (rs17231506)—which have been shown to interact together to determine global promoter activity, thus contributing to interindividual variability of plasma CETP concentration [88,90,91]. To date very few studies have evaluated the impact of CETP gene polymorphisms on HDL structure and function. Rare alleles of CETPTaq1B and CETP-629C>A are associated with reduced CETP concentrations and increases in HDL-C levels, which mostly reflect an elevation in circulating levels of large HDL particles [92,93]. Carriers of the rare allele for CETP -629C>A SNP, despite higher plasma HDL-C levels, display a reduced plasma capacity to stimulate efflux from fibroblast as compared to carriers of the most frequent allele [94]. In good agreement with this latter study, it has been recently reported that the rare allele for CETP-1337C>A SNP, despite elevated HDL-C levels, is associated with an impaired plasma efflux capacity from human THP-1 macrophages as compared to the most frequent allele [95]. These studies demonstrate that genetic elevation of plasma HDL-C levels may be associated with a defective plasma efflux capacity. Such observations might explain why functional CETP gene

Table 4.1 Impact of Genetic Variants on HDL Level, Structure, and Function

| Gene | SNP | Impact of Rare Allele on Plasma HDL-C Levels | Impact of Rare Allele on HDL Particles | | | Impact of Rare Allele on Macrophage Efflux |
			Structure Mean Size	Large HDL	Small HDL	
CETP	Taq1B G>A rs708272	↘	↘	↘	↘	↕
	-629 C>A rs18000775	↘	↘	↘	↘	↗
	-1337 C>T rs17231506	↘	n.d	n.d	n.d	↗
LIPC	-514C>T rs1800588	↘	↘	↘	n.d.	↗
LPL	D9N rs 1801177	↗	n.d.	n.d.	↘	n.d.
	N291S rs268	↗	n.d.	↗	n.d.	n.d.
	T495G rs 320	↘	n.d.	↘	↗	n.d.
	S447X rs 328	↘	↕	n.d.	n.d.	n.d.

Continued

Table 4.1 Impact of Genetic Variants on HDL Level, Structure, and Function—*cont'd*

Gene	SNP	Impact of Rare Allele on Plasma HDL-C Levels	Impact of Rare Allele on HDL Particles Structure			Impact of Rare Allele on Macrophage Efflux
			Mean Size	Large HDL	Small HDL	
ABCA1	R219K rs2230806	↕ ↘	n.d.	n.d.	↗	↘
	R230C rs 9282541	↗	↗	↗	↘	n.d.
APOA1	−75G>A rs1799837	↘ ↕	n.d.	n.d.	↗	↘
APOAII	−265T>C rs5086	↕	n.d.	n.d.	n.d.	↘
SCARB1	Exon 1 (c.4G>A) rs 4238001	↗ ↘ ↗ ↕	↗	↗	n.d.	n.d.
	Intron 5 (c.726+54C>T)	↕	↗	n.d.	n.d.	n.d.
	Exon 8 c.1050C>T	↘ ↘	↘	↘	↘	n.d.
	A350A rs5888	↗				

n.d.: Not Determined

polymorphisms, which are related to lower plasma CETP concentrations and elevated HDL-C levels predict an increased cardiovascular risk [96,97].

5.2. LIPC

HL activity is a strong negative predictor for HDL-C and HDL size. Patients with HL deficiency display elevated HDL-C levels and are characterized by the presence of circulating large HDL-TG enriched particles; however, their plasma efflux capacity is similar to control subjects [23]. Several polymorphisms located within the *LIPC* gene associated with variation of plasma HDL-C levels have been described [98]. The rare allele for LIPC-514C>T (rs1800588) is associated with reduced HL activity and increased plasma HDL-C levels [99]. The increase in plasma HDL-C levels in carriers of the rare allele for LIPC-514C>T occurs concomitantly with a specific elevation in HDL2-C levels, while those of HDL3-C remain unchanged, thus indicating that this genetic variant represents a strong predictor of HDL particle size. Interestingly, it has been demonstrated that the relationship between the LIPC-514C>T variant and HDL particle size is modulated by ethnicity and sex [100,101]. In this context it is relevant to consider that in female subjects, the rare allele of the LIPC-514C>T polymorphism is associated with a significant increase in plasma efflux capacity, whereas plasma HDL-C level is not affected [95].

5.3. LPL

The hydrolysis of TRL by LPL releases protein and lipid components that are then sequestered to HDL. By this action LPL activity is closely linked to HDL formation. In this context, several polymorphisms located on the *LPL* gene have been shown to impact both quantitative and qualitative features of HDL particles. Data concerning the impact of polymorphisms located on the *LPL* gene on circulating levels of HDL-C are controversial. Meta-analyses based on 20,000 subjects revealed that carriers of rare alleles for D9N, N291S SNP are associated with a reduction in HDL-C levels, while others, T495G (rs320) and S447X (rs328), are responsible for elevated HDL-C levels [102]. In the meantime, several independent studies have not reproduced these latter associations [103–105]. Analysis of individual HDL subfractions in association with LPL polymorphisms reveals a slight reduction in large HDL2 levels in carriers of a rare allele of N291S SNP. Carriers of the rare allele of D9N polymorphism display an increase in smaller HDL3 concentrations, concomitant with an elevation in cholesterol content of remnant-like particles and a slight increase in plasma TG levels [92]. The S447X variant has an effect on VLDL particle size and is associated with an increase in triglyceride content in VLDL particles but has no impact on HDL particle size [106,107]. The T495G polymorphism is equally associated with HDL particle subspecies profile, with the rare allele associated with reduced TG plasma levels which induces a reduction in the relative proportion of small HDL particles concomitantly with an increase in those of large HDL particles [108]. Such impact of the T495G polymorphism on the HDL profile is mostly observed in hyperlipidemic subjects displaying concomitant elevation in both plasma triglycerides and cholesterol levels. The consequence of LPL polymorphisms on HDL function in RCT has not yet been explored.

5.4. SCARB1

The P297S variant represents the first mutation identified in the human *SCARB1* gene associated with a reduction in HDL-CE liver uptake, a reduction in macrophage cholesterol efflux capacity, and an increase in HDL-C levels [109]. SCARB1 polymorphisms represent independent predictors of CLA1/SRBI protein levels and by this way are responsible for modifications in the interaction between cellular CLA-1/SRBI and HDL particles [110]. Several single-nucleotide polymorphisms (SNPs) located in the gene of SCARB1 have been described in both the coding and noncoding regions: exon 1 (c.4G>A; rs4238001), intron 5 (c.726+54C>T), and exon 8 (c.1050C>T; rs5888). The relationship between these latter SNPs and HDL-C is inconsistent from one study to another, primarily as a result of interactions with sex and diabetic status [111–114]. Indeed, a rare allele for rs4238001 has been found to be associated with a decrease in HDL-C and particularly with reduction of HDL2-C in diabetic subjects [113]. A positive association has been demonstrated between a rare allele for rs4238001 and HDL-C plasma levels in men [111]. The rare allele of rs5888 is positively associated with HDL-C levels in men [113] and negatively associated with HDL-C levels in the Chinese population [115], while several studies reveal no significant association with rs5888 and HDL-C levels [112,114]. These conflicting observations might, at least in part, result from differential sex hormone regulation of *SCARB1* gene between men and women [116]. Few studies have reported a reduction in HDL particle size in association with the rare allele of exon 1 and intron 5 variants in diabetic subjects [113]. By contrast, the rare allele for the exon 8 SNP is responsible for an increase in HDL particle size and the formation of TG-enriched HDL3a [113,117]. Finally, to our knowledge, no information is currently available on the putative association between SCARB1 polymorphisms and HDL functionality in RCT. Large genetic association studies are required to demonstrate a link between SCARB1 polymorphisms, HDL structure, and function.

5.5. ABCA1

ABCA1 mutations are frequently associated with reduced plasma HDL-C levels. The well known ABCA1 mutation called Tangier disease is character-ized by low HDL-C and reduction of HDL particle size as compared to controls [118]. Interestingly the ABCA1 efflux capacity of fibroblast isolated from Tangier disease patients, ranging according to the gravity of ABCA1 mutation, is positively associated with HDL-C levels and HDL particle size. Mutations in *ABCA1* gene affect HDL-associated proteins' activity and HDL functionality. HDL from ABCA1-mutated subjects displayed reduced anti-oxidant and anti-inflammatory capacity, concomitantly with elevation of SAA and reduction of PON1 activity that could also impair HDL efflux functionality [119]. However, the functionality of HDL from ABCA1-mutated patients in RCT has not yet been explored.

The relationship between genetic variants located in the *ABCA1* gene and HDL-C levels has been described in many studies; however, conclu-sions reached remain controversial [98]. The most frequently studied ABCA1+1051G>A; R219K; (rs2230806) variant has been shown to be associated with plasma HDL-C in some studies [120–122] but not in others [93]. Carriers of the rare allele for ABCA1+1051G>A SNP display an increase of small HDL particles [93,123]. To date, the only study that has explored efflux functionality in relation with the ABCA1+1051G>A SNP demonstrated a significant increase of plasma efflux capacity from human macrophages in carriers of the rare allele, while HDL-C plasma levels are not significantly changed, thus suggesting that genetic modulation of ABCA1 results in a shift toward small HDL particles that improve the first step of RCT [95].

5.6. ApoA-I

ApoA-I is the major structural apolipoprotein of the HDL particle. Function-disrupting mutations within the *APOAI* gene are invariably asso-ciated with decreased HDL-C levels [124]. ApoA-I mutations correspond to nonsense mutations or structural alterations of mutant apo-A-I. Mutated apoA-I fail to coactivate LCAT leading to defective maturation of HDL particles and to an acceleration of apoA-I catabolism. Mutations in the

APOAI gene induce change in HDL size and distribution with reduction of spherical HDL2 and HDL3 particles [119]. The capacity of serum to stimulate CLA-1/SR-BI dependent efflux, as evaluated using rat hepatoma Fu5AH cells, is significantly reduced in subjects carrying a mutation in the *APOAI* gene. Natural variants of apoA-I, located between residues 121 and 186, are often associated with low apoA-I levels. Heterozygous individuals for the apoA-I Milano (apoA-I$_{Milano}$) mutation display very low circulating levels of apoA-I and HDL cholesterol and a moderate elevation in plasma triglyceride. Low levels of apoA-I$_{Milano}$ in plasma appear to result from a reduction in LCAT activation, which is associated with an accelerated apoA-I clearance. Despite an extremely low HDL-C phenotype that is usually linked with an elevated premature cardiovascular risk, apoA-I$_{Milano}$ carriers display no increase in cardiovascular events, thus suggesting that apoA-I$_{Milano}$ represents a gain-of-function mutation with cardioprotective properties. Cimmino and colleagues [125] demonstrated that recombinant apoA-I$_{Milano}$ administration enhances RCT pathway and reduces atherosclerotic disease in New Zealand rabbits. The mutant form of apoA-I seems to be more efficient in RCT than the apoA-I wild type. The enhanced ABCA1-mediated efflux capacity of apoA-I$_{Milano}$ might result from the coexistence in serum of normal amounts of apoA-I-containing preβ-HDL, together with a unique small HDL particle containing the apoA-I$_{Milano}$ dimer, both effective in removing cell cholesterol via ABCA1 [126]. However, it has been recently demonstrated that, using a gene transfer approach in LDL receptor-deficient mice, systemic overexpression of apoA-I is as effective as apoA-I$_{Milano}$ in reducing atherosclerosis progression [127].

Conflicting data on the impact of apoA-I polymorphism on HDL-C levels have been reported [98]. Some studies have reported that the -75G/A rare variant located in the promoter region of the *APOAI* gene is associated with an increase in *APOAI* gene transcription activity and an elevation in HDL-C levels [128,129], while other studies failed to demonstrate such association [130,131]. The rare allele of the -75G/A polymorphism is with significant modification of HDL particle size. In addition to this latter structural modification, carriers of the rare allele for APOA1-75 display an enhanced plasma efflux capacity [95].

5.7. ApoA-II

ApoA-II is the second most important protein of HDL particles; however, apoA-II deficiency has been reported to have no significant effect on HDL-C levels. Studies on apoA-II polymorphisms also failed to demonstrate any association between apoA-II genetic variants and plasma HDL-C levels [132,133]. The precise role of apoA-II in modulation of HDL efflux capacity in humans is not well defined. Serum from mice overexpressing human apoA-II display a reduced efflux capacity as compared to control mice [134]. In good agreement with animal studies, human subjects carrying the rare allele C for the -265T>C (rs5086) variant, known to be associated with a reduced transcriptional activity of the *APOAII* gene [135], display a higher plasma efflux capacity than noncarriers [95].

5.8. LCAT

Elevated preβ–HDL plasma levels and reduction of HDL size have been reported in LCAT deficiency [119]. However, despite severe HDL deficiency cardiovascular risk does not appear to be excessively increased in classical LCAT deficiency or in the fish-eye disease. Sera from LCAT-deficient subjects are as effective as sera from control subjects for removal of cholesterol from macrophages, as a result of increased ABCA1-dependent efflux which compensates for the reduced ABCG1 and SR–B1-dependent efflux [136]. Five major polymorphisms of LCAT have been investigated for their association with HDL-C levels; however, findings are inconsistent [98]. Even though LCAT is an important gene of HDL metabolism, no study has demonstrated the impact of LCAT polymorphisms on HDL efflux function.

5.9. Multi-SNP Studies and Quantitative Relevance of HDL Genetic

In addition to single-SNP studies, multi-SNP analyses afforded a better evaluation of global impact of genetics on HDL metabolism, HDL-C, and HDL function. From genome-wide association studies (GWAS), it has been concluded that combined effects of multiple SNPs explained only 10% of the total variance of HDL-C levels [137,138]. By using such a multi-gene approach, metabolic pathways and new genes (*GALNT2*, *MVK-MMAB*) have been related to HDL-C levels [138–140]. However, the direct causality between genetic determination of HDL-C levels and cardiovascular outcome remains controversial [141,142]. Similar studies on HDL size and function are very sparse. Indeed, a recent study has reported that common variation of 43 loci explained only 12% of HDL particles size [143]. In addition, seven SNPs localized in a gene encoding for key proteins of HDL metabolism have been shown to explain only 5% of the variability of plasma to mediate cholesterol efflux from human macrophages [95].

A plethora of studies address the relationship between HDL-C levels and genetic mutations or polymorphisms. However, the lack of a strong link between inherited HDL-C and a causal link to cardiovascular diseases highlights the need to further explore HDL structure and function. Only a few studies revealed associations between the genetics of HDL metabolism and HDL structure and function. Based on these rare studies, it seems that genetically determined HDL efflux functionality is independent of HDL-C plasma levels; however, large-scale studies are needed to provide strong genetic determination of HDL function in RCT and on metabolic pathways involved.

6. CONCLUSION

A low HDL-C phenotype represents an independent risk factor for premature atherosclerosis and coronary artery disease. However, therapeutic strategies based on elevation of circulating levels of HDL-C failed to demonstrate efficacy to reduce significantly atherosclerosis progression or occurrence of cardiovascular events. These results support the contention that the level of plasma HDL-C must not represent the unique therapeutic goal and that it is necessary to equally consider structural and functional features of HDL particles. Thus, improvement of HDL function should represent the future clinical therapeutic goal; however, it is necessary to consider the genetic background of individuals as well as environmental factors and the metabolic context which are now recognized to strongly influence quantitative, qualitative, and functional properties of HDL particles.

REFERENCES

[1] Kobayashi A, Takanezawa Y, Hirata T, Shimizu Y, Misasa K, Kioka N, et al. Efflux of sphingomyelin, cholesterol, and phosphatidylcholine by ABCG1. J Lipid Res 2006;47:1791–802.

[2] Fournier N, Atger V, Cogny A, Vedie B, Giral P, Simon A, et al. Analysis of the relationship between triglyceridemia and HDL-phospholipid concentrations: consequences on the efflux capacity of serum in the Fu5AH system. Atherosclerosis 2001;157:315–23.

[3] Davidson WS, Lund-Katz S, Johnson WJ, Anantharamaiah GM, Palgunachari MN, Segrest JP, et al. The influence of apolipoprotein structure on the efflux of cellular free cholesterol to high density lipoprotein. J Biol Chem 1994;269:22975–82.

[4] Sola R, Motta C, Maille M, Bargallo MT, Boisnier C, Richard JL, et al. Dietary monounsaturated fatty acids enhance cholesterol efflux from human fibroblasts. Relation to fluidity, phospholipid fatty acid composition, overall composition, and size of HDL3. Arterioscler Thromb 1993;13:958–66.

[5] Subbaiah PV, Liu M. Role of sphingomyelin in the regulation of cholesterol esterification in the plasma lipoproteins. Inhibition of lecithin-cholesterol acyltransferase reaction. J Biol Chem 1993;268:20156–63.

[6] Greene DJ, Skeggs JW, Morton RE. Elevated triglyceride content diminishes the capacity of high density lipoprotein to deliver cholesteryl esters via the scavenger receptor class B type I (SR-BI). J Biol Chem 2001;276:4804–11.

[7] Silva RA, Hilliard GM, Li L, Segrest JP, Davidson WS. A mass spectrometric determination of the conformation of dimeric apolipoprotein A-I in discoidal high density lipoproteins. Biochemistry 2005;44:8600–7.

[8] Horowitz BS, Goldberg IJ, Merab J, Vanni TM, Ramakrishnan R, Ginsberg HN. Increased plasma and renal clearance of an exchangeable pool of apolipoprotein A-I in subjects with low levels of high density lipoprotein cholesterol. J Clin Invest 1993;91:1743–52.

[9] Curtiss LK, Bonnet DJ, Rye KA. The conformation of apolipoprotein A-I in high-density lipoproteins is influenced by core lipid composition and particle size: a surface plasmon resonance study. Biochemistry 2000;39:5712–21.

[10] Deeb SS, Takata K, Peng RL, Kajiyama G, Albers JJ. A splice-junction mutation responsible for familial apolipoprotein A-II deficiency. Am J Hum Genet 1990;46:822–7.

[11] Tanigawa H, Billheimer JT, Tohyama J, Zhang Y, Rothblat G, Rader DJ. Expression of cholesteryl ester transfer protein in mice promotes macrophage reverse cholesterol transport. Circulation 2007;116:1267–73.

[12] Tchoua U, D'Souza W, Mukhamedova N, Blum D, Niesor E, Mizrahi J, et al. The effect of cholesteryl ester transfer protein overexpression and inhibition on reverse cholesterol transport. Cardiovasc Res 2008;77(4):732–9.

[13] Ishigami M, Yamashita S, Sakai N, Arai T, Hirano K, Hiraoka H, et al. Large and cholesteryl ester-rich high-density lipoproteins in cholesteryl ester transfer protein (CETP) deficiency can not protect macrophages from cholesterol accumulation induced by acetylated low-density lipoproteins. J Biochem 1994;116:257–62.

[14] Matsuura F, Wang N, Chen W, Jiang XC, Tall AR. HDL from CETP-deficient subjects shows enhanced ability to promote cholesterol efflux from macrophages in an apoE- and ABCG1-dependent pathway. J Clin Invest 2006;116:1435–42.

[15] Plengpanich W, Le Goff W, Poolsuk S, Julia Z, Guerin M, Khovidhunkit W. CETP deficiency due to a novel mutation in the CETP gene promoter and its effect on cholesterol efflux and selective uptake into hepatocytes. Atherosclerosis 2011;216:370–3.

[16] van Haperen R, van Tol A, Vermeulen P, Jauhiainen M, van Gent T, van den Berg P, et al. Human plasma phospholipid transfer protein increases the antiatherogenic potential of high density lipoproteins in transgenic mice. Arterioscler Thromb Vasc Biol 2000;20:1082–8.

[17] Lie J, de Crom R, Jauhiainen M, van Gent T, van Haperen R, Scheek L, et al. Evaluation of phospholipid transfer protein and cholesteryl ester transfer protein as contributors to the generation of pre beta-high-density lipoproteins. Biochem J 2001;360:379–85.

[18] Qin S, Kawano K, Bruce C, Lin M, Bisgaier C, Tall AR, et al. Phospholipid transfer protein gene knock-out mice have low high density lipoprotein levels, due to hypercatabolism, and accumulate apoA-IV-rich lamellar lipoproteins. J Lipid Res 2000;41:269–76.

[19] Moerland M, Samyn H, van Gent T, Jauhiainen M, Metso J, van Haperen R, et al. Atherogenic, enlarged, and dysfunctional HDL in human PLTP/apoA-I double transgenic mice. J Lipid Res 2007;48:2622–31.

[20] Guerin M, Dachet C, Goulinet S, Chevet D, Dolphin PJ, Chapman MJ, et al. Familial lecithin:cholesterol acyltransferase deficiency: molecular analysis of a compound heterozygote: LCAT (Arg147 --> Trp) and LCAT (Tyr171 --> Stop). Atherosclerosis 1997;131:85–95.

[21] Francone OL, Haghpassand M, Bennett JA, Royer L, McNeish J. Expression of human lecithin:cholesterol acyltransferase in transgenic mice: effects on cholesterol efflux, esterification, and transport. J Lipid Res 1997;38:813–22.

[22] Goldberg IJ. Lipoprotein lipase and lipolysis: central roles in lipoprotein metabolism and atherogenesis. J Lipid Res 1996;37:693–707.

[23] Ruel IL, Couture P, Cohn JS, Bensadoun A, Marcil M, Lamarche B. Evidence that hepatic lipase deficiency in humans is not associated with proatherogenic changes in HDL composition and metabolism. J Lipid Res 2004;45:1528–37.

[24] Rashid S, Trinh DK, Uffelman KD, Cohn JS, Rader DJ, Lewis GF. Expression of human hepatic lipase in the rabbit model preferentially enhances the clearance of triglyceride-enriched versus native high-density lipoprotein apolipoprotein A-I. Circulation 2003;107:3066–72.

[25] Ma K, Cilingiroglu M, Otvos JD, Ballantyne CM, Marian AJ, Chan L. Endothelial lipase is a major genetic determinant for high-density lipoprotein concentration, structure, and metabolism. Proc Natl Acad Sci USA 2003;100:2748–53.

[26] Jin W, Millar JS, Broedl U, Glick JM, Rader DJ. Inhibition of endothelial lipase causes increased HDL cholesterol levels in vivo. J Clin Invest 2003;111:357–62.

[27] Maugeais C, Tietge UJ, Broedl UC, Marchadier D, Cain W, McCoy MG, et al. Dose-dependent acceleration of high-density lipoprotein catabolism by endothelial lipase. Circulation 2003;108:2121–6.

[28] Marduel M, Carrie A, Sassolas A, Devillers M, Carreau V, Di Filippo M, et al. Molecular spectrum of autosomal dominant hypercholesterolemia in France. Hum Mutat 2010;31:E1811–24.

[29] Guerin M. Reverse cholesterol transport in familial hypercholesterolemia. Curr Opin Lipidol 2012;23:377–85.

[30] Bellanger N, Orsoni A, Julia Z, Fournier N, Frisdal E, Duchene E, et al. Atheroprotective reverse cholesterol transport pathway is defective in familial hypercholesterolemia. Arterioscler Thromb Vasc Biol 2011;31:1675–81.

[31] Ottestad IO, Halvorsen B, Balstad TR, Otterdal K, Borge GI, Brosstad F, et al. Triglyceride-rich HDL3 from patients with familial hypercholesterolemia are less able to inhibit cytokine release or to promote cholesterol efflux. J Nutr 2006;136:877–81.

[32] Mosig S, Rennert K, Buttner P, Krause S, Lutjohann D, Soufi M, et al. Monocytes of patients with familial hypercholesterolemia show alterations in cholesterol metabolism. BMC Med Genomics 2008;1:60.

[33] Soro-Paavonen A, Naukkarinen J, Lee-Rueckert M, Watanabe H, Rantala E, Soderlund S, et al. Common ABCA1 variants, HDL levels, and cellular cholesterol efflux in subjects with familial low HDL. J Lipid Res 2007;48:1409–16.

[34] Nakanishi S, Vikstedt R, Soderlund S, Lee-Rueckert M, Hiukka A, Ehnholm C, et al. Serum, but not monocyte macrophage foam cells derived from low HDL-C subjects, displays reduced cholesterol efflux capacity. J Lipid Res 2009;50:183–92.

[35] Watanabe H, Soderlund S, Soro-Paavonen A, Hiukka A, Leinonen E, Alagona C, et al. Decreased high-density lipoprotein (HDL) particle size, prebeta-, and large HDL subspecies concentration in Finnish low-HDL families: relationship with intima-media thickness. Arterioscler Thromb Vasc Biol 2006;26:897–902.

[36] Kiss RS, Kavaslar N, Okuhira K, Freeman MW, Walter S, Milne RW, et al. Genetic etiology of isolated low HDL syndrome: incidence and heterogeneity of efflux defects. Arterioscler Thromb Vasc Biol 2007;27:1139–45.

[37] Linsel-Nitschke P, Jansen H, Aherrarhou Z, Belz S, Mayer B, Lieb W, et al. Macrophage cholesterol efflux correlates with lipoprotein subclass distribution and risk of obstructive coronary artery disease in patients undergoing coronary angiography. Lipids Health Dis 2009;8:14.

[38] Huxley R, Barzi F, Woodward M. Excess risk of fatal coronary heart disease associated with diabetes in men and women: meta-analysis of 37 prospective cohort studies. BMJ 2006;332:73–8.

[39] Kontush A, Chapman MJ. Antiatherogenic small, dense HDL—guardian angel of the arterial wall? Nat Clin Pract Cardiovasc Med 2006;3:144–53.

[40] Baynes C, Henderson AD, Anyaoku V, Richmond W, Hughes CL, Johnston DG, et al. The role of insulin insensitivity and hepatic lipase in the dyslipidaemia of type 2 diabetes. Diabet Med 1991;8:560–6.

[41] Farbstein D, Levy AP. HDL dysfunction in diabetes: causes and possible treatments. Expert Rev Cardiovasc Ther 2012;10:353–61.

[42] Syvanne M, Castro G, Dengremont C, De Geitere C, Jauhiainen M, Ehnholm C, et al. Cholesterol efflux from Fu5AH hepatoma cells induced by plasma of subjects with or without coronary artery disease and non-insulin-dependent diabetes: importance of LpA-I: A-II particles and phospholipid transfer protein. Atherosclerosis 1996;127:245–53.

[43] Dullaart RP, Groen AK, Dallinga-Thie GM, de Vries R, Sluiter WJ, van Tol A. Fibroblast cholesterol efflux to plasma from metabolic syndrome subjects is not defective despite low high-density lipoprotein cholesterol. Eur J Endocrinol 2008;158:53–60.

[44] de Vries R, Groen AK, Perton FG, Dallinga-Thie GM, van Wijland MJ, Dikkeschei LD, et al. Increased cholesterol efflux from cultured fibroblasts to plasma from hypertriglyceridemic type 2 diabetic patients: roles of pre beta-HDL, phospholipid transfer protein and cholesterol esterification. Atherosclerosis 2008;196:733–41.

[45] Tan HC, Tai ES, Sviridov D, Nestel PJ, Ng C, Chan E, et al. Relationships between cholesterol efflux and high-density lipoprotein particles in patients with type 2 diabetes mellitus. J Clin Lipidol 2011;5:467–73.

[46] Low H, Hoang A, Forbes J, Thomas M, Lyons JG, Nestel P, et al. Advanced glycation end-products (AGEs) and functionality of reverse cholesterol transport in patients with type 2 diabetes and in mouse models. Diabetologia 2012;55:2513–21.

[47] Curtiss LK, Witztum JL. Plasma apolipoproteins AI, AII, B, CI, and E are glucosylated in hyperglycemic diabetic subjects. Diabetes 1985;34:452–61.

[48] Ferretti G, Bacchetti T, Marchionni C, Caldarelli L, Curatola G. Effect of glycation of high density lipoproteins on their physicochemical properties and on paraoxonase activity. Acta Diabetol 2001;38:163–9.

[49] Hedrick CC, Thorpe SR, Fu MX, Harper CM, Yoo J, Kim SM, et al. Glycation impairs high-density lipoprotein function. Diabetologia 2000;43:312–20.

[50] Lemkadem B, Loiseau D, Larcher G, Malthiery Y, Foussard F. Effect of the nonenzymatic glycosylation of high density lipoprotein-3 on the cholesterol ester transfer protein activity. Lipids 1999;34:1281–6.

[51] Nakhjavani M, Esteghamati A, Esfahanian F, Ghanei A, Rashidi A, Hashemi S. HbA1c negatively correlates with LCAT activity in type 2 diabetes. Diabetes Res Clin Pract 2008;81:38–41.

[52] Matsuki K, Tamasawa N, Yamashita M, Tanabe J, Murakami H, Matsui J, et al. Metformin restores impaired HDL-mediated cholesterol efflux due to glycation. Atherosclerosis 2009;206:434–8.

[53] Murakami H, Tanabe J, Tamasawa N, Matsumura K, Yamashita M, Matsuki K, et al. Reduction of paraoxonase-1 activity may contribute the qualitative impairment of HDL particles in patients with type 2 diabetes. Diabetes Res Clin Pract 2013;99:30–8.

[54] Gowri MS, Van der Westhuyzen DR, Bridges SR, Anderson JW. Decreased protection by HDL from poorly controlled type 2 diabetic subjects against LDL oxidation may be due to the abnormal composition of HDL. Arterioscler Thromb Vasc Biol 1999;19:2226–33.

[55] Nobecourt E, Tabet F, Lambert G, Puranik R, Bao S, Yan L, et al. Nonenzymatic gly-cation impairs the antiinflammatory properties of apolipoprotein A-I. Arterioscler Thromb Vasc Biol 2010;30:766–72.

[56] Sorrentino SA, Besler C, Rohrer L, Meyer M, Heinrich K, Bahlmann FH, et al. Endothelial-vasoprotective effects of high-density lipoprotein are impaired in patients with type 2 diabetes mellitus but are improved after extended-release niacin therapy. Circulation 2010;121:110–22.

[57] Wang H, Peng DQ. New insights into the mechanism of low high-density lipoprotein cholesterol in obesity. Lipids Health Dis 2011;10:176.

[58] Rashid S, Patterson BW, Lewis GF. Thematic review series: patient-oriented research. What have we learned about HDL metabolism from kinetics studies in humans? J Lipid Res 2006;47:1631–42.

[59] Verges B, Petit JM, Duvillard L, Dautin G, Florentin E, Galland F, et al. Adiponectin is an important determinant of apoA-I catabolism. Arterioscler Thromb Vasc Biol 2006;26:1364–9.

[60] Chung S, Sawyer JK, Gebre AK, Maeda N, Parks JS. Adipose tissue ATP binding cas-sette transporter A1 contributes to high-density lipoprotein biogenesis in vivo. Circu-lation 2011;124:1663–72.

[61] Navarro E, Mijac V, Ryder HF. [Ultrasonography measurement of intrabdominal vis-ceral fat in obese men. Association with alterations in serum lipids and insulinemia. Arch Latinoam Nutr 2010;60:160–7.

[62] Attia N, Fournier N, Vedie B, Cambillau M, Beaune P, Ziegler O, et al. Impact of android overweight or obesity and insulin resistance on basal and postprandial SR-BI and ABCA1-mediated serum cholesterol efflux capacities. Atherosclerosis 2009;209:422–9.

[63] Aicher BO, Haser EK, Freeman LA, Carnie AV, Stonik JA, Wang X, et al. Diet-induced weight loss in overweight or obese women and changes in high-density lipoprotein levels and function. Obesity 2012;20:2057–62.

[64] Aron-Wisnewsky J, Julia Z, Poitou C, Bouillot JL, Basdevant A, Chapman MJ, et al. Effect of bariatric surgery-induced weight loss on SR-BI-, ABCG1-, and ABCA1-mediated cellular cholesterol efflux in obese women. J Clin Endocrinol Metab 2011;96:1151–9.

[65] Asztalos BF, Schaefer EJ. HDL in atherosclerosis: actor or bystander? Atheroscler Suppl 2003;4:21–9.

[66] Nestel P, Hoang A, Sviridov D, Straznicky N. Cholesterol efflux from macrophages is influenced differentially by plasmas from overweight insulin-sensitive and -resistant subjects. Int J Obes 2011;36:407–13.

[67] Wang Y, Snel M, Jonker JT, Hammer S, Lamb HJ, de Roos A, et al. Prolonged caloric restriction in obese patients with type 2 diabetes mellitus decreases plasma CETP and increases apolipoprotein AI levels without improving the cholesterol efflux properties of HDL. Diabetes Care 2011;34:2576–80.

[68] Barter P, McPherson YR, Song K, Kesaniemi YA, Mahley R, Waeber G, et al. Serum insulin and inflammatory markers in overweight individuals with and without dyslip-idemia. J Clin Endocrinol Metab 2007;92:2041–5.

[69] Pineda J, Marin F, Marco P, Roldan V, Valencia J, Ruiz-Nodar JM, et al. Premature coronary artery disease in young (age <45) subjects: interactions of lipid profile, thrombophilic and haemostatic markers. Int J Cardiol 2009;136:222–5.

[70] Bondia-Pons I, Ryan L, Martinez JA. Oxidative stress and inflammation interactions in human obesity. J Physiol Biochem 2012;68:701–11.

[71] Visser M, Bouter LM, McQuillan GM, Wener MH, Harris TB. Elevated C-reactive protein levels in overweight and obese adults. JAMA 1999;282:2131–5.

[72] Hardardottir I, Moser AH, Memon R, Grunfeld C, Feingold KR. Effects of TNF, IL-1, and the combination of both cytokines on cholesterol metabolism in Syrian hamsters. Lymphokine Cytokine Res 1994;13:161–6.

[73] Orlov SV, Mogilenko DA, Shavva VS, Dizhe EB, Ignatovich IA, Perevozchikov AP. Effect of TNFalpha on activities of different promoters of human apolipoprotein A-I gene. Biochem Biophys Res Commun 2010;398:224–30.

[74] Haas MJ, Mooradian AD. Regulation of high-density lipoprotein by inflammatory cytokines: establishing links between immune dysfunction and cardiovascular disease. Diabetes Metab Res Rev 2010;26:90–9.

[75] Badimon L, Vilahur G. LDL-cholesterol versus HDL-cholesterol in the atherosclerotic plaque: inflammatory resolution versus thrombotic chaos. Ann NY Acad Sci 2012;1254:18–32.

[76] Shao B, Heinecke JW. Impact of HDL oxidation by the myeloperoxidase system on sterol efflux by the ABCA1 pathway. J Proteomics 2011;74:2289–99.

[77] Tam SP, Flexman A, Hulme J, Kisilevsky R. Promoting export of macrophage cholesterol: the physiological role of a major acute-phase protein, serum amyloid A 2.1. J Lipid Res 2002;43:1410–20.

[78] van der Westhuyzen DR, de Beer FC, Webb NR. HDL cholesterol transport during inflammation. Curr Opin Lipidol 2007;18:147–51.

[79] McGillicuddy FC, de la Llera Moya M, Hinkle CC, Joshi MR, Chiquoine EH, Billheimer JT, et al. Inflammation impairs reverse cholesterol transport in vivo. Circulation 2009;119:1135–45.

[80] Marsche G, Frank S, Raynes JG, Kozarsky KF, Sattler W, Malle E. The lipidation status of acute-phase protein serum amyloid A determines cholesterol mobilization via scavenger receptor class B, type I. Biochem J 2007;402:117–24.

[81] Zhao Y, He X, Shi X, Huang C, Liu J, Zhou S, et al. Association between serum amyloid A and obesity: a meta-analysis and systematic review. Inflamm Res 2010;59:323–34.

[82] Shao B, Tang C, Heinecke JW, Oram JF. Oxidation of apolipoprotein A-I by myeloperoxidase impairs the initial interactions with ABCA1 required for signaling and cholesterol export. J Lipid Res 2010;51:1849–58.

[83] de la Llera Moya M, McGillicuddy FC, Hinkle CC, Byrne M, Joshi MR, Nguyen V, et al. Inflammation modulates human HDL composition and function in vivo. Atherosclerosis 2012;222:390–4.

[84] Jahangiri A, de Beer MC, Noffsinger V, Tannock LR, Ramaiah C, Webb NR, et al. HDL remodeling during the acute phase response. Arterioscler Thromb Vasc Biol 2009;29:261–7.

[85] Goode EL, Cherny SS, Christian JC, Jarvik GP, de Andrade M. Heritability of longitudinal measures of body mass index and lipid and lipoprotein levels in aging twins. Twin Res Hum Genet 2007;10:703–11.

[86] Chantepie S, Bochem AE, Chapman MJ, Hovingh GK, Kontush A. High-density lipoprotein (HDL) particle subpopulations in heterozygous cholesteryl ester transfer protein (CETP) deficiency: maintenance of antioxidative activity. PLoS One 2012;7:26.

[87] Boekholdt SM, Sacks FM, Jukema JW, Shepherd J, Freeman DJ, McMahon AD, et al. Cholesteryl ester transfer protein TaqIB variant, high-density lipoprotein cholesterol levels, cardiovascular risk, and efficacy of pravastatin treatment: individual patient meta-analysis of 13,677 subjects. Circulation 2005;111:278–87.

[88] McCaskie PA, Beilby JP, Chapman CM, Hung J, McQuillan BM, Thompson PL, et al. Cholesteryl ester transfer protein gene haplotypes, plasma high-density lipoprotein levels and the risk of coronary heart disease. Hum Genet 2007;121:401–11.

[89] Thompson A, Di Angelantonio E, Sarwar N, Erqou S, Saleheen D, Dullaart RP, et al. Association of cholesteryl ester transfer protein genotypes with CETP mass and activity, lipid levels, and coronary risk. JAMA 2008;299:2777–88.

[90] Frisdal E, Klerkx AH, Le Goff W, Tanck MW, Lagarde JP, Jukema JW, et al. Functional interaction between -629C/A, -971G/A and -1337C/T polymorphisms in the CETP gene is a major determinant of promoter activity and plasma CETP concentration in the REGRESS Study. Hum Mol Genet 2005;14:2607–18.

[91] Klerkx AH, Tanck MW, Kastelein JJ, Molhuizen HO, Jukema JW, Zwinderman AH, et al. Haplotype analysis of the CETP gene: not TaqIB, but the closely linked -629C--> A polymorphism and a novel promoter variant are independently associated with CETP concentration. Hum Mol Genet 2003;12:111–23.

[92] Lamon-Fava S, Asztalos BF, Howard TD, Reboussin DM, Horvath KV, Schaefer EJ, et al. Association of polymorphisms in genes involved in lipoprotein metabolism with plasma concentrations of remnant lipoproteins and HDL subpopulations before and after hormone therapy in postmenopausal women. Clin Endocrinol 2010;72:169–75.

[93] Tsai MY, Li N, Sharrett AR, Shea S, Jacobs Jr DR, Tracy R, et al. Associations of genetic variants in ATP-binding cassette A1 and cholesteryl ester transfer protein and differences in lipoprotein subclasses in the multi-ethnic study of atherosclerosis. Clin Chem 2009;55:481–8.

[94] Borggreve SE, de Vries R, Dallinga-Thie GM, Wolffenbuttel BH, Groen AK, van Tol A, et al. The ability of plasma to stimulate fibroblast cholesterol efflux is associated with the -629C-->A cholesteryl ester transfer protein promoter polymorphism: role of lecithin: cholesterol acyltransferase activity. Biochim Biophys Acta 2008;1781:10–5.

[95] Villard EF, Khoury PE, Frisdal E, Bruckert E, Clement K, Bonnefont-Rousselot D, et al. Genetic determination of plasma cholesterol efflux capacity is gender-specific and independent of HDL-cholesterol levels. Arterioscler Thromb Vasc Biol 2013;33:822–8.

[96] Borggreve SE, Hillege HL, Wolffenbuttel BH, de Jong PE, Zuurman MW, van der Steege G, et al. An increased coronary risk is paradoxically associated with common cholesteryl ester transfer protein gene variations that relate to higher high-density lipoprotein cholesterol: a population-based study. J Clin Endocrinol Metab 2006;91:3382–8.

[97] Takata M, Inazu A, Katsuda S, Miwa K, Kawashiri MA, Nohara A, et al. (cholesteryl ester transfer protein) promoter -1337 C>T polymorphism protects against coronary atherosclerosis in Japanese patients with heterozygous familial hypercholesterolaemia. Clin Sci (Lond) 2006;111:325–31.

[98] Boes E, Coassin S, Kollerits B, Heid IM, Kronenberg F. Genetic-epidemiological evidence on genes associated with HDL cholesterol levels: a systematic in-depth review. Exp Gerontol 2009;44:136–60.

[99] Hodoglugil U, Williamson DW, Mahley RW. Polymorphisms in the hepatic lipase gene affect plasma HDL-cholesterol levels in a Turkish population. J Lipid Res 2010;51:422–30.

[100] Feitosa MF, Myers RH, Pankow JS, Province MA, Borecki IB. LIPC variants in the promoter and intron 1 modify HDL-C levels in a sex-specific fashion. Atherosclerosis 2009;204:171–7.

[101] Miljkovic-Gacic I, Bunker CH, Ferrell RE, Kammerer CM, Evans RW, Patrick AL, et al. Lipoprotein subclass and particle size differences in Afro-Caribbeans, African Americans, and white Americans: associations with hepatic lipase gene variation. Metabolism 2006;55:96–102.

[102] Sagoo GS, Tatt I, Salanti G, Butterworth AS, Sarwar N, van Maarle M, et al. Seven lipoprotein lipase gene polymorphisms, lipid fractions, and coronary disease: a HuGE association review and meta-analysis. Am J Epidemiol 2008;168:1233–46.

[103] Agirbasli M, Sumerkan MC, Eren F, Agirbasli D. The S447X variant of lipoprotein lipase gene is inversely associated with severity of coronary artery disease. Heart Vessels 2011;26:457–63.

[104] Garcia-Rios A, Delgado-Lista J, Perez-Martinez P, Phillips CM, Ferguson JF, Gjelstad IM, et al. Genetic variations at the lipoprotein lipase gene influence plasma lipid concentrations and interact with plasma n-6 polyunsaturated fatty acids to modulate lipid metabolism. Atherosclerosis 2011;218:416–22.

[105] van Bockxmeer FM, Liu Q, Mamotte C, Burke V, Taylor R. Lipoprotein lipase D9N, N291S and S447X polymorphisms: their influence on premature coronary heart disease and plasma lipids. Atherosclerosis 2001;157:123–9.

[106] Humphries SE, Berglund L, Isasi CR, Otvos JD, Kaluski D, Deckelbaum RJ, et al. Loci for CETP, LPL, LIPC, and APOC3 affect plasma lipoprotein size and subpopulation distribution in Hispanic and non-Hispanic white subjects: the Columbia University BioMarkers Study. Nutr Metab Cardiovasc Dis 2002;12:163–72.

[107] Wood AC, Glasser S, Garvey WT, Kabagambe EK, Borecki IB, Tiwari HK, et al. Lipoprotein lipase S447X variant associated with VLDL, LDL and HDL diameter clustering in the MetS. Lipids Health Dis 2011;10:10–43.

[108] Long S, Tian Y, Zhang R, Yang L, Xu Y, Jia L, et al. Relationship between plasma HDL subclasses distribution and lipoprotein lipase gene HindIII polymorphism in hyperlipidemia. Clin Chim Acta 2006;366:316–21.

[109] Vergeer M, Korporaal SJ, Franssen R, Meurs I, Out R, Hovingh GK, et al. Genetic variant of the scavenger receptor BI in humans. N Engl J Med 2011;364:136–45.

[110] West M, Greason E, Kolmakova A, Jahangiri A, Asztalos B, Pollin TI, et al. Scavenger receptor class B type I protein as an independent predictor of high-density lipoprotein cholesterol levels in subjects with hyperalphalipoproteinemia. J Clin Endocrinol Metab 2009;94:1451–7.

[111] Acton S, Osgood D, Donoghue M, Corella D, Pocovi M, Cenarro A, et al. Association of polymorphisms at the SR-BI gene locus with plasma lipid levels and body mass index in a white population. Arterioscler Thromb Vasc Biol 1999;19:1734–43.

[112] Cerda A, Genvigir FD, Arazi SS, Hirata MH, Dorea EL, Bernik MM, et al. Influence of SCARB1 polymorphisms on serum lipids of hypercholesterolemic individuals treated with atorvastatin. Clin Chim Acta 2010;411:631–7.

[113] Osgood D, Corella D, Demissie S, Cupples LA, Wilson PW, Meigs JB, et al. Genetic variation at the scavenger receptor class B type I gene locus determines plasma lipoprotein concentrations and particle size and interacts with type 2 diabetes: the Framingham study. J Clin Endocrinol Metab 2003;88:2869–79.

[114] Perez-Martinez P, Ordovas JM, Lopez-Miranda J, Gomez P, Marin C, Moreno J, et al. Fernandez de la Puebla RA, Perez-Jimenez F. Polymorphism exon 1 variant at the locus of the scavenger receptor class B type I gene: influence on plasma LDL cholesterol in healthy subjects during the consumption of diets with different fat contents. Am J Clin Nutr 2003;77:809–13.

[115] Wu DF, Yin RX, Hu XJ, Aung LH, Cao XL, Miao L, et al. Association of rs5888 SNP in the scavenger receptor class B type 1 gene and serum lipid levels. Lipids Health Dis 2012;11:11–50.

[116] Chiba-Falek O, Nichols M, Suchindran S, Guyton J, Ginsburg GS, Barrett-Connor E, et al. Impact of gene variants on sex-specific regulation of human Scavenger receptor class B type 1 (SR-BI) expression in liver and association with lipid levels in a population-based study. BMC Med Genet 2010;11:1471–2350.

[117] Juarez-Meavepena M, Carreon-Torres E, Lopez-Osorio C, Garcia-Sanchez C, Gamboa R, Torres-Tamayo M, et al. The Srb1+1050T allele is associated with metabolic syndrome in children but not with cholesteryl ester plasma concentrations of high-density lipoprotein subclasses. Metab Syndr Relat Disord 2012;10:110–6.

[118] Brooks-Wilson A, Marcil M, Clee SM, Zhang LH, Roomp K, van Dam M, et al. Mutations in ABC1 in Tangier disease and familial high-density lipoprotein deficiency. Nat Genet 1999;22:336–45.

[119] Daniil G, Phedonos AA, Holleboom AG, Motazacker MM, Argyri L, Kuivenhoven JA, et al. Characterization of antioxidant/anti-inflammatory properties and apoA-I-containing subpopulations of HDL from family subjects with monogenic low HDL disorders. Clin Chim Acta 2011;412:1213–20.

[120] Frikke-Schmidt R, Nordestgaard BG, Jensen GB, Steffensen R, Tybjaerg-Hansen A. Genetic variation in ABCA1 predicts ischemic heart disease in the general population. Arterioscler Thromb Vasc Biol 2008;28:180–6.

[121] Mantaring M, Rhyne J, Ho Hong S, Miller M. Genotypic variation in ATP-binding cassette transporter-1 (ABCA1) as contributors to the high and low high-density lipoprotein-cholesterol (HDL-C) phenotype. Transl Res 2007;149:205–10.

[122] Yamakawa-Kobayashi K, Yanagi H, Yu Y, Endo K, Arinami T, Hamaguchi H. Associations between serum high-density lipoprotein cholesterol or apolipoprotein AI levels and common genetic variants of the ABCA1 gene in Japanese school-aged children. Metabolism 2004;53:182–6.

[123] Tsai MY, Ordovas JM, Li N, Straka RJ, Hanson NQ, Arends VL, et al. Effect of feno-fibrate therapy and ABCA1 polymorphisms on high-density lipoprotein subclasses in the Genetics of Lipid Lowering Drugs and Diet Network. Mol Genet Metab 2010;100:118–22.

[124] Hovingh GK, de Groot E, van der Steeg W, Boekholdt SM, Hutten BA, Kuivenhoven JA, et al. Inherited disorders of HDL metabolism and atherosclerosis. Curr Opin Lipidol 2005;16:139–45.

[125] Cimmino G, Ibanez B, Vilahur G, Speidl WS, Fuster V, Badimon L, et al. Up-regulation of reverse cholesterol transport key players and rescue from global inflammation by ApoA-I(Milano). J Cell Mol Med 2009;13:3226–35.

[126] Favari E, Gomaraschi M, Zanotti I, Bernini F, Lee-Rueckert M, Kovanen PT, et al. A unique protease-sensitive high density lipoprotein particle containing the apolipoprotein A-I(Milano) dimer effectively promotes ATP-binding Cassette A1-mediated cell cholesterol efflux. J Biol Chem 2007;282:5125–32.

[127] Lebherz C, Sanmiguel J, Wilson JM, Rader DJ. Gene transfer of wild-type apoA-I and apoA-I Milano reduce atherosclerosis to a similar extent. Cardiovasc Diabetol 2007;6:15.

[128] Chen ES, Mazzotti DR, Furuya TK, Cendoroglo MS, Ramos LR, Araujo LQ, et al. de Arruda Cardoso Smith M. Apolipoprotein A1 gene polymorphisms as risk factors for hypertension and obesity. Clin Exp Med 2009;9:319–25.

[129] Morcillo S, Cardona F, Rojo-Martinez G, Almaraz MC, Esteva I, Ruiz-De-Adana MS, et al. Effect of the combination of the variants -75G/A APOA1 and Trp64Arg ADRB3 on the risk of type 2 diabetes (DM2). Clin Endocrinol (Oxf) 2008;68: 102–7.

[130] Carmena-Ramon RF, Ordovas JM, Ascaso JF, Real J, Priego MA, Carmena R. Influence of genetic variation at the apo A-I gene locus on lipid levels and response to diet in familial hypercholesterolemia. Atherosclerosis 1998;139:107–13.

[131] De Oliveira e Silva ER, Kong M, Han Z, Starr C, Kass EM, Juo SH, et al. Metabolic and genetic determinants of HDL metabolism and hepatic lipase activity in normo-lipidemic females. J Lipid Res 1999;40:1211–21.

[132] Delgado-Lista J, Perez-Jimenez F, Tanaka T, Perez-Martinez P, Jimenez-Gomez Y, Marin C, et al. An apolipoprotein A-II polymorphism (-265T/C, rs5082) regulates postprandial response to a saturated fat overload in healthy men. J Nutr 2007;137: 2024–8.

[133] Xiao J, Zhang F, Wiltshire S, Hung J, Jennens M, Beilby JP, et al. The apolipoprotein AII rs5082 variant is associated with reduced risk of coronary artery disease in an Australian male population. Atherosclerosis 2008;199:333–9.

[134] Castro G, Nihoul LP, Dengremont C, de Geitere C, Delfly B, Tailleux A, et al. Cholesterol efflux, lecithin-cholesterol acyltransferase activity, and pre-beta particle formation by serum from human apolipoprotein A-I and apolipoprotein A-I/apolipoprotein A-II transgenic mice consistent with the latter being less effective for reverse cholesterol transport. Biochemistry 1997;36:2243–9.

[135] van't Hooft FM, Ruotolo G, Boquist S, de Faire U, Eggertsen G, Hamsten A. Human evidence that the apolipoprotein a-II gene is implicated in visceral fat accumulation and metabolism of triglyceride-rich lipoproteins. Circulation 2001;104:1223–8.

[136] Calabresi L, Pisciotta L, Costantin A, Frigerio I, Eberini I, Alessandrini P, et al. The molecular basis of lecithin:cholesterol acyltransferase deficiency syndromes: a comprehensive study of molecular and biochemical findings in 13 unrelated Italian families. Arterioscler Thromb Vasc Biol 2005;25:1972–8.

[137] Kathiresan S, Willer CJ, Peloso GM, Demissie S, Musunuru K, Schadt EE, et al. Common variants at 30 loci contribute to polygenic dyslipidemia. Nat Genet 2009;41:56–65.

[138] Teslovich TM, Musunuru K, Smith AV, Edmondson AC, Stylianou IM, Koseki M, et al. Biological, clinical and population relevance of 95 loci for blood lipids. Nature 2010;466:707–13.

[139] Wang K, Edmondson AC, Li M, Gao F, Qasim AN, Devaney JM, et al. Pathway-Wide Association Study implicates multiple sterol transport and metabolism genes in HDL cholesterol regulation. Front Genet 2011;2:41.

[140] Willer CJ, Sanna S, Jackson AU, Scuteri A, Bonnycastle LL, Clarke R, et al. Newly identified loci that influence lipid concentrations and risk of coronary artery disease. Nat Genet 2008;40:161–9.

[141] Kathiresan S, Melander O, Anevski D, Guiducci C, Burtt NP, Roos C, et al. Polymorphisms associated with cholesterol and risk of cardiovascular events. N Engl J Med 2008;358:1240–9.

[142] Voight BF, eloso GM, Orho-Melander M, Frikke-Schmidt R, Barbalic M, Jensen MK, et al. Plasma HDL cholesterol and risk of myocardial infarction: a Mendelian randomisation study. Lancet 2012;380:572–80.

[143] Chasman DI, Pare G, Mora S, Hopewell JC, Peloso G, Clarke R, et al. Forty-three loci associated with plasma lipoprotein size, concentration, and cholesterol content in genome-wide analysis. PLoS Genet 2009;5:e1000730.

CHAPTER 5

Role of ATP-Binding Cassette Transporters A1 and G1 in Reverse Cholesterol Transport and Atherosclerosis

Makoto Ayaori, Katsunori Ikewaki

Division of Anti-aging and Vascular Medicine, Department of Internal Medicine,
National Defense Medical College, Tokorozawa, Japan

Contents

Abstract

Reverse cholesterol transport (RCT) is a pivotal pathway involved in the return of excess cholesterol from peripheral tissues to the liver for excretion in the bile and eventually the feces. RCT from macrophages in atherosclerotic plaques (macrophage RCT) is a critical mechanism of anti-atherogenecity of high-density lipoproteins (HDL). In this paradigm, cholesterol is transferred from arterial macrophages to extracellular HDL through the action of transporters such as ATP-binding cassette transporter A1 (ABCA1) and ATP-binding cassette transporter G1 (ABCG1). Cholesterol efflux from macrophages is the first and one of the most critical mechanisms underlying macrophage RCT. Recent research has provided important insights into the molecular mechanisms of RCT, which may identify targets suitable to develop novel therapies based on pharmacologic enhancement of RCT. This chapter discusses therapeutic strategies for augmenting macrophage RCT via improved macrophage cholesterol efflux and cholesterol efflux acceptor functionality of circulating HDL.

1. INTRODUCTION

Accumulating evidence has demonstrated that treatment with statins reduces cardiovascular events by around 30% in association with a similar degree of low-density lipoprotein (LDL) reduction [1]. However, this partial response leaves a large burden of residual risk for cardiovascular diseases (CVD). A recent study demonstrated that even in patients treated with statins high-density lipoprotein cholesterol (HDL-C) levels remained an independent predictor of the likelihood of suffering cardiovascular events [2]. However, despite intense efforts to develop new pharmacological strategies to increase HDL-C levels, its robust associations with improved clinical outcomes remain sparse [3,4].

The metabolism of HDL involves a number of factors regulating HDL synthesis, intravascular remodeling, and catabolism. As proposed by Glomset [5], HDL plays a key role in the process of reverse cholesterol transport (RCT), in which it promotes the removal of excess cholesterol from peripheral tissues and returns it to the liver for biliary and fecal excretion. Thus, RCT has been recognized as a major anti-atherogenic mechanism of HDL. A major breakthrough in our understanding of the mechanism of RCT came with the discovery of the ATP-binding cassette transporter A1 (ABCA1) in which there is a molecular defect in Tangier disease, a rare genetic disease characterized by low-plasma HDL and increased CVD [6–8]. The low HDL characteristic of Tangier disease patients is caused by decreased cellular cholesterol efflux resulting from ABCA1 transporter mutations and increased catabolism of poorly lipidated apolipoprotein A-I (apoA-I) [9]. In contrast to ABCA1, ABCG1 promotes cholesterol efflux from macrophages to HDL particles but not to lipid-poor apoA-I [10]. Therefore, ABCA1 and ABCG1 are assumed to act in a sequential manner with ABCA1 generating nascent HDL particles, which then facilitate cholesterol efflux via ABCG1, resulting in the formation of mature HDL particles.

In this chapter, we will focus on the current view of the role of ABCA1 and ABCG1 in RCT and HDL metabolism, the implications for atherosclerosis, and potential therapeutic strategies for CVD.

2. THE ROLE OF REVERSE CHOLESTEROL TRANSPORT IN LIPID METABOLISM

Among the various mechanisms of cholesterol homeostasis, accumulating evidence has characterized the regulation of cholesterol biosynthesis and receptor-mediated lipoprotein uptake, which involves the sterol regulatory-element binding protein (SREBP) system at the molecular level. Cholesterol removal from the cells is also important for cholesterol homeostasis both at the cellular level and in the whole body. Although our knowledge of cellular cholesterol removal is still relatively limited, recent studies in this area have contributed significantly to our understanding of its molecular mechanism.

Extracellular cholesterol transport is mainly undertaken by plasma lipoproteins. LDL and other apoB-containing lipoproteins transport cholesterol from the liver to the peripheral cells and direct tissue uptake via receptor-mediated endocytosis. Lipoprotein transport back to the liver from the peripheral cells is also mediated by HDL. The metabolism of HDL is more complex than that of the other major lipoprotein fractions. The individual lipid and apolipoprotein components of HDL are mostly assembled after secretion, and are actively exchanged with or transferred to other lipoproteins mediated by cholesteryl ester (CE) transfer protein (CETP) and

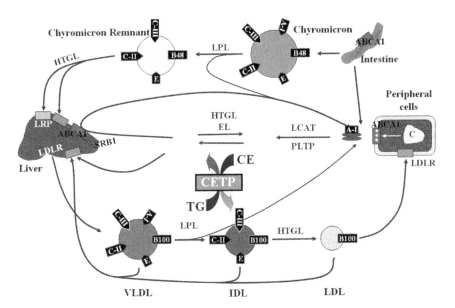

Figure 5.1 *Schematic Overview of Lipoprotein and Cholesterol Metabolism.* ABCA1, ATP binding cassette transporter A1; CE, cholesteryl ester; CETP, cholesteryl ester transfer protein; EL, endothelial lipase; HDL, high-density lipoproteins; HTGL, hepatic triglyceride lipase; IDL, intermediate lipoproteins; LCAT, lecithin-cholesterol acyltransferase; LDL, low-density lipoproteins; LDLR, LDL receptor; LPL, lipoprotein lipase; LRP, LDLR-related protein; PLTP, phospholipids transfer protein; SR-BI, scavenger receptor class B, type I; TG, triglyceride; VLDL, very low-density lipoproteins.

phospholipid transfer protein (PLTP) [11] (Figure 5.1). Further, HDL is actively remodeled within the plasma compartment by lecithin:cholesterol acyltransferase (LCAT) and various lipases [11]. However, the most critical step is the release of cholesterol from cells, a key component of cellular cholesterol homeostasis as discussed later in this chapter.

A mechanism involving RCT from macrophages to the liver and ultimately biliary and fecal excretion is the most popular one for explaining the ability of HDL to inhibit atherosclerosis. This is based on two lines of evidence—that plasma HDL cholesterol level is negatively correlated to the risk of atherosclerotic vascular disease and incubation of cells with HDL results in the reduction of cellular cholesterol in vitro. To systematically elucidate the relationship between HDL and atherosclerotic risk, it is necessary to understand the key regulatory factors that determine the net plasma concentration of HDL particles and the flux of lipids through the HDL and RCT pathways.

3. MAJOR PATHWAYS FOR CELLULAR CHOLESTEROL EFFLUX

The maintenance of cellular cholesterol levels is essential for cell function and viability, since excess free cholesterol (FC) is toxic to cells. Macrophages can protect against cholesterol toxicity by converting FC to CE or by effluxing cholesterol to extracellular acceptors, generally HDL or its related subfractions. Four major pathways have been proposed for the release of cellular cholesterol. One is nonspecific cholesterol efflux from the cellular surface by physicochemical cholesterol exchange between the cell membrane and extracellular acceptors, perhaps mediated by its passive diffusion in the aqueous phase. The net release of cellular cholesterol is driven by the cholesterol concentration gradient, and facilitated through extracellular cholesterol esterification by LCAT. HDL is a major cholesterol acceptor in this reaction because of its capacity for cholesterol accommodation and because it provides an optimum site for the LCAT reaction.

Other important pathways involve the assembly of new HDL particles with cellular phospholipid and cholesterol by means of direct interaction of helical HDL apolipoproteins with cells. This reaction involves specific cellular functions, including: 1) efflux to lipid-poor apolipoproteins, particularly apoA-I, mediated by ABCA1; 2) efflux to mature HDL particles mediated by ABCG1; and 3) efflux to mature HDL particles by other pathways, including scavenger receptor class B type I (SR-BI). While it was once assumed that passive diffusion was the primary process, recent research has demonstrated that macrophage cholesterol efflux occurs largely via ABCA1 and ABCG1.

4. MECHANISMS FOR CHOLESTEROL EFFLUX AND GENERATION OF NASCENT HDL: ROLE OF ABCA1

Several studies have shown that cholesterol-enriched fibroblasts [12] and macrophages [13] from patients with Tangier disease lacking the ability to release phospholipid and cholesterol to lipid-free apolipoproteins though efflux to mature HDL is normal. Since impaired lipidation and subsequent rapid catabolism of lipid-poor apoA-I has been observed in Tangier patients [14] in mediating the efflux of cholesterol and phospholipid from cells, ABCA1 also mediates the lipidation of apoA-I and the formation of nascent HDL. In contrast to passive diffusion and SR-BI-mediated cholesterol flux, the movement of cholesterol by ABCA1 is unidirectional, and net efflux of cellular cholesterol would always occur by this mechanism.

The mechanism for ABCA1-mediated cholesterol/phospholipid efflux involves binding of apoA-I to the membrane, followed by apoA-I-mediated solubilization of the plasma membrane domain to create discoidal nascent HDL particles. Several studies have suggested that the direct interaction of the transporter and apolipoprotein plays a role in ABCA1-mediated efflux [15]. Many models of the ABCA1-mediated efflux of phospholipid and cholesterol to apolipoproteins have been proposed, among them molecular efflux in which individual phospholipid molecules are transferred to apolipoproteins followed by acquisition of cholesterol [16], and membrane solubilization [17] in which phospholipid and cholesterol are mobilized simultaneously as discrete units. However, these models have yet to be verified with solid evidence.

Though the preferred phospholipid substrate(s) for ABCA1 has not been clearly determined, phosphatidylserine (PS) and phosphatidylcholine (PC) are major candidates. Modified lipid distribution in the cell membrane, for which the evidence is PS exofacial flopping, appears to create the biophysical microenvironment required for the docking of apoA-I at the cell surface [18]. In addition, PC-enriched domains that serve as a primary source of phospholipids and cholesterol for ABCA1-mediated efflux have been identified [19]. Although abundant evidence suggests that ABCA1 acts as a phospholipid translocase, no rigorous proof has been presented. In this regard, Vedhachalam and colleagues [20] proposed an interesting model that membrane binding induced by ABCA1 lipid translocase activity created the conditions required for nascent HDL assembly by apoA-I. Hassan and colleagues [21] also demonstrated evidence for a new cellular apoA-I binding site with higher capacity to bind apoA-I compared with the ABCA1 site in fibroblasts stimulated with liver X receptor (LXR) agonists. This new cellular apoA-I binding site was designated as a "high-capacity binding site", supporting a two binding site model for ABCA1-mediated nascent HDL genesis. In association with phospholipid efflux from the cells, two papers demonstrated the role of PLTP in nascent HDL generation. Oram and colleagues [22] reported that the N-terminal domain of PLTP promoted ABCA1-dependent cholesterol efflux and stabilized ABCA1 protein. These findings are supported by the observation by Lee-Rueckert and colleagues [23] that endogenous PLTP produced by macrophages contributes to the optimal function of the ABCA1-mediated cholesterol efflux–promoting machinery in these cells.

A question that should be asked is how lipid-free or lipid-poor stages of apolipoproteins feature in HDL assembly. This stems from the hypothesis that lipid-poor HDL is an efficient "acceptor" of cellular cholesterol, based on its appearance in so-called "preβ-HDL" fractions when the cells are incubated with plasma, though another interpretation of this observation is that these HDL fractions are newly created by apolipoprotein-cell interactions in an ABCA1-dependent fashion [24] (Figure 5.1). Preβ-HDL generally contains two copies of apoA-I per particle and less than 10% by mass of lipid (free cholesterol and phospholipids). Similar particles are regenerated during the catabolism of triglyceride-rich lipoproteins and mature HDL. However, the presence of lipid-poor and lipid-free apoA-I has not been definitively established in vivo, because it is either rapidly reincorporated into mature HDL or relipidated to form preβ-HDL, and irreversibly cleared by the kidneys. It is likely that preβ-HDL particles are formed rapidly when monomeric apoA-I is exposed to cells, and lipid efflux from cells to such acceptor lipoprotein particles can occur by various mechanisms, including regulated transporter-facilitated processes and passive diffusion [25].

5. MECHANISMS FOR MATURATION OF HDL PARTICLES AND MAINTENANCE OF CIRCULATING HDL LEVELS: ROLE OF ABCA1

Many questions as to the site and mechanism of the assembly of HDL particles remain unanswered. The major sites for the synthesis of helical apolipoproteins, especially for the main apolipoproteins of HDL apoA-I, are believed to be the liver and intestines. Unlike apoB-containing lipoproteins, however, no HDL particles (not even of premature HDL) have been identified in secretory pathways such as the endoplasmic reticulum and the Golgi apparatus of liver and intestinal cells. Nevertheless, HDL particles have been found in hepatocyte culture media [26] and in liver perfusates [27]. This is principally nascent HDL composed mainly of surface lipids, phospholipids, and cholesterol, but it does not contain much core lipid and consequently has a disc-like shape. The question is thus: how and where are these particles formed? If the apolipoproteins–cell interaction is a major mechanism for the production of HDL, it is possible that HDL is assembled by an autocrine mechanism, such that apoA-I or E are first secreted by cells and then interact with the cell surface to generate HDL [28].

ABCA1 knockout mice have an extremely low HDL-C phenotype similar to that of humans with Tangier disease [29]. Interestingly, in mice a macrophage-specific deficiency of ABCA1 had a minimal effect in reducing plasma HDL-C levels [30], while a liver-specific deficiency of ABCA1 dramatically reduced HDL levels [31]. Further, Brunham and colleagues [32] demonstrated that approximately 30% and 80% of the steady-state plasma HDL pool is contributed by, respectively, intestinal and hepatic ABCA1 in mice, and suggested that HDL derived from intestinal ABCA1 was secreted directly into the circulation and that HDL in lymph was predominantly derived from the plasma compartment. Thus, hepatic and intestinal ABCA1 appear to be critical for the initial lipidation of newly secreted lipid-poor apoA-I, protecting it from rapid degradation and allowing it to go on to form mature HDL. As described earlier, hepatic ABCA1 predominantly contributes to circulating HDL levels as compared to intestinal levels; however, several reports have demonstrated that, in particular settings, intestinal ABCA1 generated HDL particles independent of the liver. Brunham and colleagues [33] observed that increased HDL levels induced by an LXR agonist were comparable between wild-type mice and those lacking hepatic ABCA1, and this effect of the LXR agonist on HDL levels was completely abrogated in mice lacking intestinal ABCA1. Singaraja and colleagues [34] reported that infusion of human apoA-I/phospholipid reconstituted HDL particles, normalized plasma HDL-C levels in liver-specific ABCA1 knockout mice, but had no effect on HDL-C levels in whole body knockout mice. Although these studies have clearly demonstrated the role of ABCA1 in HDL metabolism, the data from mice studies must be accepted with caution because mice lack CETP expression.

In human settings, it is clear that defective ABCA1 has resulted in marked reductions in circulating HDL levels as was extensively observed in patients with Tangier disease [6–8]. Marcil and colleagues [35] demonstrated that, despite the absence of typical physical findings in Tangier disease (e.g., orange tonsil), mutations in ABCA1 were the major cause of familial HDL deficiency associated with defective cholesterol efflux. This identification was made by using skin fibroblasts derived from the patients. Clee and colleagues also reported that each 8% change in ABCA1-mediated efflux due to its mutations was predicted to be associated with a 0.1 mmol/L change in HDL-C [36]. The patients with heterozygous ABCA1 deficiency have an approximately 40–45% decrease in HDL-C and apoA-I compared with unaffected family members. Several researchers have reported that single nucleotide polymorphisms in the *ABCA1* gene are significantly associated with low HDL-C levels [37–41]. Recent genome-wide association studies have also revealed the association of ABCA1 locus and low HDL levels [42–45].

6. THE ROLE OF ABCG1 IN CHOLESTEROL EFFLUX AND HDL METABOLISM

ABCG1 has been reported to mediate the efflux of cholesterol to HDL [10], but the mechanism by which ABCG1 does this still remains unclear. In contrast to ABCA1, which specifically promotes efflux of cholesterol to the acceptor apoA-I [20], the efflux activity of ABCG1 is relatively nonspecific, as it can promote efflux not only to HDL but also to LDL and to cyclodextrin [10]. Moreover, although ABCG1 can traffic to the plasma membrane, several observations in various cell types have demonstrated that most of the transporter is intracellular [46]. It remains unclear whether ABCG1 is mainly mobilized to the cell surface to support cholesterol efflux [46] or whether its main function is to regulate intracellular cholesterol distribution. Recently, Tarling and Edwards [47] reported that ABCG1 is primarily an intracellular sterol transporter that localizes to endocytic vesicles, facilitating the redistribution of specific intracellular sterols away from the endoplasmic reticulum.

Regarding the role of ABCG1 in circulating HDL levels in mice, ABCG1 knockout mice demonstrated low HDL-C levels under conditions of dietary and pharmacological (LXR) activation as compared to wild-type mice. However, though there is no study showing a relationship between the genomic ABCG1 locus and circulating HDL-C levels in humans,

Furuyama and colleagues [48] reported that in Japanese men a novel ABCG1 -257T>G promoter polymorphism influences coronary artery diseases (CAD) severity, but not lipid or HDL levels. These findings were confirmed in a Chinese population [49]. Association analysis in a cohort containing extremes of HDL-C (case-control, $n = 1733$), and replication cohorts in three additional populations consisting of 7857 individuals revealed that HDL-C levels were independently associated with the ABCG1 locus [50]. However, its association was much less remarkable as compared to CETP, endothelial lipase, lipoprotein lipase, and hepatic lipase. Thus, to date, further studies are needed to conclude that ABCG1 contributes to circulating HDL-C levels.

7. TRANSCRIPTIONAL REGULATION OF ABCA1/G1

As described earlier, ABCA1 and ABCG1 play important roles in cellular cholesterol homeostasis and their expression levels are upregulated by loading cells with cholesterol and downregulated by depleting cholesterol. A ligand-dependent nuclear receptor, liver receptor X (LXR), is involved in such a mechanism. Oxysterols and 9-cis-retinoic acid (9cRA) activate transcription by binding to LXR and retinoid X receptor (RXR), respectively, to form heterodimers that bind to conserved LXR-responsive elements (LXREs). These LXREs contain direct repeats spaced by four nucleotides (DR4), and are found in the ABCA1 [51] and ABCG1 [52–54] promoter regions. Endogenous oxysterols are derived from intracellular cholesterol [55].

Other examples of transcriptional regulation of ABCA1 have been discovered. The suppression of acyl-CoA: cholesterol acyltransferase activity increases cellular unesterified cholesterol and transactivates ABCA1, whether this is done pharmacologically [56] or genetically [57]. ABCA1/G1 in macrophages are transcriptionally regulated by ligand-dependent nuclear receptors, such as the peroxisome proliferator–activated receptor γ (PPARγ)-LXR pathway [58]. Induction of ABCA1 by PPARγ agonists does not occur in the absence of LXRα/β expression in mice [58], indicating that the effects of PPARγ are dependent on LXRs. LXRβ compensated for LXR-mediated transactivation of the respective target genes, including ABCA1/G1, in LXRα-deficient mice [59]. In our study [60], however, LXRβ was not affected by PPARγ agonists, consistent with previous studies [58]. Indeed, a PPARγ-responsive element has been reported only in the LXRα [61], but not the LXRβ promoter. Retinoic acid receptor (RAR)/RXR heterodimer can also bind LXRE in ABCA1/G1 promoters and transactivate ABCA1 [62] and ABCG1 [63] in macrophages.

Tamehiro and colleagues [64] reported that the dual promoter system driven by SREBP-2 and LXR regulates hepatic ABCA1 expression. In this study a liver type promoter located upstream of exon 2 was activated via its sterol-responsive element under cholesterol depletion; in sharp contrast, a peripheral type promoter located upstream of exon 1 was activated by its LXRE under cholesterol overload. Finally, HMG-CoA reductase inhibitors (statins) are widely known to not only reduce LDL-C levels, but also raise HDL-C levels [65], a finding that may be due in part to increased liver-specific transcripts via activation of liver type promoter.

Other transcriptional regulation of ABCA1 has been observed. Analysis of the ABCA1 promoter revealed several promoter elements, including an E-box, as well as binding motifs for Sp1, Sp3, and AP1 [66]. Both USF1 and USF2 can bind to the E-box in the promoter of the ABCA1 transporter gene [66] and SREBP-2 also binds to E-box, resulting in decreased ABCA1 promoter activity [67]. Arakawa and colleagues [68] reported that PPARα agonists and fibrates increase transcription of ABCA1 in an LXR-dependent manner. In murine (but not human) cells, cyclic AMP (cAMP) strongly induces ABCA1 gene transcription and this increases ABCA1 expression and apoA-I-mediated cell lipid release [69]. Phosphorylated cAMP-responsive element (CRE) binding protein 1 controls the cAMP-mediated induction of murine ABCA1 gene expression through a CRE site [70]. Iwamoto and colleagues [71] reported that the antihypertensive drug, doxazosin, inhibits AP2α activity independent of α1-adrenoceptor blockade and increases ABCA1 expression and HDL biogenesis. They also demonstrated that protein kinase D phosphorylates AP2α to negatively regulate expression of the ABCA1 gene [72]. Ohoka and colleagues [73] identified a regulatory enhancer element for expression of a liver-specific transcript lying within intron 3 of the human *ABCA1* gene, to which hepatocyte nuclear factor 4α binds in response to cholesterol depletion.

Multiple promoters for the gene encoding ABCG1 have been reported by several investigators. Figure 5.2 shows the genomic organization and reported transcripts of human ABCG1. Lorkowski and colleagues [74] identified several transcripts containing variable combinations of exons 1–6 in cDNA derived from THP-1 macrophages. These transcripts were assumed to result from alternative splicing and/or the use of four putative promoters postulated to be upstream of exons 1, 4, 5, or 6. Four additional transcripts were identified by Kennedy and colleagues [52] using cDNA that derived

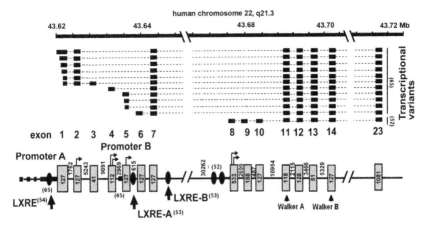

Figure 5.2 *Genomic Organization of Human ABCG1 and LXREs.* A, the 23 exons (indicated by shaded boxes, numbers indicate size of each exon) and the introns (numbers indicate size of each intron) of the human ABCG1 gene are illustrated (window from the Ensembl Genome Browser: http://uswest.ensembl.org/index.html). Putative promoters (bold lines), LXRE (closed ellipses), and translation initiation sites (horizontal arrows) are also indicated. Reported transcripts are indicated by closed boxes connected by dashed lines with the corresponding reference numbers in parentheses.

from THP-1 macrophages. These new transcripts resulted in the identification of exons 8–10, a putative promoter, and two LXREs upstream of exon 8. However, these authors observed that the activity of this promoter (assayed by luciferase constructs) was only two-fold greater than in a promoterless luciferase vector, implying that the exon 8–10 promoter did not greatly contribute to overall ABCG1 expression. Uehara and colleagues [54] also reported an LXRE in the promoter upstream of exon 1, which was designated as promoter A. Sabol and colleagues [53] reported that the promoter upstream of exon 5, designated as promoter B, was LXR ligand-responsive in the presence of three LXREs located in intron 5 or 7. To date, the promoter activity of the 5'-flanking region upstream of exons 4 and 6 has not been examined. Jackobsson and colleagues [75] confirmed the role of these LXREs in ABCG1 transcription, and observed that upon ligand activation a cofactor for nuclear receptors, G protein pathway suppressor 2, facilitated LXR recruitment to an ABCG1 promoter/enhancer (LXREs downstream of *ABCG1* gene) unit, thus identifying functional links to histone H3K9 demethylation. By performing promoter assays and measuring each transcript driven by promoters A and B, Ozasa and colleagues [60] demonstrated that promoter A activity was 20-fold lower than that of promoter B, indicating ABCG1 expression is mainly regulated by promoter B in macrophages.

8. POSTTRANSCRIPTIONAL/POSTTRANSLATIONAL REGULATION OF ABCA1/G1

ABCA1 is also regulated by posttranslational mechanisms. Recently, miR-33, an intronic microRNA (miRNA) located within the gene encoding SREBP-2, was found to modulate ABCA1/G1 expression in a posttranscriptional level [76–78] in mice. Peritoneal macrophages derived from miR-33 null mice demonstrated a marked increase in ABCA1 levels and higher cholesterol efflux than those from WT mice. Hepatic ABCA1 expression was also greater in miR-33 null mice, resulting in raised circulation HDL levels [78]. Moreover, anti-miR33 oligonucleotide treatment accompanied by raising HDL levels promotes RCT and inhibits atherosclerosis development in mice [79,80]. More recently, miR-758 was reported to be a novel miRNA that controls ABCA1 levels and regulates macrophage cellular cholesterol efflux to apoA-I [81]. Coffee polyphenols enhance ABCG1 expression by stabilizing mRNA in a 3'-untranslated lesion-dependent manner [82].

ABCA1 is regulated by posttranslational protein modification and protein–protein interaction. ApoA-I interaction with ABCA1 prevented phosphorylation of the PEST sequence in a cytoplasmic domain of ABCA1, resulting in less degradation by calpain proteolysis and increased surface expression of ABCA1 [83]. PEST sequence is also important for internalization of ABCA1 protein and cholesterol efflux from late endosomal cholesterol pools [84].

Palmitoylation can regulate ABCA1 localization at the plasma membrane, and regulate its cholesterol efflux capacity [85]. ABCA1 was robustly palmitoylated at cysteines in various positions, and prevention of palmitoylation of ABCA1 by mutation of the cysteines decreased both ABCA1 localization at the plasma membranes and ABCA1-dependent cholesterol efflux. On the other hand, Tamehiro and colleagues [86] observed that a serine palmitoyltransferase enzyme, SPTLC1, bound to ABCA1 and attenuated cholesterol efflux.

ABCA1 interacts with several molecules involving cytoskeletal formation and to adapter proteins that localize a variety of signaling molecules to specific intracellular locations. This mechanism to regulate intracellular cholesterol localization is mediated by a PDZ protein-binding motif located in the terminal four residues of the cytoplasmic tail of ABCA1 transporter. The syntrophin family contains intracellular membrane–associated proteins that interact with both ion channels and signaling proteins. β1-syntrophin acts through a class-I PDZ interaction with the C terminus of ABCA1 to regulate the cellular distribution and activity of the transporter [87]. β2-syntrophin may also play a pivotal role in the retaining of ABCA1 in cytoplasmic vesicles and for the targeting of ABCA1 to plasma membrane microdomains [88]. α1-syntrophin is involved in intracellular signaling, which determines the stability of ABCA1 and modulates cellular cholesterol efflux [89]. Syntrophins associate with utrophin, a large scaffolding molecule that links to actin cytoskeleton. Thus, cell surface ABCA1 could be stabilized by being anchored to the actin cytoskeleton.

Okuhira and colleagues [90] identified a physical and functional interaction between ABCA1 and PDZ-RhoGEF/LARG, which activated RhoA, resulting in ABCA1 stabilization, and promoted cholesterol efflux. Their observation that ABCA1 interacted with PDZ-RhoGEF and that apoA-I provoked RhoA activation suggests that ABCA1 plays an important role not only in transporting cholesterol but also in transmitting signals from apoA-I to activate RhoA and downstream molecules. This model is reasonable because apoA-I directly associates with ABCA1 [15], and apoA-I stimulates signaling of Cdc42, another Rho GTPase [91]. Similarly, protein kinase C and JAK2 kinases are also stimulated by apoA-I in an ABCA1-dependent fashion [92]. Thus, ABCA1 may act as an apoA-I receptor to transmit signals to the cytoplasm by associating with RhoGEFs. This pathway might further modulate cholesterol efflux in a feed-forward fashion by inhibiting the degradation of ABCA1.

LXRs also contribute to ABCA1 transactivation in response to cellular cholesterol content. Hozoji and colleagues [93] identified a novel posttranslational mechanism of LXR-dependent ABCA1 regulation. In their study, the LXRβ/RXR complex bound directly to ABCA1 on the plasma membrane of macrophages and modulated cholesterol efflux. When intracellular cholesterol levels were low, LXRβ/RXR complex anchored ABCA1 to the plasma membrane; once cholesterol accumulated, oxysterols bound to LXRβ, leading to dissociation of the LXRβ/RXR complex from ABCA1. Thus, ABCA1 activity is restored, allowing apoA-I-dependent cholesterol efflux to proceed. This is the first report that observed a nongenomic function of LXR.

The ubiquitin-proteasome system (UPS) mediates nonlysosomal pathways for protein degradation and is known to be involved in atherosclerosis. Wang and colleagues [83] have reported that ABCA1 protein is ubiquitinated in the presence of lactacystin. Moreover, Azuma and colleagues [94] observed proteasomal degradation of ABCA1 and interaction between ABCA1 and the COP9 signalosome, a key molecule controlling protein ubiquitination and deubiquitination. These reports led to the findings observed by Ogura and colleagues [95], who demonstrated that ABCA1/G1 was degraded via polyubiquitination in the UPS. Furthermore, inhibition of the UPS using proteasome inhibitors enhanced both apoA-I- and HDL-mediated cholesterol efflux from macrophages by increasing ABCA1/G1 expressions. Finally, bortezomib, a proteasome inhibitor, promoted overall RCT in vivo in mice. These findings may ultimately provide the basis for a novel therapeutic strategy for atherosclerotic diseases.

Compared with ABCA1, there is scarce evidence of posttranscriptional regulation or protein–protein interactions involving ABCG1. Hori and colleagues [96] reported that calpain promoted ABCG1 degradation by cleaving cell surface–resident ABCG1, and consequently attenuated cholesterol efflux. Nagelin and colleagues [97] demonstrated that ABCG1 serine phosphorylation was induced during degradation by eicosanoids. They also observed that proteasomal inhibition by eicosanoids and lactacystin prevented ABCG1 degradation and induced the accumulation of phosphorylated ABCG1 [98]. Since phosphorylation of ABCA1 also accelerates its own degradation, there might be similar, if not identical, machinery involving ABCA1 and ABCG1 degradation via the UPS.

9. THE ROLE OF ABCA1/G1 IN REVERSE CHOLESTEROL TRANSPORT AND ATHEROSCLEROSIS IN ANIMAL MODELS

The RCT system consists of various steps between the efflux of cholesterol from macrophage foam cells and the final excretion of cholesterol into bile either as cholesterol or bile acids (Figure 5.3) [99]. Cholesterol efflux occurs from macrophages as FC via ABCA1 with poorly lipidated apoA-I, resulting in formation of nascent HDL particles, which further take up cholesterol via ABCG1 and SR-BI. Within HDL, cholesterol is esterified by LCAT to form CE, which moves to the core of HDL particles thereby allowing HDL to take up additional FC. Cholesterol is then transported back to the liver via the systemic circulation. HDL is mainly taken up by the liver through SR-BI, which positively regulates RCT despite reduction in circulating HDL-C levels. HDL-derived CE is de-esterified and secreted into the bile as FC or as bile acids. Humans, but not rodents, demonstrate CETP expression, which provides a shunt between HDL-mediated RCT

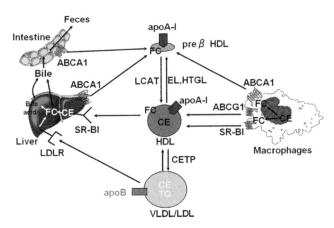

Figure 5.3 *Overview of Reverse Cholesterol Transport System.* ABCA1, ATP-binding cassette transporter A1; ABCG1, ATP-binding cassette transporter G1; apo, apolipoproteins; CE, cholesteryl ester; CETP, cholesteryl ester transfer protein; EL, endothelial lipase; FC, free cholesterol; HDL, high-density lipoproteins; HTGL, hepatic triglyceride lipase; LCAT, lecithin:cholesterol acyltransferase; LDL, low-density lipoproteins; LDLR, LDL receptor; SR-BI, scavenger receptor class B type I; TG, triglyceride; VLDL, very low-density lipoproteins.

and metabolic pathways of apoB–containing lipoproteins, transporting cholesterol to liver via the LDL receptor. Finally, fecal excretion of macrophage-derived cholesterol can be also affected by absorption via an enterohepatic circulation [100].

To evaluate RCT in vivo, centripetal cholesterol flow from extrahepatic organs to the liver were initially determined [101]. However, macrophage foam cell–derived cholesterol is not able to be specifically traced by this method. Although macrophage-specific RCT represents only a small fraction of total flux through the pathway, it may warrant particular focus given the critical importance of cholesterol-laden macrophage foam cells in the development of atheromatous lesions. In this perspective, Zhang and colleagues [102] introduced a novel in vivo method to specifically trace the movement of cholesterol from macrophages to plasma, liver, and feces (Figure 5.4). In brief, mice were intraperitoneally injected with macrophages loaded with radiolabeled cholesterol, which was supposed to be transferred to the systemic circulation by HDL and eventually to the liver. RCT in vivo was evaluated by counting tracer levels in plasma, liver, bile, and feces. Indeed, this murine assay that quantifies macrophage RCT has proven to be a better predictor of atherosclerosis in mice than HDL cholesterol concentration [103].

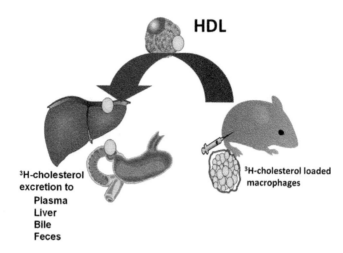

HDL

^{3}H-cholesterol
excretion to
    Plasma
    Liver
    Bile
    Feces

^{3}H-cholesterol loaded
macrophages

Figure 5.4 *A Method Evaluating Reverse Cholesterol Transport In Vivo.*

Macrophages deficient in ABCA1 expression have significantly impaired cholesterol efflux to lipid-poor apoA-I in vitro. Consistent with this observation, mice that were transplanted with bone marrow from ABCA1 knockout mice increased the development of atherosclerotic lesions despite maintaining normal plasma HDL-C levels [104], whereas mice that were transplanted with bone marrow from ABCA1-overexpressing mice showed reduced atherosclerosis [105]. Further supporting these findings, ABCA1-deficient macrophages also demonstrated significantly reduced RCT in vivo [106].

ABCA1 and ABCG1 also act in concert to promote macrophage cellular cholesterol efflux. For example, ABCA1 can lipidate lipid-poor apoA-I to generate nascent HDL particles, which can then serve as substrates for ABCG1-mediated cholesterol efflux. In fact, knockdown of both ABCA1 and ABCG1 in macrophages reduced cholesterol efflux ex vivo and macrophage RCT in vivo to a greater extent than loss of function of either transporter alone [106]. Moreover, transplantation of ABCA1/G1 double-knockout bone marrow into atherosclerosis-prone mice resulted in substantially greater atherosclerosis than bone marrow from either single knockout [107]. Supporting these observations, a synthetic LXR agonist, GW3965, enhanced macrophage RCT in mice accompanied with robust increase in ABCA1/G1 expression [108].

ABCG1 facilitated cholesterol efflux to mature HDL, but not to lipid-poor apoA-I [10,109]. Mice that were deficient in ABCG1 demonstrated lipid accumulation in macrophages within multiple tissues when they were fed a high-fat, high-cholesterol diet [109]. Macrophages lacking ABCG1 expression had impaired cholesterol efflux to mature HDL ex vivo and demonstrated significantly reduced RCT in vivo [106], which translated into inhibition of atherosclerosis development in mice [107] and rabbits [110]. In contrast, several investigators have reported opposite results; macrophage deficiency of ABCG1 is paradoxically associated with decreased atherosclerosis [111,112], possibly due to compensatory upregulation of macrophage ABCA1 and apoE or increased susceptibility of ABCG1-deficient macrophages to induce oxidized LDL-mediated apoptosis [111]. Meurs and colleagues [113] reported that the effect of ABCG1 deficiency on lesion development in LDL receptor–null mice depends on the stage of atherogenesis, whereby the absence of ABCG1 led to increased lesions while in more advanced stages of atherosclerosis enhanced apoptosis and/or compensatory mechanisms led to retarded lesion progression. It is possible that these observations provide an explanation for the discrepant results in ABCG1 deficiency.

10. THE ROLE OF ABCA1/G1 IN ATHEROSCLEROSIS IN HUMANS

In humans, initial studies on obligate Tangier heterozygotes have reported conflicting findings [114]. However, several recent reports have suggested that ABCA1 mutations were associated with increased cardiovascular events and atherosclerosis burden. Clee and colleagues reported that symptomatic vascular disease was more than three times as frequent in the adult heterozygotes as in unaffected family members. Interestingly, the presentation of vascular disease was generally more severe in the heterozygotes than in their unaffected family members. van Dam and colleagues [115] observed that subjects with ABCA1 mutations had lower amounts of cholesterol efflux from their own skin fibroblasts, lower HDL cholesterol concentrations, and greater intima-media thicknesses than controls. Further, intima-media thickness in the carotid artery at the upper limit of normal was reached by age 55 years in the ABCA1 heterozygotes, but not until age 80 years in unaffected controls. Besides ABCA1 mutations causing Tangier disease and familial HDL deficiency, single-nucleotide polymorphisms in ABCA1 locus have been associated with CVD risk and severity of atherosclerosis. Carriers with a common ABCA1 variant, R219K (frequency, 46% in Europeans) have a reduced severity of CAD and fewer coronary events as compared to non-carriers [116]. Moreover, Cenarro and colleagues [117] confirmed the above observations in patients with familial hypercholesterolemia (FH); the K allele of the R219K variant was significantly more frequent in FH subjects without premature CAD than in FH subjects with premature CAD. Kyriakou [118] also demonstrated that common ABCA1 functional polymorphisms affected atherosclerosis development; ABCA1 gene -565C>T polymorphism was associated with severity of coronary atherosclerosis. They further observed that age of symptom onset was associated with the promoter -407G>C polymorphism, which yield higher promoter activity, being 2.82 years higher in C allele homozygotes than in G allele homozygotes and intermediate in heterozygotes [119]. They also found a nonsynonymous variant, V825I, had a higher activity in mediating cholesterol efflux than the wild-type, and a trend toward higher symptom onset age in 825I allele carriers was observed. On the other hand, a negative result has been recently reported [120].

In contrast to ABCA1, there is scant evidence demonstrating whether ABCG1 affects atherosclerotic development in humans. Furuyama and colleagues [48] reported, however, that a novel ABCG1 -257T>G promoter polymorphism, which causes lower ABCG1 transcription activity, influences CAD severity in Japanese men.

11. THERAPEUTIC STRATEGIES AGAINST ATHEROSCLEROSIS INVOLVING RCT MODIFICATION

There have been accumulating trials to treat and prevent atherosclerotic diseases based on the strategy of promoting RCT, in particular, mediated by increasing ABCA1/G1. LXRs serve as the major regulators of macrophage ABCA1/G1 expression, and synthetic LXR agonists have been shown to promote macrophage RCT [108] and decrease atherosclerosis in mouse models [121]. Studies have highlighted the primary role of the macrophage to exert anti-atherogenic activity of LXR agonists [122]. However, development of first-generation LXR agonists has been hampered by the induction of hepatic steatosis, which might be due to increased fatty acid biosynthesis via increased expression of SREBP 1-c [123].

A proof-of-mechanism study was performed in human cells using LXR-623, a nonselective small molecule agonist of LXR [124]. This compound increased expression of ABCA1/G1 in blood leukocytes obtained from human subjects. However, adverse effects related to the central nervous system were noted in more than half of the patients, leading to termination of the study.

Because LXRα is the predominant subtype in the liver, selective agonism of LXRβ may help overcome the unfavorable hepatic effects observed in nonselective stimulation. Indeed, LXRβ-selective agonists have been developed to show the ability to promote macrophage cholesterol efflux [125]. Alternatively, restricting LXR activation to the small intestine might also result in an increase in intestinal HDL formation via ABCA1, without developing fatty liver [33]. Thus, LXR agonists remain a highly plausible and conceptually attractive therapeutic target, particularly if it can be accomplished with selective targeting of the macrophage or intestine.

Pioglitazone, a PPARγ agonist, is an oral antidiabetic agent that positively affects blood glucose and lipids by improving insulin sensitivity. Recently, it has been reported to have an anti-atherogenic effect in Type 2 diabetic patients, in comparison with a sulfonylurea agent [126,127]. These two trials demonstrated that pioglitazone treatment achieved significant regression of atherosclerotic plaques as detected by carotid ultrasonography [127] and coronary intravascular ultrasonography, as compared with glimepiride, though both drugs provided comparable glycemic control [126]. Overall, these favorable effects of pioglitazone on atherosclerosis can be interpreted as nonmetabolic, pleiotropic effects on the vasculature. Ozasa and colleagues [60] reported that the vitro findings were borne out by an ex vivo experiment which showed that ABCA1/G1 expressions and cholesterol efflux from macrophages were enhanced when cultured with sera obtained from diabetic patients after administration of a clinical dose of pioglitazone. Based on these observations, it is conceivable that pioglitazone would provide macrophage-associated anti-atherogenesis in humans, in vivo. It has yet to be demonstrated, however, that these effects on carotid plaques are sufficiently robust to result in improved clinical outcomes.

PPARα regulates gene expression in response to the binding of fatty acids and their metabolites. Although PPARα stimulation has multiple effects on lipid metabolism, including increased apoA-I production [128], we recently found that a specific PPARα agonist GW7647 promoted macrophage RCT in vivo, and this promotion was due to the enhanced macrophage cholesterol efflux by increasing ABCA1 and ABCG1 expression through the PPARα-LXR pathway, even under the condition of increased human apoA-I levels [129]. However, disappointing results of two recent large clinical trials using the weak PPARα agonist fenofibrate [130,131] have reduced the enthusiasm for this approach. While fibrates are weak PPARα agonists, several agents with substantially increased potency and selectivity for PPARα have been developed. One of these compounds was shown to upregulate apoA-I production by 30% in humans, although it was not assessed for its effects on blood leukocyte ABCA1 upregulation [132]. Combined with the data provided earlier suggesting that more potent PPARα agonists have a greater potential to activate macrophage RCT, it remains possible that more potent PPARα agonists would have a greater effect on atherosclerotic cardiovascular disease than using fibrates has shown.

Cilostazol, a selective inhibitor of phosphodiesterase 3, inhibits platelet aggregation and has been widely used to treat peripheral artery diseases and ischemic stroke worldwide. Collective clinical data [133,134] further demonstrate that cilostazol increases plasma concentrations of HDL, an antiatherogenic lipoprotein inversely associated with risk of atherosclerotic diseases [135]. Nakaya and colleagues [136] showed that cilostazol increased ABCA1/G1 gene expression and stimulated macrophage ABCA1/G1 expression in a cAMP/protein kinase A–independent manner. It also enhanced cholesterol efflux mediated by apoA-I and HDL, resulting in promotion of RCT in vivo. These findings and previous human studies on HDL metabolism indicate that cilostazol might provide anti-atherosclerotic effects by promoting RCT in addition to raising HDL-C levels.

Natural nutrients have been a focus for the strategy of promoting RCT. A recent meta-analysis demonstrated that coffee consumption was inversely associated with total and CVD mortality [137]. Uto-Kondo and colleagues [82] showed that coffee's phenolic acids induced ABCG1 and SR-BI expressions, leading to the increased HDL-mediated cholesterol efflux from macrophages. Furthermore, a cross-over human study revealed that single oral administration of coffee raised plasma levels of phenolic acids and their metabolites, and induced ABCG1 and SR-BI expression, thus leading to the enhanced cholesterol efflux. In line with the in vitro finding, ferulic acid was found to promote overall RCT in vivo in mice. These comparable results obtained from in vitro, ex vivo, and in vivo studies indicate that the potential anti-atherogenic properties of coffee proposed from previous studies could be explained, at least in part, by the upregulated cholesterol efflux from macrophages mediated through ABCG1 and SR-BI pathways.

12. CONCLUSION

In summary, the metabolism of HDL involves a complex interaction of factors that regulate the synthesis, intravascular remodeling, and catabolism of HDL. Observations in Tangier patients and the findings of intensive studies on ABCA1 indicate that the most important step in RCT is cholesterol efflux from peripheral cells. Also, although HDL promotes RCT from extrahepatic tissues to the liver, the liver itself is a major site for the lipidation of nascent HDL, and the cholesterol pool in macrophage that is effluxed to HDL and returned to the liver is probably the most important aspect in protection from atherosclerosis. New therapeutic strategies against atherosclerosis will be developed as more progress is made in our understanding of cholesterol efflux, HDL metabolism, and RCT in conjunction with ABCA1/G1 regulation.

REFERENCES

[1] Kearney PM, Blackwell L, Collins R, Keech A, Simes J, Peto R, et al. Efficacy of cholesterol-lowering therapy in 18,686 people with diabetes in 14 randomised trials of statins: a meta-analysis. Lancet 2008;371:117–25.

[2] Mureddu GF, Brandimarte F, De Luca L. High-density lipoprotein levels and risk of cardiovascular events: a review. J Cardiovasc Med (Hagerstown) 2012;13:575–86.

[3] Singh IM, Shishehbor MH, Ansell BJ. High-density lipoprotein as a therapeutic target: a systematic review. JAMA 2007;298:786–98.

[4] Briel M, Ferreira-Gonzalez I, You JJ, Karanicolas PJ, Akl EA, Wu P, et al. Association between change in high density lipoprotein cholesterol and cardiovascular disease morbidity and mortality: systematic review and meta-regression analysis. BMJ 2009;338; b92.

[5] Glomset JA. The plasma lecithins:cholesterol acyltransferase reaction. J Lipid Res 1968;9:155–67.

[6] Brooks-Wilson A, Marcil M, Clee SM, Zhang LH, Roomp K, van Dam M, et al. Mutations in ABC1 in Tangier disease and familial high-density lipoprotein deficiency. Nat Genet 1999;22:336–45.

[7] Bodzioch M, Orso E, Klucken J, Langmann T, Bottcher A, Diederich W, et al. The gene encoding ATP-binding cassette transporter 1 is mutated in Tangier disease. Nat Genet 1999;22:347–51.

[8] Rust S, Rosier M, Funke H, Real J, Amoura Z, Piette JC, et al. Tangier disease is caused by mutations in the gene encoding ATP-binding cassette transporter 1. Nat Genet 1999;22:352–5.

[9] Tall AR, Yvan-Charvet L, Terasaka N, Pagler T, Wang N HDL. ABC transporters, and cholesterol efflux: implications for the treatment of atherosclerosis. Cell Metab 2008;7:365–75.

[10] Wang N, Lan D, Chen W, Matsuura F, Tall AR. ATP-binding cassette transporters G1 and G4 mediate cellular cholesterol efflux to high-density lipoproteins. Proc Natl Acad Sci USA 2004;101:9774–9.

[11] Brewer Jr HB. Clinical review: the evolving role of HDL in the treatment of high-risk patients with cardiovascular disease. J Clin Endocrinol Metab 2011;96:1246–57.

[12] Cheung MC, Mendez AJ, Wolf AC, Knopp RH. Characterization of apolipoprotein A-I- and A-II-containing lipoproteins in a new case of high density lipoprotein deficiency resembling Tangier disease and their effects on intracellular cholesterol efflux. J Clin Invest 1993;91:522–9.

[13] Schmitz G, Assmann G, Robenek H, Brennhausen B. Tangier disease: a disorder of intracellular membrane traffic. Proc Natl Acad Sci USA 1985;82:6305–9.

[14] Schaefer EJ, Anderson DW, Zech LA, Lindgren FT, Bronzert TB, Rubalcaba EA, et al. Metabolism of high density lipoprotein subfractions and constituents in Tangier disease following the infusion of high density lipoproteins. J Lipid Res 1981;22:217–28.

[15] Oram JF, Lawn RM, Garvin MR, Wade DP. ABCA1 is the cAMP-inducible apolipo-protein receptor that mediates cholesterol secretion from macrophages. J Biol Chem 2000;275:34508–11.

[16] Fielding PE, Nagao K, Hakamata H, Chimini G, Fielding CJ. A two-step mechanism for free cholesterol and phospholipid efflux from human vascular cells to apolipopro-tein A-1. Biochemistry 2000;39:14113–20.

[17] Gillotte KL, Zaiou M, Lund-Katz S, Anantharamaiah GM, Holvoet P, Dhoest A, et al. Apolipoprotein-mediated plasma membrane microsolubilization. Role of lipid affin-ity and membrane penetration in the efflux of cellular cholesterol and phospholipid. J Biol Chem 1999;274:2021–8.

[18] Chambenoit O, Hamon Y, Marguet D, Rigneault H, Rosseneu M, Chimini G. Spe-cific docking of apolipoprotein A-I at the cell surface requires a functional ABCA1 transporter. J Biol Chem 2001;276:9955–60.

[19] Drobnik W, Borsukova H, Bottcher A, Pfeiffer A, Liebisch G, Schutz GJ, et al. Apo AI/ ABCA1-dependent and HDL3-mediated lipid efflux from compositionally distinct cholesterol-based microdomains. Traffic 2002;3:268–78.

[20] Vedhachalam C, Duong PT, Nickel M, Nguyen D, Dhanasekaran P, Saito H, et al. Mechanism of ATP-binding cassette transporter A1-mediated cellular lipid efflux to apolipoprotein A-I and formation of high density lipoprotein particles. J Biol Chem 2007;282:25123–30.

[21] Hassan HH, Denis M, Lee DY, Iatan I, Nyholt D, Ruel I, et al. Identification of an ABCA1-dependent phospholipid-rich plasma membrane apolipoprotein A-I binding site for nascent HDL formation: implications for current models of HDL biogenesis. J Lipid Res 2007;48:2428–42.

[22] Oram JF, Wolfbauer G, Tang C, Davidson WS, Albers JJ. An amphipathic helical region of the N-terminal barrel of phospholipid transfer protein is critical for ABCA1-dependent cholesterol efflux. J Biol Chem 2008;283:11541–9.

[23] Lee-Rueckert M, Vikstedt R, Metso J, Ehnholm C, Kovanen PT, Jauhiainen M. Absence of endogenous phospholipid transfer protein impairs ABCA1-dependent efflux of cholesterol from macrophage foam cells. J Lipid Res 2006;47:1725–32.

[24] Duong PT, Collins HL, Nickel M, Lund-Katz S, Rothblat GH, Phillips MC. Charac-terization of nascent HDL particles and microparticles formed by ABCA1-mediated efflux of cellular lipids to apoA-I. J Lipid Res 2006;47:832–43.

[25] Yancey PG, Bortnick AE, Kellner-Weibel G, de la Llera-Moya M, Phillips MC, Roth-blat GH. Importance of different pathways of cellular cholesterol efflux. Arterioscler Thromb Vasc Biol 2003;23:712–9.

[26] Bell-Quint J, Forte T. Time-related changes in the synthesis and secretion of very low density, low density and high density lipoproteins by cultured rat hepatocytes. Bio-chim Biophys Acta 1981;663:83–98.

[27] Sorci-Thomas M, Prack MM, Dashti N, Johnson F, Rudel LL, Williams DL. Apolipo-protein (apo) A-I production and mRNA abundance explain plasma apoA-I and high density lipoprotein differences between two nonhuman primate species with high and low susceptibilities to diet-induced hypercholesterolemia. J Biol Chem 1988;263:5183–9.

[28] Smith JD, Miyata M, Ginsberg M, Grigaux C, Shmookler E, Plump AS. Cyclic AMP induces apolipoprotein E binding activity and promotes cholesterol efflux from a macrophage cell line to apolipoprotein acceptors. J Biol Chem 1996;271:30647–55.

[29] Aiello RJ, Brees D, Francone OL. ABCA1-deficient mice: insights into the role of monocyte lipid efflux in HDL formation and inflammation. Arterioscler Thromb Vasc Biol 2003;23:972–80.

[30] Haghpassand M, Bourassa PA, Francone OL, Aiello RJ. Monocyte/macrophage expression of ABCA1 has minimal contribution to plasma HDL levels. J Clin Invest 2001;108:1315–20.

[31] Lee JY, Parks JS. ATP-binding cassette transporter AI and its role in HDL formation. Curr Opin Lipidol 2005;16:19–25.

[32] Brunham LR, Kruit JK, Iqbal J, Fievet C, Timmins JM, Pape TD, et al. Intestinal ABCA1 directly contributes to HDL biogenesis in vivo. J Clin Invest 2006;116: 1052–62.

[33] Brunham LR, Kruit JK, Pape TD, Parks JS, Kuipers F, Hayden MR. Tissue-specific induction of intestinal ABCA1 expression with a liver X receptor agonist raises plasma HDL cholesterol levels. Circ Res 2006;99:672–4.

[34] Singaraja RR, Van Eck M, Bissada N, Zimetti F, Collins HL, Hildebrand RB, et al. Both hepatic and extrahepatic ABCA1 have discrete and essential functions in the maintenance of plasma high-density lipoprotein cholesterol levels in vivo. Circulation 2006;114:1301–9.

[35] Marcil M, Brooks-Wilson A, Clee SM, Roomp K, Zhang LH, Yu L, et al. Mutations in the ABC1 gene in familial HDL deficiency with defective cholesterol efflux. Lancet 1999;354:1341–6.

[36] Clee SM, Kastelein JJ, van Dam M, Marcil M, Roomp K, Zwarts KY, et al. Age and residual cholesterol efflux affect HDL cholesterol levels and coronary artery disease in ABCA1 heterozygotes. J Clin Invest 2000;106:1263–70.

[37] Wang J, Burnett JR, Near S, Young K, Zinman B, Hanley AJ, et al. Common and rare ABCA1 variants affecting plasma HDL cholesterol. Arterioscler Thromb Vasc Biol 2000;20:1983–9.

[38] Lutucuta S, Ballantyne CM, Elghannam H, Gotto Jr AM, Marian AJ. Novel polymorphisms in promoter region of ATP binding cassette transporter gene and plasma lipids, severity, progression, and regression of coronary atherosclerosis and response to therapy. Circ Res 2001;88:969–73.

[39] Harada T, Imai Y, Nojiri T, Morita H, Hayashi D, Maemura K, et al. A common Ile 823 Met variant of ATP-binding cassette transporter A1 gene (ABCA1) alters high density lipoprotein cholesterol level in Japanese population. Atherosclerosis 2003; 169:105–12.

[40] Shioji K, Nishioka J, Naraba H, Kokubo Y, Mannami T, Inamoto N, et al. A promoter variant of the ATP-binding cassette transporter A1 gene alters the HDL cholesterol level in the general Japanese population. J Hum Genet 2004;49:141–7.

[41] Frikke-Schmidt R, Nordestgaard BG, Jensen GB, Tybjaerg-Hansen A. Genetic variation in ABC transporter A1 contributes to HDL cholesterol in the general population. J Clin Invest 2004;114:1343–53.

[42] Willer CJ, Sanna S, Jackson AU, Scuteri A, Bonnycastle LL, Clarke R, et al. Newly identified loci that influence lipid concentrations and risk of coronary artery disease. Nat Genet 2008;40:161–9.

[43] Kathiresan S, Melander O, Guiducci C, Surti A, Burtt NP, Rieder MJ, et al. Six new loci associated with blood low-density lipoprotein cholesterol, high-density lipoprotein cholesterol or triglycerides in humans. Nat Genet 2008;40:189–97.

[44] Talmud PJ, Drenos F, Shah S, Shah T, Palmen J, Verzilli C, et al. Gene-centric association signals for lipids and apolipoproteins identified via the HumanCVD BeadChip. Am J Hum Genet 2009;85:628–42.

[45] Acuna-Alonzo V, Flores-Dorantes T, Kruit JK, Villarreal-Molina T, Arellano-Campos O, Hunemeier T, et al. A functional ABCA1 gene variant is associated with low HDL-cholesterol levels and shows evidence of positive selection in Native Americans. Hum Mol Genet 2010.

[46] Wang N, Ranalletta M, Matsuura F, Peng F, Tall AR. LXR-induced redistribution of ABCG1 to plasma membrane in macrophages enhances cholesterol mass efflux to HDL. Arterioscler Thromb Vasc Biol 2006;26:1310–6.

[47] Tarling EJ, Edwards PA. ATP binding cassette transporter G1 (ABCG1) is an intracellular sterol transporter. Proc Natl Acad Sci USA 2011;108:19719–24.

[48] Furuyama S, Uehara Y, Zhang B, Baba Y, Abe S, Iwamoto T, et al. Genotypic effect of ABCG1 gene promoter -257T>G polymorphism on coronary artery disease severity in Japanese men. J Atheroscler Thromb 2009;16:194–200.

[49] Xu Y, Wang W, Zhang L, Qi LP, Li LY, Chen LF, et al. A polymorphism in the ABCG1 promoter is functionally associated with coronary artery disease in a Chinese Han population. Atherosclerosis 2011;219:648–54.

[50] Edmondson AC, Braund PS, Stylianou IM, Khera AV, Nelson CP, Wolfe ML, et al. Dense genotyping of candidate gene loci identifies variants associated with high-density lipoprotein cholesterol. Circ Cardiovasc Genet 2011;4:145–55.

[51] Costet P, Luo Y, Wang N, Tall AR. Sterol-dependent transactivation of the ABC1 promoter by the liver X receptor/retinoid X receptor. J Biol Chem 2000;275:28240–5.

[52] Kennedy MA, Venkateswaran A, Tarr PT, Xenarios I, Kudoh J, Shimizu N, et al. Characterization of the human ABCG1 gene: liver X receptor activates an internal promoter that produces a novel transcript encoding an alternative form of the protein. J Biol Chem 2001;276:39438–47.

[53] Sabol SL, Brewer Jr HB, Santamarina-Fojo S. The human ABCG1 gene: identification of LXR response elements that modulate expression in macrophages and liver. J Lipid Res 2005;46:2151–67.

[54] Uehara Y, Engel T, Li Z, Goepfert C, Rust S, Zhou X, et al. Polyunsaturated fatty acids and acetoacetate downregulate the expression of the ATP-binding cassette transporter A1. Diabetes 2002;51:2922–8.

[55] Janowski BA, Willy PJ, Devi TR, Falck JR, Mangelsdorf DJ. An oxysterol signalling pathway mediated by the nuclear receptor LXR alpha. Nature 1996;383:728–31.

[56] Sugimoto K, Tsujita M, Wu CA, Suzuki K, Yokoyama S. An inhibitor of acylCoA: cholesterol acyltransferase increases expression of ATP-binding cassette transporter A1 and thereby enhances the ApoA-I-mediated release of cholesterol from macrophages. Biochim Biophys Acta 2004;1636:69–76.

[57] Yamauchi Y, Chang CC, Hayashi M, Abe-Dohmae S, Reid PC, Chang TY, et al. Intracellular cholesterol mobilization involved in the ABCA1/apolipoprotein-mediated assembly of high density lipoprotein in fibroblasts. J Lipid Res 2004;45:1943–51.

[58] Li AC, Binder CJ, Gutierrez A, Brown KK, Plotkin CR, Pattison JW, et al. Differential inhibition of macrophage foam-cell formation and atherosclerosis in mice by PPAR-alpha, beta/delta, and gamma. J Clin Invest 2004;114:1564–76.

[59] Quinet EM, Savio DA, Halpern AR, Chen L, Schuster GU, Gustafsson JA, et al. Liver X receptor (LXR)-beta regulation in LXRalpha-deficient mice: implications for therapeutic targeting. Mol Pharmacol 2006;70:1340–9.

[60] Ozasa H, Ayaori M, Iizuka M, Terao Y, Uto-Kondo H, Yakushiji E, et al. Pioglitazone enhances cholesterol efflux from macrophages by increasing ABCA1/ABCG1 expressions via PPARgamma/LXRalpha pathway: findings from in vitro and ex vivo studies. Atherosclerosis 2011;219:141–50.

[61] Oberkofler H, Schraml E, Krempler F, Patsch W. Potentiation of liver X receptor transcriptional activity by peroxisome-proliferator-activated receptor gamma coactivator 1 alpha. Biochem J 2003;371:89–96.

[62] Costet P, Lalanne F, Gerbod-Giannone MC, Molina JR, Fu X, Lund EG, et al. Retinoic acid receptor-mediated induction of ABCA1 in macrophages. Mol Cell Biol 2003;23:7756–66.

[63] Ayaori M, Yakushiji E, Ogura M, Nakaya K, Hisada T, Uto-Kondo H, et al. Retinoic acid receptor agonists regulate expression of ATP-binding cassette transporter G1 in macrophages. Biochim Biophys Acta 2012;1821:561–72.

[64] Tamehiro N, Shigemoto-Mogami Y, Kakeya T, Okuhira K, Suzuki K, Sato R, et al. Sterol regulatory element-binding protein-2- and liver X receptor-driven dual promoter regulation of hepatic ABC transporter A1 gene expression: mechanism underlying the unique response to cellular cholesterol status. J Biol Chem 2007;282:21090–9.

[65] Brewer Jr HB. Benefit-risk assessment of Rosuvastatin 10 to 40 milligrams. Am J Cardiol 2003;92:23K–9K.

[66] Yang XP, Freeman LA, Knapper CL, Amar MJ, Remaley A, Brewer Jr HB, et al. The E-box motif in the proximal ABCA1 promoter mediates transcriptional repression of the ABCA1 gene. J Lipid Res 2002;43:297–306.

[67] Zeng L, Liao H, Liu Y, Lee TS, Zhu M, Wang X, et al. Sterol-responsive element-binding protein (SREBP) 2 down-regulates ATP-binding cassette transporter A1 in vascular endothelial cells: a novel role of SREBP in regulating cholesterol metabolism. J Biol Chem 2004;279:48801–7.

[68] Arakawa R, Tamehiro N, Nishimaki-Mogami T, Ueda K, Yokoyama S. Fenofibric acid, an active form of fenofibrate, increases apolipoprotein A-I-mediated high-density lipoprotein biogenesis by enhancing transcription of ATP-binding cassette transporter A1 gene in a liver X receptor-dependent manner. Arterioscler Thromb Vasc Biol 2005;25:1193–7.

[69] Abe-Dohmae S, Suzuki S, Wada Y, Aburatani H, Vance DE, Yokoyama S. Characterization of apolipoprotein-mediated HDL generation induced by cAMP in a murine macrophage cell line. Biochemistry 2000;39:11092–9.

[70] Le Goff W, Zheng P, Brubaker G, Smith JD. Identification of the cAMP-responsive enhancer of the murine ABCA1 gene: requirement for CREB1 and STAT3/4 elements. Arterioscler Thromb Vasc Biol 2006;26:527–33.

[71] Iwamoto N, Abe-Dohmae S, Ayaori M, Tanaka N, Kusuhara M, Ohsuzu F, et al. ATP-binding cassette transporter A1 gene transcription is downregulated by activator protein 2alpha. Doxazosin inhibits activator protein 2alpha and increases high-density lipoprotein biogenesis independent of alpha1-adrenoceptor blockade. Circ Res 2007;101:156–65.

[72] Iwamoto N, Abe-Dohmae S, Lu R, Yokoyama S. Involvement of protein kinase D in phosphorylation and increase of DNA binding of activator protein 2 alpha to down-regulate ATP-binding cassette transporter A1. Arterioscler Thromb Vasc Biol 2008;28:2282–7.

[73] Ohoka N, Okuhira K, Cui H, Wu W, Sato R, Naito M, et al. HNF4alpha increases liver-specific human ATP-binding cassette transporter A1 expression and cholesterol efflux to apolipoprotein A-I in response to cholesterol depletion. Arterioscler Thromb Vasc Biol 2012;32:1005–14.

[74] Lorkowski S, Rust S, Engel T, Jung E, Tegelkamp K, Galinski EA, et al. Genomic sequence and structure of the human ABCG1 (ABC8) gene. Biochem Biophys Res Commun 2001;280:121–31.

[75] Jakobsson T, Venteclef N, Toresson G, Damdimopoulos AE, Ehrlund A, Lou X, et al. GPS2 is required for cholesterol efflux by triggering histone demethylation, LXR recruitment, and coregulator assembly at the ABCG1 locus. Mol Cell 2009;34:510–8.

[76] Rayner KJ, Suarez Y, Davalos A, Parathath S, Fitzgerald ML, Tamehiro N, et al. MiR-33 contributes to the regulation of cholesterol homeostasis. Science 2010;328:1570–3.

[77] Najafi-Shoushtari SH, Kristo F, Li Y, Shioda T, Cohen DE, Gerszten RE, et al. MicroRNA-33 and the SREBP host genes cooperate to control cholesterol homeostasis. Science 2010;328:1566–9.

[78] Horie T, Ono K, Horiguchi M, Nishi H, Nakamura T, Nagao K, et al. MicroRNA-33 encoded by an intron of sterol regulatory element-binding protein 2 (Srebp2) regulates HDL in vivo. Proc Natl Acad Sci USA 2010;107:17321–6.

[79] Rayner KJ, Sheedy FJ, Esau CC, Hussain FN, Temel RE, Parathath S, et al. Antagonism of miR-33 in mice promotes reverse cholesterol transport and regression of atherosclerosis. J Clin Invest 2011;121:2921–31.

[80] Horie T, Baba O, Kuwabara Y, Chujo Y, Watanabe S, Kinoshita M, et al. MicroRNA-33 Deficiency reduces the progression of atherosclerotic plaque in ApoE(−/−) Mice. J Am Heart Assoc 2012;1:e003376.

[81] Ramirez CM, Davalos A, Goedeke L, Salerno AG, Warrier N, Cirera-Salinas D, et al. MicroRNA-758 regulates cholesterol efflux through posttranscriptional repression of ATP-binding cassette transporter A1. Arterioscler Thromb Vasc Biol 2011;31: 2707–14.

[82] Uto-Kondo H, Ayaori M, Ogura M, Nakaya K, Ito M, Suzuki A, et al. Coffee consumption enhances high-density lipoprotein-mediated cholesterol efflux in macrophages. Circ Res 2010;106:779–87.

[83] Wang N, Chen W, Linsel-Nitschke P, Martinez LO, Agerholm-Larsen B, Silver DL, et al. A PEST sequence in ABCA1 regulates degradation by calpain protease and stabilization of ABCA1 by apoA-I. J Clin Invest 2003;111:99–107.

[84] Chen W, Wang N, Tall AR. A PEST deletion mutant of ABCA1 shows impaired internalization and defective cholesterol efflux from late endosomes. J Biol Chem 2005;280:29277–81.

[85] Singaraja RR, Kang MH, Vaid K, Sanders SS, Vilas GL, Arstikaitis P, et al. Palmitoylation of ATP-binding cassette transporter A1 is essential for its trafficking and function. Circ Res 2009;105:138–47.

[86] Tamehiro N, Zhou S, Okuhira K, Benita Y, Brown CE, Zhuang DZ, et al. SPTLC1 binds ABCA1 to negatively regulate trafficking and cholesterol efflux activity of the transporter. Biochemistry 2008;47:6138–47.

[87] Okuhira K, Fitzgerald ML, Sarracino DA, Manning JJ, Bell SA, Goss JL, et al. Purification of ATP-binding cassette transporter A1 and associated binding proteins reveals the importance of beta1-syntrophin in cholesterol efflux. J Biol Chem 2005;280:39653–64.

[88] Buechler C, Boettcher A, Bared SM, Probst MC, Schmitz G. The carboxyterminus of the ATP-binding cassette transporter A1 interacts with a beta2-syntrophin/utrophin complex. Biochem Biophys Res Commun 2002;293:759–65.

[89] Munehira Y, Ohnishi T, Kawamoto S, Furuya A, Shitara K, Imamura M, et al. Alpha1-syntrophin modulates turnover of ABCA1. J Biol Chem 2004;279:15091–5.

[90] Okuhira K, Fitzgerald ML, Tamehiro N, Ohoka N, Suzuki K, Sawada JI, et al. Binding of PDZ-RhoGEF to ATP-binding cassette transporter A1 (ABCA1) induces cholesterol efflux through RhoA activation and prevention of transporter degradation. J Biol Chem 2010;285(21):16369–77.

[91] Tsukamoto K, Hirano K, Tsujii K, Ikegami C, Zhongyan Z, Nishida Y, et al. ATP-binding cassette transporter-1 induces rearrangement of actin cytoskeletons possibly through Cdc42/N-WASP. Biochem Biophys Res Commun 2001;287:757–65.

[92] Yamauchi Y, Hayashi M, Abe-Dohmae S, Yokoyama S. Apolipoprotein A-I activates protein kinase C alpha signaling to phosphorylate and stabilize ATP binding cassette transporter A1 for the high density lipoprotein assembly. J Biol Chem 2003;278: 47890–7.

[93] Hozoji M, Munehira Y, Ikeda Y, Makishima M, Matsuo M, Kioka N, et al. Direct interaction of nuclear liver X receptor-beta with ABCA1 modulates cholesterol efflux. J Biol Chem 2008;283:30057–63.

[94] Azuma Y, Takada M, Maeda M, Kioka N, Ueda K. The COP9 signalosome controls ubiquitinylation of ABCA1. Biochem Biophys Res Commun 2009;382:145–8.

[95] Ogura M, Ayaori M, Terao Y, Hisada T, Iizuka M, Takiguchi S, et al. Proteasomal inhibition promotes ATP-binding cassette transporter A1 (ABCA1) and ABCG1 expression and cholesterol efflux from macrophages in vitro and in vivo. Arterioscler Thromb Vasc Biol 2011;31:1980–7.

[96] Hori N, Hayashi H, Sugiyama Y. Calpain-mediated cleavage negatively regulates the expression level of ABCG1. Atherosclerosis 2011;215:383–91.

[97] Nagelin MH, Srinivasan S, Lee J, Nadler JL, Hedrick CC. 12/15-Lipoxygenase activity increases the degradation of macrophage ATP-binding cassette transporter G1. Arterioscler Thromb Vasc Biol 2008;28:1811–9.

[98]   Nagelin MH, Srinivasan S, Nadler JL, Hedrick CC. Murine 12/15-lipoxygenase regulates ATP-binding cassette transporter G1 protein degradation through p38- and JNK2-dependent pathways. J Biol Chem 2009;284:31303–14.

[99]   Lewis GF, Rader DJ. New insights into the regulation of HDL metabolism and reverse cholesterol transport. Circ Res 2005;96:1221–32.

[100] van der Velde AE, Brufau G, Groen AK. Transintestinal cholesterol efflux. Curr Opin Lipidol 2012;21:167–71.

[101] Jolley CD, Woollett LA, Turley SD, Dietschy JM. Centripetal cholesterol flux to the liver is dictated by events in the peripheral organs and not by the plasma high density lipoprotein or apolipoprotein A-I concentration. J Lipid Res 1998;39:2143–9.

[102] Zhang Y, Zanotti I, Reilly MP, Glick JM, Rothblat GH, Rader DJ. Overexpression of apolipoprotein A-I promotes reverse transport of cholesterol from macrophages to feces in vivo. Circulation 2003;108:661–3.

[103] Rader DJ, Alexander ET, Weibel GL, Billheimer J, Rothblat GH. The role of reverse cholesterol transport in animals and humans and relationship to atherosclerosis. J Lipid Res 2009;50(Suppl):S189–94.

[104] van Eck M, Bos IS, Kaminski WE, Orso E, Rothe G, Twisk J, et al. Leukocyte ABCA1 controls susceptibility to atherosclerosis and macrophage recruitment into tissues. Proc Natl Acad Sci USA 2002;99:6298–303.

[105] Van Eck M, Singaraja RR, Ye D, Hildebrand RB, James ER, Hayden MR, et al. Macrophage ATP-binding cassette transporter A1 overexpression inhibits atherosclerotic lesion progression in low-density lipoprotein receptor knockout mice. Arterioscler Thromb Vasc Biol 2006;26:929–34.

[106] Wang X, Collins HL, Ranalletta M, Fuki IV, Billheimer JT, Rothblat GH, et al. Macrophage ABCA1 and ABCG1, but not SR-BI, promote macrophage reverse cholesterol transport in vivo. J Clin Invest 2007;117:2216–24.

[107] Yvan-Charvet L, Ranalletta M, Wang N, Han S, Terasaka N, Li R, et al. Combined deficiency of ABCA1 and ABCG1 promotes foam cell accumulation and accelerates atherosclerosis in mice. J Clin Invest 2007;117:3900–8.

[108] Naik SU, Wang X, Da Silva JS, Jaye M, Macphee CH, Reilly MP, et al. Pharmacological activation of liver X receptors promotes reverse cholesterol transport in vivo. Circulation 2006;113:90–7.

[109] Kennedy MA, Barrera GC, Nakamura K, Baldan A, Tarr P, Fishbein MC, et al. ABCG1 has a critical role in mediating cholesterol efflux to HDL and preventing cellular lipid accumulation. Cell Metab 2005;1:121–31.

[110] Munch G, Bultmann A, Li Z, Holthoff HP, Ullrich J, Wagner S, et al. Overexpression of ABCG1 protein attenuates arteriosclerosis and endothelial dysfunction in atherosclerotic rabbits. Heart Int 2012;7:e12.

[111] Baldan A, Pei L, Lee R, Tarr P, Tangirala RK, Weinstein MM, et al. Impaired development of atherosclerosis in hyperlipidemic Ldlr−/− and ApoE−/− mice transplanted with Abcg1−/− bone marrow. Arterioscler Thromb Vasc Biol 2006;26:2301–7.

[112] Ranalletta M, Wang N, Han S, Yvan-Charvet L, Welch C, Tall AR. Decreased atherosclerosis in low-density lipoprotein receptor knockout mice transplanted with Abcg1−/− bone marrow. Arterioscler Thromb Vasc Biol 2006;26:2308–15.

[113] Meurs I, Lammers B, Zhao Y, Out R, Hildebrand RB, Hoekstra M, et al. The effect of ABCG1 deficiency on atherosclerotic lesion development in LDL receptor knockout mice depends on the stage of atherogenesis. Atherosclerosis 2012;221:41–7.

[114] Assmann G, Simantke O, Schaefer HE, Smootz E. Characterization of high density lipoproteins in patients heterozygous for Tangier disease. J Clin Invest 1977;60:1025–35.

[115] van Dam MJ, de Groot E, Clee SM, Hovingh GK, Roelants R, Brooks-Wilson A, et al. Association between increased arterial-wall thickness and impairment in ABCA1-driven cholesterol efflux: an observational study. Lancet 2002;359:37–42.

[116] Clee SM, Zwinderman AH, Engert JC, Zwarts KY, Molhuizen HO, Roomp K, et al. Common genetic variation in ABCA1 is associated with altered lipoprotein levels and a modified risk for coronary artery disease. Circulation 2001;103:1198–205.

[117] Cenarro A, Artieda M, Castillo S, Mozas P, Reyes G, Tejedor D, et al. A common variant in the ABCA1 gene is associated with a lower risk for premature coronary heart disease in familial hypercholesterolaemia. J Med Genet 2003;40:163–8.

[118] Kyriakou T, Hodgkinson C, Pontefract DE, Iyengar S, Howell WM, Wong YK, et al. Genotypic effect of the -565C>T polymorphism in the ABCA1 gene promoter on ABCA1 expression and severity of atherosclerosis. Arterioscler Thromb Vasc Biol 2005;25:418–23.

[119] Kyriakou T, Pontefract DE, Viturro E, Hodgkinson CP, Laxton RC, Bogari N, et al. Functional polymorphism in ABCA1 influences age of symptom onset in coronary artery disease patients. Hum Mol Genet 2007;16:1412–22.

[120] Frikke-Schmidt R, Nordestgaard BG, Stene MC, Sethi AA, Remaley AT, Schnohr P, et al. Association of loss-of-function mutations in the ABCA1 gene with high-density lipoprotein cholesterol levels and risk of ischemic heart disease. JAMA 2008;299:2524–32.

[121] Terasaka N, Hiroshima A, Koieyama T, Ubukata N, Morikawa Y, Nakai D, et al. T-0901317, a synthetic liver X receptor ligand, inhibits development of atherosclerosis in LDL receptor-deficient mice. FEBS Lett 2003;536:6–11.

[122] Levin N, Bischoff ED, Daige CL, Thomas D, Vu CT, Heyman RA, et al. Macrophage liver X receptor is required for antiatherogenic activity of LXR agonists. Arterioscler Thromb Vasc Biol 2005;25:135–42.

[123] Li AC, Glass CK. PPAR- and LXR-dependent pathways controlling lipid metabolism and the development of atherosclerosis. J Lipid Res 2004;45:2161–73.

[124] Katz A, Udata C, Ott E, Hickey L, Burczynski ME, Burghart P, et al. Safety, pharmacokinetics, and pharmacodynamics of single doses of LXR-623, a novel liver X-receptor agonist, in healthy participants. J Clin Pharmacol 2009;49:643–9.

[125] Molteni V, Li X, Nabakka J, Liang F, Wityak J, Koder A, et al. N-Acylthiadiazolines, a new class of liver X receptor agonists with selectivity for LXRbeta. J Med Chem 2007;50:4255–9.

[126] Nissen SE, Nicholls SJ, Wolski K, Nesto R, Kupfer S, Perez A, et al. Comparison of pioglitazone vs glimepiride on progression of coronary atherosclerosis in patients with type 2 diabetes: the PERISCOPE randomized controlled trial. JAMA 2008;299:1561–73.

[127] Mazzone T, Meyer PM, Feinstein SB, Davidson MH, Kondos GT, D'Agostino Sr RB, et al. Effect of pioglitazone compared with glimepiride on carotid intima-media thickness in type 2 diabetes: a randomized trial. JAMA 2006;296:2572–81.

[128] Duffy D, Rader DJ. Update on strategies to increase HDL quantity and function. Nat Rev Cardiol 2009;6:455–63.

[129] Nakaya K, Tohyama J, Naik SU, Tanigawa H, MacPhee C, Billheimer JT, et al. Peroxisome proliferator-activated receptor-alpha activation promotes macrophage reverse cholesterol transport through a liver X receptor-dependent pathway. Arterioscler Thromb Vasc Biol 2011;31(6):1276–82.

[130] Keech A, Simes RJ, Barter P, Best J, Scott R, Taskinen MR, et al. Effects of long-term fenofibrate therapy on cardiovascular events in 9795 people with type 2 diabetes mellitus (the FIELD study): randomised controlled trial. Lancet 2005;366:1849–61.

[131] Ginsberg HN, Elam MB, Lovato LC, Crouse 3rd JR, Leiter LA, Linz P, et al. Effects of combination lipid therapy in type 2 diabetes mellitus. N Engl J Med 2010;362(17):1563–74.

[132] Millar JS, Duffy D, Gadi R, Bloedon LT, Dunbar RL, Wolfe ML, et al. Potent and selective PPAR-alpha agonist LY518674 upregulates both ApoA-I production and catabolism in human subjects with the metabolic syndrome. Arterioscler Thromb Vasc Biol 2009;29:140–6.

[133] Ikewaki K, Mochizuki K, Iwasaki M, Nishide R, Mochizuki S, Tada N. Cilostazol, a potent phosphodiesterase type III inhibitor, selectively increases antiatherogenic high-density lipoprotein subclass LpA-I and improves postprandial lipemia in patients with type 2 diabetes mellitus. Metabolism 2002;51:1348–54.

[134] Elam MB, Heckman J, Crouse JR, Hunninghake DB, Herd JA, Davidson M, et al. Effect of the novel antiplatelet agent cilostazol on plasma lipoproteins in patients with intermittent claudication. Arterioscler Thromb Vasc Biol 1998;18:1942–7.

[135] deGoma EM, deGoma RL, Rader DJ. Beyond high-density lipoprotein cholesterol levels evaluating high-density lipoprotein function as influenced by novel therapeutic approaches. J Am Coll Cardiol 2008;51:2199–211.

[136] Nakaya K, Ayaori M, Uto-Kondo H, Hisada T, Ogura M, Yakushiji E, et al. Cilostazol enhances macrophage reverse cholesterol transport in vitro and in vivo. Atherosclerosis 2010;213:135–41.

[137] Freedman ND, Park Y, Abnet CC, Hollenbeck AR, Sinha R. Association of coffee drinking with total and cause-specific mortality. N Engl J Med 2012;366:1891–904.

CHAPTER 6

Sphingolipids and HDL Metabolism

Xian-Cheng Jiang*,**, Zhiqiang Li*,**, Amirfarbod Yazdanyar*
*Department of Cell Biology, State University of New York, Downstate Medical Center, Brooklyn, NY, USA
**Molecular and Cellular Cardiology Program, VA New York Harbor Healthcare System, Brooklyn, NY, USA

Contents

Abstract

The major sphingolipids in the human plasma are sphingomyelin, ceramide, sphingo-sine-1-phosphate, and glycosphingolipids. Sphingolipids are found to be associated with plasma lipoproteins and their presence can influence lipoprotein, including high-density lipoprotein (HDL), metabolism. Sphingolipids are also present on cell plasma membrane and they also affect lipoprotein, including HDL, production and catabo-lism. In this chapter, we will summarize what we know about sphingolipids and HDL metabolism.

1. INTRODUCTION

Sphingolipids consist of an 18-carbon amino-alcohol backbone, sphingo-sine. Sphingolipids, especially ceramides, sphingomyelins (SM), and sphin-gosine-1-phosphates (S1P), are known to play important roles in maintaining membrane function and integrity, preserving lipoprotein structure and functions, and preventing or promoting many diseases such as atherosclerosis.

Epidemiological data from the Framingham Heart Study [1–2] and other prospective studies [3] demonstrate that high levels of high-density lipopro-tein (HDL) cholesterol in blood are inversely associated with risk for cardio-vascular disease. However, some individuals with high HDL-cholesterol and normal low-density lipoprotein (LDL) cholesterol still develop cardiovascu-lar disease [4]. Pharmacologically increased HDL cholesterol (HDL-C) levels, so far, have not demonstrated any effect on cardiovascular disease prevention [5–10]. This has lead to the hypothesis that the HDL in some individuals might be dysfunctional as an anti-atherogenic agent or perhaps even pro-atherogenic as a result of the HDL lipid content, particularly that of sphingolipids. Here, we will summarize what is known about the synthesis and metabolism of lipoprotein sphingolipids, especially that of HDL sphingolipids.

2. PLASMA HDL-SM ORIGINATION

2.1. From ABCA1 Pathway

HDL biogenesis is not entirely understood. Initially, HDL particle forma-tion was thought to occur inside the cell by a process similar to that for the formation of very low-density lipoprotein (VLDL) [11]. However, the assembly of free cholesterol and phospholipids, including SM and phospha-tidylcholine (PC), with lipid-free apoA-I to form nascent HDL particles is now thought to occur extracellularly [12–13]. ATP-binding cassette A1

(ABCA1) is a critical molecule regulating nascent HDL particle assembly [14–16]. It has been reported that ABCA1 mediates SM and PC efflux from cells [17–18]. Recently, Sorci-Thomas and colleagues [19] detailed the lipid composition of nascent HDL particles formed by the action of the ABCA1 on apoA-I. They showed that the proportions of the principal nascent HDL lipids, free cholesterol and SM, were similar to that of lipid rafts (enriched with both lipids), suggesting that the lipid originated from a raft-like region of the cell. Since only two tissues, the liver and small intestine, are quantitatively important in the synthesis and secretion of apoA-I [20], both tissues are the major sources of HDL particles. Indeed, the combined deletion of both hepatic and intestinal ABCA1 resulted in an approximately 90% decrease in plasma HDL levels [21].

2.2. From apoB-Containing Lipoprotein Particles (BLp)

BLp are SM-rich particles [22–23]. BLp-SM is one of the sources of HDL-SM. ApoB is the major protein component of VLDL and chylomicron [24]. ApoB exists in two forms, apoB48 and apoB100 [25–26]. Accumulating evidence suggests that the formation of apoB100-BLp [27] and apoB48-BLp [28–29] is accomplished sequentially. The "two-step" model postulates that the initial product is a primordial particle, formed during apoB translation in the endoplasmic reticulum (ER) [30]. It is clear that MTP is involved in the early stage (first step) of apoB lipidation. However, the mechanism involved in the later stage (second step), in which the apoB-containing primordial particle fuses with apoB-free/triglyceride/SM/PC-lipid droplets [30], is still not well understood.

More than 30 years ago, Tall and Small [31] provided a possible mechanism for the transfer of surface components, including SM, PC, and free cholesterol, from BLp to HDL fraction during lipolysis. After lipoprotein lipase (LPL) hydrolyzes BLp triglyceride, the core of BLp shrinks, and the redundant surface constituents form lipid bilayer folds projecting from the BLp, which can be the molecular basis of nascent HDL. Indeed, LPL deficiency is associated with low HDL cholesterol levels in both homozygous and heterozygous states [32].

2.3. From Phospholipid Transfer Protein (PLTP) Activity

PLTP may be involved in HDL-SM origination through BLp production pathway and ABCA1 pathway (Figure 6.1). The second step of BLp lipidation is involved in the fusion of primordial BLp and apoB-free

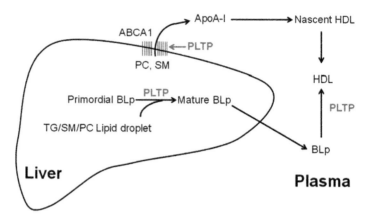

Figure 6.1 *The Potential Role of PLTP on HDL Production.* PLTP may be involved in HDL-SM origination through BLp production pathway. The 2nd step of BLp lipidation is involved in the fusion of primordial BLp and TG/SM/PC-rich lipid droplets. PLTP can also promote ABCA1-mediated nascent HDL production through stabilizing ABCA1 and shuttling lipids between cells and existing HDL particles.

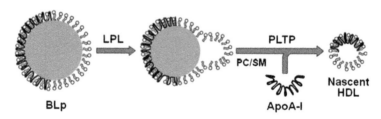

Figure 6.2 *The potential role of PLTP on BLp-derived HDL.* After mature BLp secreted into the blood and LPL-mediated triglyceride hydrolysis, the core of BLp shrinks, and the redundant surface constituents (SM and PC, as well as free cholesterol) can be the substrates of PLTP, transferring from BLp to HDL. BLp, apoB-containing lipoprotein, LPL, lipoprotein lipase.

TG/SM/PC-rich lipid droplets, a process similar to HDL remodeling [33–34]. We speculate that cellular PLTP activity could be involved in this process. After mature BLp secreted into the blood and LPL-mediated triglyceride hydrolysis, the core of BLp shrinks, and the redundant surface constituents (SM, PC, and free cholesterol) can be the substrates of PLTP, transferring from BLp to HDL (Figure 6.2). PLTP can also promote ABCA1-mediated nascent HDL production (Figure 6.1).

2.3.1. PLTP

PLTP belongs to a family of lipid transfer/lipopolysaccharide-binding proteins, including cholesteryl ester transfer protein (CETP), lipopolysaccharide-binding protein (LBP), and bactericidal/permeability-increasing protein (BPI) [35]. It is a monomeric protein of 81 kDa [36]. Besides transferring PC, PLTP also efficiently transfers SM, cholesterol, diacylglycerol, α-tocopherol, cerebroside, and lipopolysaccharide [37–38]. Although CETP can also transfer phospholipids, there is no redundancy in the functions of PLTP and CETP [39]. The liver [36,40] and small intestine [41] are two important sites of PLTP expression. It was also shown that PLTP is highly expressed in macrophages [42–44] and in atherosclerotic lesions [45–46].

2.3.2. PLTP and BLp Production

We have unexpectedly found that PLTP deficiency causes a significant impairment in hepatic secretion of VLDL [47]. Likewise, it has been reported that animals overexpressing PLTP exhibit hepatic VLDL overproduction [48].

Associations of plasma PLTP activity with elevated apoB levels have been found in humans as well [49]. Dr. Largrost's group [50] found that human PLTP transgenic rabbits showed a significant increase of BLp cholesterol in the circulation. Nevertheless, the surprising finding that PLTP affects BLp secretion from the liver has remained unexplained. We believe that PLTP activity is involved in promoting BLp lipidation, since PLTP activity and triglyceride enrichment are two factors for PLTP-mediated HDL enlargement [33–34], a process similar to the second step of BLp lipidation [30]. PLTP-mediated BLp production influences plasma HDL levels. PLTP gene knockout (KO) mice have: 1) reduced BLp production [47] and reduced SM levels on BLp [51]; and 2) no lipid, including SM and PC, transfer activity from VLDL to HDL [37,39]. To address the impact of liver-expressed PLTP on lipoprotein metabolism, we created a mouse model that expresses PLTP acutely and specifically in the liver on a PLTP-null background. We found that liver PLTP expression dramatically increases plasma non–HDL-cholesterol, non–HDL-phospholipid, and triglyceride levels, but has no significant influence on plasma HDL-lipids, compared with controls [52]. We believe that 1) the major function of liver PLTP is driving VLDL production, and 2) liver-generated PLTP makes a small contribution to plasma PLTP activity (20–25%) which is not sufficient to influence HDL levels by transferring phospholipid and free cholesterol from BLp to HDL.

Recently, we created liver-specific PLTP KO mice that have 75–80% PLTP activity in the circulation. More importantly, we found that these mice not only displayed significant decrease in BLp levels, but also HDL levels. Reduction of BLp is probably one of the mechanisms for HDL reduction (Yazdanyar and Jiang, unpublished observation).

2.3.3. PLTP and HDL Metabolism

In human studies, PLTP activity is inversely associated with HDL levels [53–54]. PLTP KO mice demonstrated a complete loss of phospholipid transfer activities [37]. The deficient mice showed a marked decrease in HDL cholesterol and apoA-I levels [37,51,55]. PLTP transgenic mice showed a 2.5- to 4.5-fold increase in PLTP activity in plasma compared with controls. This resulted in a 30–40% reduction of plasma HDL cholesterol levels [56]. Overall, PLTP over-expression or deficiency causes a significant reduction of HDL levels in the circulation, and we still do not have a good explanation for that.

Oram and colleagues [57] reported that exogenous PLTP can promote removing cholesterol and phospholipids from cells by the ABCA1 pathway. In contrast, PLTP had no effect on lipid efflux from fibroblasts isolated from a patient with Tangier disease [58], an HDL deficiency syndrome caused by mutations in ABCA1. The same group of researchers also indicated that an amphipathic helical region (aa144–aa163) of PLTP is critical for ABCA1-dependent cholesterol efflux [59], PLTP functions to: 1) stabilize ABCA1 [57,59], and 2) shuttle lipids between cells and existing HDL particles (formed first through ABCA1 action) [57].

Recently, we isolated primary hepatocytes from PLTP KO mice and labeled them with 3H-cholesterol. The cells were then incubated with human recombinant apoA-I with or without recombinant PLTP (rPLTP). We then isolated HDL by ultracentrifugation and found that rPLTP significantly promoted hepatocyte nascent 3H-HDL production. Furthermore, we found that this PLTP-mediated effect requires the presence of ABCA1, since rPLTP has no effect on ABCA1 KO hepatocytes (Yazdanyar and Jiang, unpublished observation). Therefore, reduction of HDL production in the liver may be another mechanism for HDL decreasing in liver-specific PLTP KO mice.

2.3.4. PLTP-Mediated HDL Remodeling

PLTP has a considerable impact on the heterogeneity of circulating HDL. The regulatory role of PLTP in HDL metabolism is achieved through its transfer activity [60–61]. In plasma, PLTP mediates the rearrangement and

conversion of HDL particles in favor of larger sized HDL with concomitant generation of preβ-HDL particles [62–63]. It has been shown that
addition of apoA-I to cells expressing ABCA1 results in the formation of
up to five distinct preβ-HDL subfractions [64]. These nascent HDLs
undergo rapid intravascular remodeling which appears to be dependent
on PLTP activity [65].

PLTP and CETP have been shown to have opposite effects on HDL size
distribution. Treatment of isolated plasma lipoproteins with purified PLTP
or CETP leads to production of different HDL spices. PLTP promotes the
formation of HDL2b particles at the expense of HDL3a, whereas CETP
increases the relative proportion of both HDL3b and HDL3c at the expense
of HDL2a [66–67]. It has been reported that the initial step in PLTP-mediated HDL conversion is surface phospholipid (PC and SM) transfer followed by release and fusion of unstable particles [68]. Moreover, nuclear
magnetic resonance (NMR) spectroscopy study has shown that particle
fusion, but not aggregation, accounts for HDL particle enlargement [69]. It
has been shown that the process of HDL remodeling by PLTP is dependent
on efficient phospholipid transfer [33], triglyceride enrichment [34,19],
oxidative modifications of HDL [70], and also lipid/protein ratio of the
particles. It is conceivable that HDL surface lipid, including SM and PC,
levels may play an important role in HDL remodeling, thus influencing
HDL metabolism.

3. SPHINGOLIPID DE NOVO SYNTHESIS AND HDL METABOLISM

3.1. Major Sphingolipid Biosynthesis Pathway

Sphingolipid biosynthesis starts in the endoplasmic reticulum (ER) using
non-sphingolipid hydrophilic precursor molecules serine and palmitoyl-
CoA (Figure 6.3). Condensation of L-serine and palmitoyl-CoA into
3-ketodihydrosphingosine is facilitated by ER membrane–associated serine
palmitoyltransferases (SPT). Next step in sphingolipid biosynthesis is the
reduction of 3-ketodihydrosphingosine to dihydrosphingosine by a reductase. N-acylation of dihydrosphingosine gives rise to dihydroceramide, a
product that is still relatively hydrophilic. Conversion of dihydroceramide to
ceramides is facilitated by ceramide synthases and involves a desaturation
step. Ceramides are hydrophobic and therefore become membrane associated. Sphingomyelin synthase–related protein (SMSr) converts some of the
ceramide into ceramide phosphatidylethanolamine (CPE) whose function

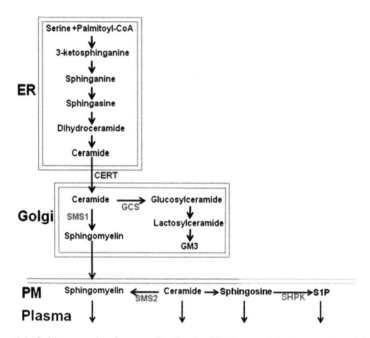

Figure 6.3 *Sphingomyelin de novo Synthesis.* CERT, ceramide transport protein; PM, plasma membrane; GCS, glucosylceramide; SMS, sphingomyelin synthase; SPHK, sphingosine kinase.

is still unknown. However, the majority of ceramides are transported from the ER to the Golgi by ceramide transport protein (CERT) [71].

In the Golgi, ceramides are further converted to sphingomyelin by sphingomyelin synthase (SMS) 1 and SMS2 [72–73], to glucosylceramide by glucosylceramide synthase, and then to more complex sphingolipids such as GM3, to galactosylceramide by galactosylceramide synthase, or to ceramide-1-phosphate by ceramide kinase (Figure 6.3). These products are then transported to plasma membrane, which is the major cellular reservoir for these lipids.

Very little is known about the transport of different sphingolipids out of the Golgi. It is assumed that they are trafficked to plasma membrane through the vesicular transport pathway that carries secretory and plasma membrane–destined proteins.

Ceramide plays a pivotal role in sphingosine-1-phosphate (S1P) formation, but a key enzyme sphingosine kinase (SphK) can phosphorylate its substrate sphingosine to generate S1P [74]. There are two forms of Sphk— Sphk1 and Sphk2. Sphk1 is found in the cytosol of eukaryotic cells, and it migrates to the plasma membrane on activation.

Plasma membrane is enriched in SMS2 that synthesizes sphingomyelin from ceramides [75] as well as in sphingomyelinase (SMase) that hydrolyzes sphingomyelin to ceramides [76]. Thus, plasma membrane ceramide levels are balanced by the activities of these two enzymes.

The major sphingolipids in the human plasma are sphingomyelin (SM) (~90%), ceramides (~7%), and S1P (~1%) [77]. These sphingolipids are found associated with plasma lipoproteins. SM is the second most abundant phospholipid in mammalian plasma. Atherogenic lipoproteins such as VLDL and LDL are the major carriers of SM and ceramide [22–23]. The major carriers of ceramide in the blood are also non-HDL [77]. Plasma S1P, which is derived from several cellular sources [78], is associated with HDL (~65%) and albumin (~35%) [79–80]

There is paucity of knowledge about the metabolism of sphingolipids in the plasma compartment. Due to their structural similarities and localization on the surface of lipoproteins, sphingolipid catabolism is expected to be very similar to that of phospholipids and free cholesterol. Nascent plasma lipoproteins are hydrolyzed at endothelial cell surfaces by the action of lipoprotein lipases resulting in the hydrolysis of triglycerides and phospholipids and shedding of some of the surface components. It is not known whether sphingolipids remain associated with lipoproteins during and after lipase action. Similarly, very little is known about the exchange of sphingolipids in the plasma compartment by plasma transfer proteins such as PLTP and CETP. Hydrolyzed lipoprotein remnants are removed from plasma through endocytosis involving members of the LDL receptor family [21]. It is likely that some of the sphingolipids are taken by cells during endocytosis of BLp. Thus, sphingolipid catabolism might follow the path of their lipoprotein carriers; however, experimental evidence for this is lacking.

3.2. Serine Palmitoyltransferase (SPT)

SPT is the rate-limiting enzyme involved in the biosynthesis of SM [81]. Mammalian SPT holoenzyme is primarily a heterodimer, composed of two protein subunits, SPTLC1 (53kDa) and SPTLC2 (63 kDa), with a 20% sequence homology [82–83]. However, more recent studies indicate the existence of a third subunit, SPTLC3, which shows a 68% homology to SPTLC2 [84]. Interestingly, the expression of SPTLC3 is almost negligible in hematopoietic tissues like peripheral blood cells, macrophages, bone marrow, or spleen. In fact, SPTLC3 levels are compensated by increased expression of SPTLC2 in these tissues [85]. Additionally, there are two low molecular weight proteins, ssSPTa and ssSPTb, that enhance enzyme activity and confer distinct acyl-CoA substrate specificities to mammalian SPT like the yeast Tsc3p subunit [86]. Compelling evidence from multiple studies has established the fact that inhibition of SPT by myriocin results in lower plasma sphingolipid levels, leading to reduced atherosclerosis in apoE KO mice [87–89]. Mechanisms that link sphingolipid metabolism and atherogenicity include: 1) reduction of sphingolipid levels, including SM, ceramide, and S1P in the circulation, mainly on non-HDL [88]; and 2) reduction of macrophage plasma membrane sphingolipid levels, especially those enriched in lipid rafts, thus influencing lipid raft–associated cellular functions—inflammation and cholesterol efflux.

3.2.1. SPT Inhibitors

Myriocin, sphingofungins, and lipoxamycin are potent and highly selective naturally occurring inhibitors of SPT, inhibiting fungal and mammalian SPT in cell-free preparations with IC^{50} values in the nanomolar range [90–92]. Structurally, they resemble the transient intermediate postulated to form in the condensation of L-serine and palmitoyl CoA. Myriocin-linked resins bind the SPTLC1/SPTLC2 complex tightly [93]. All the inhibitors significantly inhibit SM accumulation in cultured cells and in vivo [90–92].

3.2.2. The Effect of SPT Inhibition on Plasma Lipoproteins

Park and colleagues [87] and we[88] have reported that myriocin treatment (oral administration and intraperitoneal injection, respectively) decreases plasma SM levels and atherosclerosis in apoE knockout mice. Oral administration also decreases plasma non-HDL-C and increases HDL-C [87].

Heterozygous SPTLC1 knockout mice also absorbed significantly less cholesterol than controls [94]. To understand the mechanism, the protein levels of Niemann-Pick C1-like 1 (NPC1L1), ABCG5, and ABCA1—three key factors involved in intestinal cholesterol absorption—were measured. NPC1L1 and ABCA1 were significantly decreased, whereas ABCG5 was increased in the SPT-deficient small intestine. SM levels on the apical membrane were also measured and they were significantly decreased in SPT deficient mice, compared with controls [94]. These result suggested that SPT deficiency might reduce intestinal cholesterol absorption by altering NPC1L1 and ABCG5 levels in the apical and reducing ABCA1 levels in the basal membranes of enterocytes. Thus, manipulation of SPT activity could provide a novel alternative treatment for dyslipidemia.

Mice totally lacking SPT are embryonic lethal [95]. Since the liver is the major site for plasma lipoprotein biosynthesis, secretion, and degradation, we utilized a liver-specific knockout approach to evaluate liver SPT activity and its role in plasma SM and lipoprotein metabolism [96]. We found liver-specific SPTLC2 deficiency decreases liver SPT protein mass and activity by 95% and 92%, respectively, while there are no effects on other tissues. Liver SPTLC2 deficiency significantly decreased plasma SM levels (on both HDL and non-HDL fractions), compared with controls. The deficiency also significantly decreased SM levels in the liver and on the hepatocyte plasma membrane. Moreover, plasma from liver-specific SPTLC2 KO mice has significantly stronger potential to promote cholesterol efflux from macrophages than that from control mice.

3.3. Sphingomyelin Synthase (SMS) and Plasma Lipoprotein Metabolism

Sphingomyelin synthase (SMS) sits at the crossroads of sphingolipid biosynthesis. Blockage of SMS activity should influence not only SM and ceramide levels, but also those of other related sphingolipids, including glycosphingolipid, sphingosine, and S1P. SMS has two isoforms: SMS1 and SMS2. Although both enzymes catalyze the same reactions, their subcellular localizations are different: SMS1 is found in the trans-Golgi apparatus, while SMS2 is predominantly found in the plasma membranes [73]. Also, we found that the liver is the major site for SMS2 expression [97]. To evaluate the in vivo role of SMS2 in SM metabolism, we prepared SMS2 knockout (KO) and SMS2 liver-specific transgenic (LTg) mice and studied plasma SM and lipoprotein metabolism in mice. On a chow diet, SMS2 KO mice showed significant decrease of plasma SM levels, mainly on HDL, but no significant changes in total cholesterol, total phospholipids, and triglyceride, compared with wild-type (WT) littermates. On a high-fat diet, SMS2 KO mice showed further decrease of plasma SM levels, on non-HDL and HDL, whereas, SMS2LTg mice showed significant increase of plasma SM levels, mainly on HDL, but no significant changes in other lipids, compared with WT littermates [98]. We also found that SMS1 deficiency significantly decreased plasma SM levels, but had only a marginal effect on plasma ceramide levels. Surprisingly, we found that SMS1 deficiency dramatically increased glucosylceramide and GM3 levels in plasma, while SMS2 deficiency had no such effect [97].

3.4. Sphingomyolinase and Lipoprotein Metabolism

Sphingomyelinase (SMase) can degrade sphingomyelin and generate phosphocholine and ceramide. There are several forms of SMase that differ by pH optima and subcellular localizations. Acid SMase can be secreted and it functions equally well at neutral pH [99–100]. Deficiency of acid SMase is associated with Niemann-Pick disease types A and B, a severe neurodegenerative disease [101]. Acid SMase–deficient mice develop a severe neurodegenerative course leading to death at 8 months of age. Blood cholesterol, mainly on HDL, is elevated in this mouse model [102]. The patients with Niemann-Pick disease A and B present increased LDL-C and decreased HDL-C [103].

Alkaline SMase is present in the intestinal tract [104] and additionally in human bile [105]. It hydrolyses sphingomyelin in both intestinal lumen and the mucosal membrane in a specific bile salt–dependent manner [104–105]. In the small intestine, alkaline SMase is the key enzyme for sphingomyelin digestion [106]. The hydrolysis of sphingomyelin may affect the cholesterol uptake and have an impact on sphingomyelin levels in plasma lipoproteins [107].

4. PLASMA SM AND HDL METABOLISM

Although SM is the second most abundant phospholipid in the plasma lipoproteins, its physiological function in plasma has not attracted much attention. HDL particles are involved in the process of reverse cholesterol transport (RCT), which removes cholesterol from peripheral tissues and cells. The cholesterol on HDL becomes cholesterol ester (CE) through the activity of lecithin:cholesterol acyltransferase (LCAT). The CE on HDL may then be delivered to liver or steroidogenic tissues by the scavenger receptor B1 (SR-B1) [108–109] (primary fate) or transferred through plasma CETP to LDL and TG-rich lipoproteins in exchange for TG [110].

These processes may be affected by the presence of SM in HDL. SM inhibits LCAT by decreasing its binding to HDL [111]. A negative correlation between the SM content of HDL and LCAT activity was observed in studies with proteoliposomes or reconstituted HDL [112]. Rye, Hime, and Barter [113] found that SM influences the structure of discoidal and spherical HDL and confirmed that SM inhibits the LCAT reaction. We recently reported that plasma SM is a physiological modulator of LCAT activity and plasma CE composition [114], and this may contribute to the previously reported pro-atherogenic effect of high plasma SM levels.

SR–BI is the first molecularly defined receptor for HDL and can mediate the selective uptake of CE into cells. Subbaiah and colleagues [115] investigated the effect of SM in lipoproteins on the selective uptake in three different cell lines: SR–BI-transfected CHO cells, hepatocytes (HepG2), and adrenocortical cells (Y1BS1). They found that SM in the lipoproteins regulates the SR–BI-mediated selective uptake of CE, possibly by interacting with the sterol ring or with SR–BI itself.

A key function of HDL-SM may be to regenerate HDL during normal lipid metabolism such as after liver SR-B1 extracts CE from spherical HDL and releases some lipid-free apoA-I protein. Addition of SM-enriched phospholipids and cholesterol to apoA-I produces a new lipoprotein species optimized to accept cholesterol from cells, thus restarting HDL growth [62].

5. PLASMA S1P AND HDL METABOLISM

S1P, a biologically active lipid mediator, is generated by phosphorylation of sphingosine by sphingosine kinases [74]. It has been reported that the concentration of S1P in the plasma is much higher than that of tissues [116]; we confirmed this phenomenon in our mouse studies [97–98]. It has been reported that S1P is a mediator in many of the cardiovascular effects of HDL [80]. Likewise, the composition of S1P in HDL defines HDL function [117–118]. A recent study indicated that apoM, as a carrier of S1P in HDL, possesses a binding site of S1P [119].

6. PLASMA MEMBRANE SPHINGOLIPID LEVELS AND CHOLESTEROL EFFLUX

6.1. The Effect of Membrane SM on Cholesterol Efflux

ABCA1 transport of cholesterol and phospholipids to nascent HDL particles plays a central role in lipoprotein metabolism and macrophage cholesterol homeostasis. SPTLC1, but not SPTLC2, binds ABCA1 and negatively regulates ABCA1-mediated cholesterol efflux [120]. Plasma membrane SM levels also play a critical role in this process.

The interaction of SM, cholesterol, and glycosphingolipid drives the formation of plasma membrane rafts [121]. Lipid rafts have been shown to be involved in cell signaling, lipid and protein sorting, and membrane trafficking [121–123]. It is well known that both class A and B scavenger receptors are located in lipid rafts [124–126]. Lipoprotein metabolism–related proteins, including LDL receptor–related protein [127], ABCA1 [128–129], and ABCG1[129] are also associated with membrane rafts. Reductions in cholesterol efflux from SMase-deficient macrophages and induction of cholesterol efflux by ABCA1 from SM-deficient Chinese hamster ovary (CHO) cells have been reported [130–131].

We have analyzed cholesterol efflux and reverse cholesterol transport SPTLC2 haploinsufficient (SPTLC2$^{+/-}$) macrophages. We found that SPTLC2$^{+/-}$ macrophages have significantly lower SM levels in plasma membrane and lipid rafts. This reduction enhanced ex vivo cholesterol efflux and in vivo reverse cholesterol transport mediated by ABC transporters [132].

We found that SMS2 deficiency significantly decreases SM levels in SM-rich microdomain (lipid rafts) on hepatocytes, red blood cells, embryo fibroblast cells, and macrophages, while hepatocytes overexpressing SMS2 have significantly higher SM levels on these membrane microdomains [98]. Moreover, we found that SMS2 deficiency caused significant induction of macrophage cholesterol efflux in vitro and in vivo [133].

6.2. The Effect of Other Sphingolipids on Cholesterol Efflux

Witting and colleagues [134] reported that ceramide enhances cholesterol efflux to apoA-I by increasing the cell surface presence of ABCA. Further, the same group of researchers characterized the structural features of ceramide required for this effect, and concluded that the overall ceramide shape and the amide bond are critical for the cholesterol efflux effect, and suggested that ceramide acts through a protein-mediated pathway to affect ABCA1 activity [135].

Glycosphingolipids have been implicated as potentially atherogenic [89]. It has been reported that glycosphingolipid synthesis inhibitor stimulates ABCA1/apoA-I-mediated cholesterol efflux [136].

7. SPHINGOLIPIDS AND ATHEROSCLEROSIS

We found that LDL receptor KO mice with myeloid cell-specific SPTLC2 haploinsufficiency exhibited significantly less atherosclerosis than that of controls [132]. We also found that SMS2 total deficiency and hematopoietically derived cell SMS1 or SMS2 deficiency decrease atherosclerosis in mice [97,133,137].

7.1. SM and Atherosclerosis

The relationship between SM and atherosclerosis has been intensively reviewed [138–140].

7.2. S1P and Atherosclerosis

S1P is a signaling sphingolipid as a ligand for a family of G-protein-coupled receptors that are divided into five subtypes: $S1P_1$, $S1P_2$, $S1P_3$, $S1P_4$, and $S1P_5$ [99,141–144]. These S1P receptors are differentially expressed, coupled to different G proteins to regulate cell proliferation, migration, adhesion, and inflammation in endothelial cells, smooth muscle cells (SMCs), and macrophages [145–148], all of which are central to the development of atherosclerosis, but the role of S1P in atherosclerosis is hard to define as pro- or anti-atherogenic.

FTY720, a synthetic analogue of S1P, functions as a potent agonist of four of the five G protein–coupled S1P receptors except $S1P_2$. Two groups reported that FTY720 dramatically reduced atherosclerotic lesion volume in both ApoE KO mice [149] and LDL receptor KO mice [150]. The possible mechanism is that the drug stimulates S1P3-mediated nitric oxide (NO) production, thus inhibiting the release of the monocyte chemokine MCP-1 by smooth muscle cells, resulting in suppressing recruitment of monocyte/macrophage into atherosclerotic lesions [149]. Another group indicated that FTY720 treatment inhibited atherosclerosis through reducing splenocyte proliferation, interferon-γ levels, and proinflammatory cytokines in plasma [150]. The ratio between inflammatory M1- and anti-inflammatory M2-macrophages was decreased in LDL receptor KO mice treated with FTY720 [150]. In conclusion, though the role of S1P is implicated in the development of atherosclerosis, and $S1P_2$ and $S1P_3$ are both pro-atherogenic factors.

7.3. Ceramide and Atherosclerosis

Roles have been proposed for ceramide in atherogenesis, and it has been shown to induce apoptosis [151]. Ceramide mediates an inflammatory response initiated by cytokines or oxidized LDL, a response that uregulates adhesion molecule expression and induces adhesion and migration of monocytes, both important events in the initiation and progression of atherogenesis [152–153]. Plasma ceramide may contribute to maladaptive inflammation in patients with coronary heart disease [154]. It has been reported that plasma ceramide levels in apoE KO mice are higher than in WT mice [155]. Plasma ceramides may possibly correlate with an increase in LDL oxidation, becoming a risk factor for atherosclerosis [155]. In general, ceramide appears to be a pro-atherogenic factor.

7.4. Glycosphingolipids and Atherosclerosis

Glycosphingolipids induce monocyte adhesion to endothelial cells [156], stimulate vascular smooth muscle cell proliferation [157], and accelerate LDL uptake by macrophages [158]. One report showed that inhibition of glucosylceramide synthase reduces atherosclerosis in apoE KO mice [159].

On the other hand, glycosphingolipids may have anti-inflammatory functions. Dietary gangliosides decrease cholesterol content, caveolin expression, and inflammatory mediators in rat intestinal cell lipid rafts [160], and oral glucosylceramide reduces 2,4-dinitrofluorobenzene-induced inflammatory response in mice by lowering TNFα levels and leukocyte infiltration [161]. Another report has indicated that inhibition of glucosylceramide synthase activity does not inhibit atherosclerosis in a mouse model [162]. Thus, the relationship between glycosphingolipid metabolism and atherosclerosis is still largely unknown.

8. CONCLUSION

Although HDL levels are generally known to be a negative risk factor in human populations, recent studies indicate that the characteristics of HDL are just as important as its plasma concentration in leading to atherosclerotic lesion development. This conclusion is also supported by human studies. Among patients with premature coronary heart disease in the Framingham Study population, ~25% did not have a detectable abnormality in their lipid profiles [163]. In the original Framingham Study, approximately 44% of clinical events occurred in patients with normal HDL-C levels [4]. Although CETP inhibitors can significantly increase HDL levels [164–165], two phase III clinical trials have failed. A most recent study indicated that some genetic mechanisms that raise plasma HDL cholesterol do not seem to lower the risk of myocardial infarction [166]. These data challenge the concept that raising of plasma HDL cholesterol will uniformly translate into reductions in heart diseases. Thus, we should pay more attention to HDL heterogeneity, in terms of compositions and functions.

To explore the effect of sphingolipids on HDL metabolism is a big challenge. On the one hand HDL is a carrier for many sphingolipids, on the other hand, sphingolipids also define and modulate many HDL functions. Much progress has been made from the description of sphingolipids in HDL production and catabolism. However, sphingolipid studies in the lipoprotein field, especially in HDL, are still in the infant stage. Fundamental questions, including the delineation of molecular pathways, remain to be addressed.

Atherosclerosis is the major cause of mortality in the developed countries. Therapy aimed at lowering LDL cholesterol reduces only a small fraction (roughly 30%) of the burden of atherosclerotic disease [167–168]. It is extremely important to find new approaches for a better understanding of the disease. Exploration of the relationship between sphingolipids and HDL metabolism is one of these approaches. Although it is very promising, we still have a long way to go.

REFERENCES

[1] Gordon T, Castelli WP, Hjortland MC, Kannel WB, Dawber TR. High density lipoprotein as a protective factor against coronary heart disease. The Framingham Study. Am J Med 1977;62:707–14.

[2] Gordon T, Kannel WB, Castelli WP, Dawber TR. Lipoproteins, cardiovascular disease, and death. The Framingham study. Arch Intern Med 1981;141:1128–31.

[3] Gordon DJ, Probstfield JL, Garrison RJ, Neaton JD, Castelli WP, Knoke JD, et al. High-density lipoprotein cholesterol and cardiovascular disease. Four prospective American studies. Circulation 1989;79:8–15.

[4] Ansell BJ, Navab M, Hama S, Kamranpour N, Fonarow G, Hough G, et al. Inflammatory/antiinflammatory properties of high-density lipoprotein distinguish patients from control subjects better than high-density lipoprotein cholesterol levels and are favorably affected by simvastatin treatment. Circulation 2003;108:2751–6.

[5] Lincoff AM, Wolski K, Nicholls SJ, Nissen SE. Pioglitazone and risk of cardiovascular events in patients with type 2 diabetes mellitus: a meta-analysis of randomized trials. JAMA 2007;298:1180–8.

[6] Kastelein JJ, van Leuven SI, Burgess L, Evans GW, Kuivenhoven JA, et al. Effect of torcetrapib on carotid atherosclerosis in familial hypercholesterolemia. N Engl J Med 2007;356:1620–30.

[7] Bots ML, Visseren FL, Evans GW, Riley WA, Revkin JH, Tegeler CH, et al. Torcetrapib and carotid intima-media thickness in mixed dyslipidaemia (RADIANCE 2 study): a randomised, double-blind trial. Lancet 2007;370:153–60.

[8] Fayad ZA, Mani V, Woodward M, Kallend D, Abt M, Burgess T, et al. Safety and efficacy of dalcetrapib on atherosclerotic disease using novel non-invasive multimodality imaging (dal-PLAQUE): a randomised clinical trial. Lancet 2011;378:1547–59.

[9] Barter PJ, Rye KA. Cholesteryl ester transfer protein inhibition as a strategy to reduce cardiovascular risk. J Lipid Res 2012;53:1755–66.

[10] Voight BF, Peloso GM, Orho-Melander M, Frikke-Schmidt R, Barbalic M, Jensen MK, et al. Plasma HDL cholesterol and risk of myocardial infarction: a Mendelian randomisation study. Lancet 2012;380:572–80.

[11] Hamilton RL, Moorehouse A, Havel RJ. Isolation and properties of nascent lipoproteins from highly purified rat hepatocytic Golgi fractions. J Lipid Res 1991;32:529–43.

[12] Hamilton RL, Guo LS, Felker TE, Chao YS, Havel RJ. Nascent high density lipoproteins from liver perfusates of orotic acid-fed rats. J Lipid Res 1986;27:967–78.

[13] Oram JF, Yokoyama S. Apolipoprotein-mediated removal of cellular cholesterol and phospholipids. J Lipid Res 1996;37:2473–91.

[14] Brooks-Wilson A, Marcil M, Clee SM, Zhang LH, Roomp K, van Dam M, et al. Mutations in ABC1 in Tangier disease and familial high-density lipoprotein deficiency. Nat Genet 1999;22:336–45.

[15] Bodzioch M, Orso E, Klucken J, Langmann T, Bottcher A, Diederich W, et al. The gene encoding ATP-binding cassette transporter 1 is mutated in Tangier disease. Nat Genet 1999;22:347–51.

[16] Rust S, Rosier M, Funke H, Real J, Amoura Z, Piette JC, et al. Tangier disease is caused by mutations in the gene encoding ATP-binding cassette transporter 1. Nat Genet 1999;22:352–5.

[17] Wang N, Silver DL, Thiele C, Tall AR. ATP-binding cassette transporter A1 (ABCA1) functions as a cholesterol efflux regulatory protein. J Biol Chem 2001;276:23742–7.

[18] Schifferer R, Liebisch G, Bandulik S, Langmann T, Dada A, Schmitz G. ApoA-I induces a preferential efflux of monounsaturated phosphatidylcholine and medium chain sphingomyelin species from a cellular pool distinct from HDL(3) mediated phospholipid efflux. Biochim Biophys Acta 2007;1771:853–63.

[19] Sorci-Thomas MG, Owen JS, Fulp B, Bhat S, Zhu X, Parks JS, et al. Nascent high density lipoproteins formed by ABCA1 resemble lipid rafts and are structurally organized by three apoA-I monomers. J Lipid Res 2012;53:1890–909.

[20] Wu AL, Windmueller HG. Relative contributions by liver and intestine to individual plasma apolipoproteins in the rat. J Biol Chem 1979;254:7316–22.

[21] Brunham LR, Kruit JK, Iqbal J, Fievet C, Timmins JM, Pape TD, et al. Intestinal ABCA1 directly contributes to HDL biogenesis in vivo. J Clin Invest 2006;116:1052–62.

[22] Nilsson A, Duan RD. Absorption and lipoprotein transport of sphingomyelin. J Lipid Res 2006;47:154–71.

[23] Rodriguez JL, Ghiselli GC, Torreggiani D, Sirtori CR. Very low density lipoproteins in normal and cholesterol-fed rabbits: lipid and protein composition and metabolism. Part 1. Chemical composition of very low density lipoproteins in rabbits. Atherosclerosis 1976;23:73–83.

[24] Young SG. Recent progress in understanding apolipoprotein B. Circulation 1990;82:1574–94.

[25] Chen SH, Habib G, Yang CY, Gu ZW, Lee BR, Weng SA, et al. Apolipoprotein B-48 is the product of a messenger RNA with an organ-specific in-frame stop codon. Science 1987;238:363–6.

[26] Powell LM, Wallis SC, Pease RJ, Edwards YH, Knott TJ, Scott J. A novel form of tissue-specific RNA processing produces apolipoprotein-B48 in intestine. Cell 1987;50:831–40.

[27] Rustaeus S, Stillemark P, Lindberg K, Gordon D, Olofsson SO. The microsomal triglyceride transfer protein catalyzes the post-translational assembly of apolipoprotein B-100 very low density lipoprotein in McA-RH7777 cells. J Biol Chem 1998;273:5196–203.

[28] Boren J, Rustaeus S, Olofsson SO. Studies on the assembly of apolipoprotein B-100- and B-48-containing very low density lipoproteins in McA-RH7777 cells. J Biol Chem 1994;269:25879–88.

[29] Wang Y, McLeod RS, Yao Z. Normal activity of microsomal triglyceride transfer protein is required for the oleate-induced secretion of very low density lipoproteins containing apolipoprotein B from McA-RH7777 cells. J Biol Chem 1997;272:12272–8.

[30] Hamilton RL, Wong JS, Cham CM, Nielsen LB, Young SG. Chylomicron-sized lipid particles are formed in the setting of apolipoprotein B deficiency. J Lipid Res 1998;39:1543–57.

[31] Tall AR, Small DM. Plasma high-density lipoproteins. N Engl J Med 1978;299: 1232–6.

[32] Ross CJ, Twisk J, Meulenberg JM, Liu G, van den Oever K, Moraal E, et al. Long-term correction of murine lipoprotein lipase deficiency with AAV1-mediated gene transfer of the naturally occurring LPL(S447X) beneficial mutation. Hum Gene Ther 2004;15:906–19.

[33] Huuskonen J, Olkkonen VM, Ehnholm C, Metso J, Julkunen I, Jauhiainen M. Phospholipid transfer is a prerequisite for PLTP-mediated HDL conversion. Biochemistry 2000;39:16092–8.

[34] Rye KA, Jauhiainen M, Barter PJ, Ehnholm C. Triglyceride-enrichment of high density lipoproteins enhances their remodelling by phospholipid transfer protein. J Lipid Res 1998;39:613–22.

[35] Bruce C, Beamer LJ, Tall AR. The implications of the structure of the bactericidal/permeability-increasing protein on the lipid-transfer function of the cholesteryl ester transfer protein. Curr Opin Struct Biol 1998;8:426–34.

[36] Day JR, Albers JJ, Lofton-Day CE, Gilbert TL, Ching AF, Grant FJ, et al. Complete cDNA encoding human phospholipid transfer protein from human endothelial cells. J Biol Chem 1994;269:9388–91.

[37] Jiang XC, Bruce C, Mar J, Lin M, Ji Y, Francone OL, et al. Targeted mutation of plasma phospholipid transfer protein gene markedly reduces high-density lipoprotein levels. J Clin Invest 1999;103:907–14.

[38] Massey JB, Hickson D, She HS, Sparrow JT, Via DP, Gotto Jr AM, et al. Measurement and prediction of the rates of spontaneous transfer of phospholipids between plasma lipoproteins. Biochim Biophys Acta 1984;794:274–80.

[39] Kawano K, Qin SC, Lin M, Tall AR, Jiang XC. Cholesteryl ester transfer protein and phospholipid transfer protein have nonoverlapping functions in vivo. J Biol Chem 2000;275:29477–81.

[40] Jiang XC, Bruce C. Regulation of murine plasma phospholipid transfer protein activity and mRNA levels by lipopolysaccharide and high cholesterol diet. J Biol Chem 1995;270:17133–8.

[41] Liu R, Iqbal J, Yeang C, Wang DQ, Hussain MM, Jiang XC. Phospholipid transfer protein-deficient mice absorb less cholesterol. Arterioscler Thromb Vasc Biol 2007;27:2014–21.

[42] Cao G, Beyer TP, Yang XP, Schmidt RJ, Zhang Y, Bensch WR, et al. Phospholipid transfer protein is regulated by liver X receptors in vivo. J Biol Chem 2002;277:39561–5.

[43] Valenta DT, Ogier N, Bradshaw G, Black AS, Bonnet DJ, Lagrost L, et al. Atheroprotective potential of macrophage-derived phospholipid transfer protein in low-density lipoprotein receptor-deficient mice is overcome by apolipoprotein AI overexpression. Arterioscler Thromb Vasc Biol 2006;26:1572–8.

[44] Lee-Rueckert M, Vikstedt R, Metso J, Ehnholm C, Kovanen PT, Jauhiainen M. Absence of endogenous phospholipid transfer protein impairs ABCA1-dependent efflux of cholesterol from macrophage foam cells. J Lipid Res 2006;47:1725–32.

[45] Desrumaux CM, Mak PA, Boisvert WA, Masson D, Stupack D, Jauhiainen M, et al. Phospholipid transfer protein is present in human atherosclerotic lesions and is expressed by macrophages and foam cells. J Lipid Res 2003;44:1453–61.

[46] O'Brien KD, Vuletic S, McDonald TO, Wolfbauer G, Lewis K, Tu AY, et al. Cell-associated and extracellular phospholipid transfer protein in human coronary atherosclerosis. Circulation 2003;108:270–4.

[47] Jiang XC, Qin S, Qiao C, Kawano K, Lin M, Skold A, et al. Apolipoprotein B secretion and atherosclerosis are decreased in mice with phospholipid-transfer protein deficiency. Nat Med 2001;7:847–52.

[48] Lie J, de Crom R, van Gent T, van Haperen R, Scheek L, Lankhuizen I, et al. Elevation of plasma phospholipid transfer protein in transgenic mice increases VLDL secretion. J Lipid Res 2002;43:1875–80.

[49] Colhoun HM, Taskinen MR, Otvos JD, Van Den Berg P, O'Connor J, Van Tol A. Relationship of phospholipid transfer protein activity to HDL and apolipoprotein B-containing lipoproteins in subjects with and without type 1 diabetes. Diabetes 2002;51:3300–5.

[50] Masson D, Deckert V, Gautier T, Klein A, Desrumaux C, Viglietta C, et al. Worsening of diet-induced atherosclerosis in a new model of transgenic rabbit expressing the human plasma phospholipid transfer protein. Arterioscler Thromb Vasc Biol 2011;2011(31):766–74.

[51] Qin S, Kawano K, Bruce C, Lin M, Bisgaier C, Tall AR, et al. Phospholipid transfer protein gene knock-out mice have low high density lipoprotein levels, due to hypercatabolism, and accumulate apoA-IV-rich lamellar lipoproteins. J Lipid Res 2000;41:269–76.

[52] Yazdanyar A, Jiang XC. Liver phospholipid transfer protein (PLTP) expression with a PLTP-null background promotes very low density lipoprotein production. Hepatology 2012;56:576–84.

[53] Chen X, Sun A, Mansoor A, Zou Y, Ge J, Lazar JM, et al. Plasma PLTP activity is inversely associated with HDL-C levels. Nutr Metab (Lond) 2009;6:49.

[54] Vergeer M, Boekholdt SM, Sandhu MS, Ricketts SL, Wareham NJ, Brown MJ, et al. Genetic variation at the phospholipid transfer protein locus affects its activity and high-density lipoprotein size and is a novel marker of cardiovascular disease susceptibility. Circulation 2010;122:470–7.

[55] Yan D, Navab M, Bruce C, Fogelman AM, Jiang XC. PLTP deficiency improves the anti-inflammatory properties of HDL and reduces the ability of LDL to induce monocyte chemotactic activity. J Lipid Res 2004;45:1852–8.

[56] van Haperen R, van Tol A, Vermeulen P, Jauhiainen M, van Gent T, van den Berg P, et al. Human plasma phospholipid transfer protein increases the antiatherogenic potential of high density lipoproteins in transgenic mice. Arterioscler Thromb Vasc Biol 2000;20:1082–8.

[57] Oram JF, Wolfbauer G, Vaughan AM, Tang C, Albers JJ. Phospholipid transfer protein interacts with and stabilizes ATP-binding cassette transporter A1 and enhances cholesterol efflux from cells. J Biol Chem 2003;278:52379–85.

[58] Wolfbauer G, Albers JJ, Oram JF. Phospholipid transfer protein enhances removal of cellular cholesterol and phospholipids by high-density lipoprotein apolipoproteins. Biochim Biophys Acta 1999;1439:65–76.

[59] Oram JF, Wolfbauer G, Tang C, Davidson WS, Albers JJ. An amphipathic helical region of the N-terminal barrel of phospholipid transfer protein is critical for ABCA1-dependent cholesterol efflux. J Biol Chem 2008;283:11541–9.

[60] Tall AR, Abreu E, Shuman J. Separation of a plasma phospholipid transfer protein from cholesterol ester/phospholipid exchange protein. J Biol Chem 1983;258:2174–80.

[61] Rao R, Albers JJ, Wolfbauer G, Pownall HJ. Molecular and macromolecular specificity of human plasma phospholipid transfer protein. Biochemistry 1997;36:3645–53.

[62] Marques-Vidal P, Jauhiainen M, Metso J, Ehnholm C. Transformation of high density lipoprotein 2 particles by hepatic lipase and phospholipid transfer protein. Atherosclerosis 1997;133:87–95.

[63] von Eckardstein A, Jauhiainen M, Huang Y, Metso J, Langer C, Pussinen P, et al. Phospholipid transfer protein mediated conversion of high density lipoproteins generates pre beta 1-HDL. Biochim Biophys Acta 1996;1301:255–62.

[64] Mulya A, Lee JY, Gebre AK, Thomas MJ, Colvin PL, Parks JS. Minimal lipidation of pre-beta HDL by ABCA1 results in reduced ability to interact with ABCA1. Arterioscler Thromb Vasc Biol 2007;27:1828–36.

[65] Mulya A, Lee JY, Gebre AK, Boudyguina EY, Chung SK, Smith TL, et al. Initial interaction of apoA-I with ABCA1 impacts in vivo metabolic fate of nascent HDL. J Lipid Res 2008;49:2390–401.

[66] Lagrost L, Athias A, Herbeth B, Guyard-Dangremont V, Artur Y, Paille F, et al. Opposite effects of cholesteryl ester transfer protein and phospholipid transfer protein on the size distribution of plasma high density lipoproteins. Physiological relevance in alcoholic patients. J Biol Chem 1996;271:19058–65.

[67] Pulcini T, Terru P, Sparrow JT, Pownall HJ, Ponsin G. Plasma factors affecting the in vitro conversion of high-density lipoproteins labeled with a non-transferable marker. Biochim Biophys Acta 1995;1254:13–21.

[68] Lusa S, Jauhiainen M, Metso J, Somerharju P, Ehnholm C. The mechanism of human plasma phospholipid transfer protein-induced enlargement of high-density lipoprotein particles: evidence for particle fusion. Biochem J 1996;313 (Pt 1):275–82.

[69] Korhonen A, Jauhiainen M, Ehnholm C, Kovanen PT, Ala-Korpela M. Remodeling of HDL by phospholipid transfer protein: demonstration of particle fusion by 1H NMR spectroscopy. Biochem Biophys Res Commun 1998;249:910–6.

[70] Pussinen PJ, Metso J, Keva R, Hirschmugl B, Sattler W, Jauhiainen M, et al. Plasma phospholipid transfer protein-mediated reactions are impaired by hypochlorite-modification of high density lipoprotein. Int J Biochem Cell Biol 2003;35: 192–202.

[71] Hanada K, Kumagai K, Yasuda S, Miura Y, Kawano M, Fukasawa M, et al. Molecular machinery for non-vesicular trafficking of ceramide. Nature 2003;426:803–9.

[72] Yamaoka S, Miyaji M, Kitano T, Umehara H, Okazaki T. Expression cloning of a human cDNA restoring sphingomyelin synthesis and cell growth in sphingomyelin synthase-defective lymphoid cells. J Biol Chem 2004;279:18688–93.

[73] Huitema K, van den Dikkenberg J, Brouwers JF, Holthuis JC. Identification of a family of animal sphingomyelin synthases. EMBO J 2004;23:33–44.

[74] Hait NC, Oskeritzian CA, Paugh SW, Milstien S, Spiegel S. Sphingosine kinases, sphingosine 1-phosphate, apoptosis and diseases. Biochim Biophys Acta 2006;1758:2016–26.

[75] Tafesse FG, Ternes P, Holthuis JC. The multigenic sphingomyelin synthase family. J Biol Chem 2006;281:29421–5.

[76] Milhas D, Clarke CJ, Hannun YA. Sphingomyelin metabolism at the plasma membrane: implications for bioactive sphingolipids. FEBS Lett 2010;584:1887–94.

[77] Hammad SM, Pierce JS, Soodavar F, Smith KJ, Al Gadban MM, et al. Blood sphingolipidomics in healthy humans: impact of sample collection methodology. J Lipid Res 2010;51:3074–87.

[78] Pappu R, Schwab SR, Cornelissen I, Pereira JP, Regard JB, Xu Y, et al. Promotion of lymphocyte egress into blood and lymph by distinct sources of sphingosine-1-phosphate. Science 2007;316:295–8.

[79] Aoki S, Yatomi Y, Ohta M, Osada M, Kazama F, Satoh K, et al. Sphingosine 1-phosphate-related metabolism in the blood vessel. J Biochem 2005;138:47–55.

[80] Argraves KM, Argraves WS. HDL serves as a S1P signaling platform mediating a multitude of cardiovascular effects. J Lipid Res 2007;48:2325–33.

[81] Merrill Jr AH. Characterization of serine palmitoyltransferase activity in Chinese hamster ovary cells. Biochim Biophys Acta 1983;754:284–91.

[82] Weiss B, Stoffel W. Human and murine serine-palmitoyl-CoA transferase–cloning, expression and characterization of the key enzyme in sphingolipid synthesis. Eur J Biochem/FEBS 1997;249:239–47.

[83] Hanada K, Hara T, Nishijima M. Purification of the serine palmitoyltransferase complex responsible for sphingoid base synthesis by using affinity peptide chromatography techniques. J Biol Chem 2000;275:8409–15.

[84] Hornemann T, Richard S, Rutti MF, Wei Y, von Eckardstein A. Cloning and initial characterization of a new subunit for mammalian serine-palmitoyltransferase. J Biol Chem 2006;281:37275–81.

[85] Hornemann T, Wei Y, von Eckardstein A. Is the mammalian serine palmitoyltransferase a high-molecular-mass complex? Biochem J 2007;405:157–64.

[86] Han G, Gupta SD, Gable K, Niranjanakumari S, Moitra P, Eichler F, et al. Identification of small subunits of mammalian serine palmitoyltransferase that confer distinct acyl-CoA substrate specificities. Proc Natl Acad Sci USA 2009; 106:8186–91.

[87] Park TS, Panek RL, Mueller SB, Hanselman JC, Rosebury WS, Robertson AW, et al. Inhibition of sphingomyelin synthesis reduces atherogenesis in apolipoprotein E-knockout mice. Circulation 2004;110:3465–71.

[88] Hojjati MR, Li Z, Zhou H, Tang S, Huan C, Ooi E, et al. Effect of myriocin on plasma sphingolipid metabolism and atherosclerosis in apoE-deficient mice. J Biol Chem 2005;280:10284–9.

[89] Glaros EN, Kim WS, Wu BJ, Suarna C, Quinn CM, Rye KA, et al. Inhibition of atherosclerosis by the serine palmitoyl transferase inhibitor myriocin is associated with reduced plasma glycosphingolipid concentration. Biochem Pharmacol 2007;73:1340–6.

[90] Zweerink MM, Edison AM, Wells GB, Pinto W, Lester RL. Characterization of a novel, potent, and specific inhibitor of serine palmitoyltransferase. J Biol Chem 1992;267:25032–8.

[91] Mandala SM, Frommer BR, Thornton RA, Kurtz MB, Young NM, Cabello MA, et al. Inhibition of serine palmitoyl-transferase activity by lipoxamycin. J Antibiot (Tokyo) 1994;47:376–9.

[92] Miyake Y, Kozutsumi Y, Nakamura S, Fujita T, Kawasaki T. Serine palmitoyltransferase is the primary target of a sphingosine-like immunosuppressant, ISP-1/myriocin. Biochem Biophys Res Commun 1995;211:396–403.

[93] Chen JK, Lane WS, Schreiber SL. The identification of myriocin-binding proteins. Chem Biol 1999;6:221–35.

[94] Li Z, Park TS, Li Y, Pan X, Iqbal J, Lu D, et al. Serine palmitoyltransferase (SPT) deficient mice absorb less cholesterol. Biochim Biophys Acta 2009;1791: 297–306.

[95] Hojjati MR, Li Z, Jiang XC. Serine palmitoyl-CoA transferase(SPT) deficiency and sphingolipid levels in mice. Biochim Biophys Acta 2005;1737:44–51.

[96] Li Z, Li Y, Chakraborty M, Fan Y, Bui HH, Peake DA, et al. Liver-specific deficiency of serine palmitoyltransferase subunit 2 decreases plasma sphingomyelin and increases apolipoprotein E levels. J Biol Chem 2009;284:27010–9.

[97] Li Z, Fan Y, Liu J, Li Y, Quan C, Bui HH, et al. Impact of sphingomyelin synthase 1 deficiency on sphingolipid metabolism and atherosclerosis in mice. Arterioscler Thromb Vasc Biol 2012.

[98] Liu J, Zhang H, Li Z, Hailemariam TK, Chakraborty M, Jiang K, et al. Sphingomyelin synthase 2 is one of the determinants for plasma and liver sphingomyelin levels in mice. Arterioscler Thromb Vasc Biol 2009;29:850–6.

[99] Jeong T, Schissel SL, Tabas I, Pownall HJ, Tall AR, Jiang X. Increased sphingomyelin content of plasma lipoproteins in apolipoprotein E knockout mice reflects combined production and catabolic defects and enhances reactivity with mammalian sphingomyelinase. J Clin Invest 1998;101:905–12.

[100] Schissel SL, Jiang X, Tweedie-Hardman J, Jeong T, Camejo EH, Najib J, et al. Secretory sphingomyelinase, a product of the acid sphingomyelinase gene, can hydrolyze atherogenic lipoproteins at neutral pH. Implications for atherosclerotic lesion development. J Biol Chem 1998;273:2738–46.

[101] Brady RO, Kanfer JN, Mock MB, Fredrickson DS. The metabolism of sphingomyelin. II. Evidence of an enzymatic deficiency in Niemann-Pick diseae. Proc Natl Acad Sci USA 1966;55:366–9.

[102] Horinouchi K, Erlich S, Perl DP, Ferlinz K, Bisgaier CL, Sandhoff K, et al. Acid sphingomyelinase deficient mice: a model of types A and B Niemann-Pick disease. Nat Genet 1995;10:288–93.

[103] McGovern MM, Pohl-Worgall T, Deckelbaum RJ, Simpson W, Mendelson D, Desnick RJ, et al. Lipid abnormalities in children with types A and B Niemann Pick disease. J Pediatr 2004;145:77–81.

[104] Nilsson A. The presence of spingomyelin- and ceramide-cleaving enzymes in the small intestinal tract. Biochim Biophys Acta 1969;176:339–47.

[105] Duan RD, Nilsson A. Purification of a newly identified alkaline sphingomyelinase in human bile and effects of bile salts and phosphatidylcholine on enzyme activity. Hepatology 1997;26:823–30.

[106] Duan RD, Nyberg L, Nilsson A. Alkaline sphingomyelinase activity in rat gastrointestinal tract: distribution and characteristics. Biochim Biophys Acta 1995;1259:49–55.

[107] Duan RD. Alkaline sphingomyelinase: an old enzyme with novel implications. Biochim Biophys Acta 2006;1761:281–91.

[108] Krieger M. Charting the fate of the "good cholesterol": identification and characterization of the high-density lipoprotein receptor SR-BI. Annu Rev Biochem 1999;68:523–58.

[109] Van Eck M, Pennings M, Hoekstra M, Out R, Van Berkel TJ. Scavenger receptor BI and ATP-binding cassette transporter A1 in reverse cholesterol transport and atherosclerosis. Curr Opin Lipidol 2005;16:307–15.

[110] Barter PJ, Brewer Jr HB, Chapman MJ, Hennekens CH, Rader DJ, Tall AR. Cholesteryl ester transfer protein: a novel target for raising HDL and inhibiting atherosclerosis. Arterioscler Thromb Vasc Biol 2003;23:160–7.

[111] Subbaiah PV, Liu M. Role of sphingomyelin in the regulation of cholesterol esterification in the plasma lipoproteins. Inhibition of lecithin-cholesterol acyltransferase reaction. J Biol Chem 1993;268:20156–63.

[112] Bolin DJ, Jonas A. Sphingomyelin inhibits the lecithin-cholesterol acyltransferase reaction with reconstituted high density lipoproteins by decreasing enzyme binding. J Biol Chem 1996;271:19152–8.

[113] Rye KA, Hime NJ, Barter PJ. The influence of sphingomyelin on the structure and function of reconstituted high density lipoproteins. J Biol Chem 1996;271:4243–50.

[114] Subbaiah PV, Jiang XC, Belikova NA, Aizezi B, Huang ZH, Reardon CA. Regulation of plasma cholesterol esterification by sphingomyelin: effect of physiological variations of plasma sphingomyelin on lecithin-cholesterol acyltransferase activity. Biochim Biophys Acta 2012;1821:908–13.

[115] Subbaiah PV, Gesquiere LR, Wang K. Regulation of the selective uptake of cholesteryl esters from high density lipoproteins by sphingomyelin. J Lipid Res 2005;46:2699–705.

[116] Liu X, Xiong SL, Yi GH. ABCA1, ABCG1, and SR-BI: Transit of HDL-associated sphingosine-1-phosphate. Clin Chim Acta 2012;413:384–90.

[117] Lee MH, Hammad SM, Semler AJ, Luttrell LM, Lopes-Virella MF, Klein RL. HDL3, but not HDL2, stimulates plasminogen activator inhibitor-1 release from adipocytes: the role of sphingosine-1-phosphate. J Lipid Res 2010;51:2619–28.

[118] Sekine Y, Suzuki K, Remaley AT. HDL and sphingosine-1-phosphate activate stat3 in prostate cancer DU145 cells via ERK1/2 and S1P receptors, and promote cell migration and invasion. Prostate 2011;71:690–9.

[119] Christoffersen C, Obinata H, Kumaraswamy SB, Galvani S, Ahnstrom J, Sevvana M, et al. Endothelium-protective sphingosine-1-phosphate provided by HDL-associated apolipoprotein M. Proc Natl Acad Sci USA 2011;108:9613–8.

[120] Tamehiro N, Zhou S, Okuhira K, Benita Y, Brown CE, Zhuang DZ, et al. SPTLC1 binds ABCA1 to negatively regulate trafficking and cholesterol efflux activity of the transporter. Biochemistry 2008;47:6138–47.

[121] Simons K, Ikonen E. Functional rafts in cell membranes. Nature 1997;387:569–72.

[122] Futerman AH, Hannun YA. The complex life of simple sphingolipids. EMBO Rep 2004;5:777–82.

[123] Holthuis JC, van Meer G, Huitema K. Lipid microdomains, lipid translocation and the organization of intracellular membrane transport (Review). Mol Membr Biol 2003;20:231–41.

[124] Kim S, Watarai M, Suzuki H, Makino S, Kodama T, Shirahata T. Lipid raft microdomains mediate class A scavenger receptor-dependent infection of Brucella abortus. Microb Pathog 2004;37:11–9.

[125] Graf GA, Connell PM, van der Westhuyzen DR, Smart EJ. The class B, type I scavenger receptor promotes the selective uptake of high density lipoprotein cholesterol ethers into caveolae. J Biol Chem 1999;274:12043–8.

[126] Rhainds D, Bourgeois P, Bourret G, Huard K, Falstrault L, Brissette L. Localization and regulation of SR-BI in membrane rafts of HepG2 cells. J Cell Sci 2004;117:3095–105.

[127] von Arnim CA, Kinoshita A, Peltan ID, Tangredi MM, Herl L, et al. The low density lipoprotein receptor-related protein (LRP) is a novel beta-secretase (BACE1) substrate. J Biol Chem 2005;280:17777–85.

[128] Landry YD, Denis M, Nandi S, Bell S, Vaughan AM, Zha X. ATP-binding cassette transporter A1 expression disrupts raft membrane microdomains through its ATPase-related functions. J Biol Chem 2006;281:36091–101.

[129] Jessup W, Gelissen IC, Gaus K, Kritharides L. Roles of ATP binding cassette transporters A1 and G1, scavenger receptor BI and membrane lipid domains in cholesterol export from macrophages. Curr Opin Lipidol 2006;17:247–57.

[130] Leventhal AR, Chen W, Tall AR, Tabas I. Acid sphingomyelinase-deficient macrophages have defective cholesterol trafficking and efflux. J Biol Chem 2001;276:44976–83.

[131] Nagao K, Takahashi K, Hanada K, Kioka N, Matsuo M, Ueda K. Enhanced apoA-I-dependent cholesterol efflux by ABCA1 from sphingomyelin-deficient Chinese hamster ovary cells. The Journal of Biologicalchemistry 2007;282:14868–74.

[132] Chakraborty M, Lou C, Huan C, Kuo MS, Park TS, Cao G, et al. Myeloid cell-specific serine palmitoyltransferase subunit 2 haploinsufficiency reduces murine atherosclerosis. J Clin Invest 2013;123:1784–97.

[133] Liu J, Huan C, Chakraborty M, Zhang H, Lu D, Kuo MS, et al. Macrophage sphingomyelin synthase 2 deficiency decreases atherosclerosis in mice. Circ Res 2009; 105:295–303.

[134] Witting SR, Maiorano JN, Davidson WS. Ceramide enhances cholesterol efflux to apolipoprotein A-I by increasing the cell surface presence of ATP-binding cassette transporter A1. J Biol Chem 2003;278:40121–7.

[135] Ghering AB, Davidson WS. Ceramide structural features required to stimulate ABCA1-mediated cholesterol efflux to apolipoprotein A-I. J Lipid Res 2006;47: 2781–8.

[136] Glaros EN, Kim WS, Quinn CM, Wong J, Gelissen I, Jessup W, et al. Glycosphingolipid accumulation inhibits cholesterol efflux via the ABCA1/apolipoprotein A-I pathway: 1-phenyl-2-decanoylamino-3-morpholino-1-propanol is a novel cholesterol efflux accelerator. J Biol Chem 2005;280:24515–23.

[137] Fan Y, Shi F, Liu J, Dong J, Bui HH, Peake DA, et al. Selective reduction in the sphingomyelin content of atherogenic lipoproteins inhibits their retention in murine aortas and the subsequent development of atherosclerosis. Arterioscler Thromb Vasc Biol 2010;30:2114–20.

[138] Worgall TS. Sphingolipid synthetic pathways are major regulators of lipid homeostasis. Adv Exp Med Biol 2011;721:139–48.

[139] Hornemann T, Worgall TS. Sphingolipids and atherosclerosis. Atherosclerosis 2013;226:16–28.

[140] Jiang XC, Goldberg IJ, Park TS. Sphingolipids and cardiovascular diseases: lipoprotein metabolism, atherosclerosis and cardiomyopathy. Adv Exp Med Biol 2011;721:19–39.

[141] Hla T, Maciag T. An abundant transcript induced in differentiating human endothelial cells encodes a polypeptide with structural similarities to G-protein-coupled receptors. J Biol Chem 1990;265:9308–13.

[142] Im DS, Clemens J, Macdonald TL, Lynch KR. Characterization of the human and mouse sphingosine 1-phosphate receptor, S1P5 (Edg-8): structure–activity relationship of sphingosine1-phosphate receptors. Biochemistry 2001;40:14053–60.

[143] Okazaki H, Ishizaka N, Sakurai T, Kurokawa K, Goto K, Kumada M, et al. Molecular cloning of a novel putative G protein-coupled receptor expressed in the cardiovascular system. Biochem Biophysical Res Commun 1993;190:1104–9.

[144] Yamaguchi F, Tokuda M, Hatase O, Brenner S. Molecular cloning of the novel human G protein-coupled receptor (GPCR) gene mapped onchromosome 9. Biochem Biophys Res Commun 1996;227:608–14.

[145] Lee MJ, Thangada S, Claffey KP, Ancellin N, Liu CH, Kluk M, et al. Vascular endothelial cell adherens junction assembly and morphogenesis induced by sphingosine-1-phosphate. Cell 1999;99:301–12.

[146] Michaud J, Im DS, Hla T. Inhibitory role of sphingosine 1-phosphate receptor 2 in macrophage recruitment during inflammation. J Immunol 2010;184:1475–83.

[147] Takuwa Y, Okamoto Y, Yoshioka K, Takuwa N. Sphingosine-1-phosphate signaling and biological activities in the cardiovascular system. Biochim Biophys Acta 2008;1781:483–8.

[148] Zhang H, Desai NN, Olivera A, Seki T, Brooker G, Spiegel S. Sphingosine-1-phosphate, a novel lipid, involved in cellular proliferation. J Cell Biol 1991;114:155–67.

[149] Keul P, Tolle M, Lucke S, von Wnuck Lipinski K, Heusch G, Schuchardt M, et al. The sphingosine-1-phosphate analogue FTY720 reduces atherosclerosis in apolipoprotein E-deficient mice. Arterioscler, Thrombosis Vasc Biol 2007;27:607–13.

[150] Nofer JR, Bot M, Brodde M, Taylor PJ, Salm P, Brinkmann V, et al. FTY720, a synthetic sphingosine 1 phosphate analogue, inhibits development of atherosclerosis in low-density lipoprotein receptor-deficient mice. Circulation 2007;115:501–8.

[151] Mallat Z, Tedgui A. Current perspective on the role of apoptosis in atherothrombotic disease. Circ Res 2001;88:998–1003.

[152] Chatterjee S. Sphingolipids in atherosclerosis and vascular biology. Arterioscler, Thrombosis Vasc Biol 1998;18:1523–33.

[153] Gulbins E, Kolesnick R. Raft ceramide in molecular medicine. Oncogene 2003;22:7070–7.

[154] de Mello VD, Lankinen M, Schwab U, Kolehmainen M, Lehto S, et al. Link between plasma ceramides, inflammation and insulin resistance: association with serum IL-6 concentration in patients with coronary heart disease. Diabetologia 2009;52:2612–5.

[155] Ichi I, Takashima Y, Adachi N, Nakahara K, Kamikawa C, Harada-Shiba M, et al. Effects of dietary cholesterol on tissue ceramides and oxidation products of apolipoprotein B-100 in ApoE-deficient mice. Lipids 2007;42:893–900.

[156] Gong N, Wei H, Chowdhury SH, Chatterjee S. Lactosylceramide recruits PKCalpha/epsilon and phospholipase A2 to stimulate PECAM-1 expression in human monocytes and adhesion to endothelial cells. Proc Natl Acad Sci USA 2004;101:6490–5.

[157] Bhunia AK, Han H, Snowden A, Chatterjee S. Redox-regulated signaling by lactosyl-ceramide in the proliferation of human aortic smooth muscle cells. J Biol Chem 1997;272:15642–9.

[158] Prokazova NV, Mikhailenko IA, Bergelson LD. Ganglioside GM3 stimulates the uptake and processing of low density lipoproteins by macrophages. Biochem Biophys Res Commun 1991;177:582–7.

[159] Bietrix F, Lombardo E, van Roomen CP, Ottenhoff R, Vos M, et al. Inhibition of glycosphingolipid synthesis induces a profound reduction of plasma cholesterol and inhibits atherosclerosis development in APOE*3 Leiden and low-density lipoprotein receptor–/– mice. Arterioscler Thromb Vasc Biol 2010;30:931–7.

[160] Park EJ, Suh M, Thomson B, Thomson AB, Ramanujam KS, Clandinin MT. Dietary ganglioside decreases cholesterol content, caveolin expression and inflammatory mediators in rat intestinal microdomains. Glycobiology 2005;15:935–42.

[161] Duan J, Sugawara T, Sakai S, Aida K, Hirata T. Oral glucosylceramide reduces 2,4-dinitrofluorobenzene induced inflammatory response in mice by reducing TNF-alpha levels and leukocyte infiltration. Lipids 2011;46:505–12.

[162] Glaros EN, Kim WS, Rye KA, Shayman JA, Garner B. Reduction of plasma glyco-sphingolipid levels has no impact on atherosclerosis in apolipoprotein E-null mice. J Lipid Res 2008;49:1677–81.

[163] Genest Jr JJ, Martin-Munley SS, McNamara JR, Ordovas JM, Jenner J, Myers RH, et al. Familial lipoprotein disorders in patients with premature coronary artery disease. Circulation 1992;85:2025–33.

[164] Clark RW, Ruggeri RB, Cunningham D, Bamberger MJ. Description of the torce-trapib series of cholesteryl ester transfer protein inhibitors, including mechanism of action. J Lipid Res 2006;47:537–52.

[165] Krishna R, Anderson MS, Bergman AJ, Jin B, Fallon M, Cote J, et al. Effect of the cholesteryl ester transfer protein inhibitor, anacetrapib, on lipoproteins in patients with dyslipidaemia and on 24-h ambulatory blood pressure in healthy individuals: two double-blind, randomised placebo-controlled phase I studies. Lancet 2007;370:1907–14.

[166] Voight BF, Peloso GM, Orho-Melander M, Frikke-Schmidt R, Barbalic M, Jensen MK, et al. Plasma HDL cholesterol and risk of myocardial infarction: a Mendelian randomisation study. Lancet 2012.

[167] Maher V, Sinfuego J, Chao P, Parekh J. Primary prevention of coronary heart disease. What has WOSCOPS told us and what questions remain? West of Scotland Coronary Prevention Study. Drugs 1997;54:1–8.

[168] Kjekshus J, Pedersen TR. Reducing the risk of coronary events: evidence from the Scandinavian Simvastatin Survival Study (4S). Am J Cardiol 1995;76:64C–8C.

CHAPTER 7

Role of Lecithin: Cholesterol Acyltransferase in HDL Metabolism and Atherosclerosis

Lusana Ahsan, Alice F. Ossoli, Lita Freeman, Boris Vaisman,
Marcelo J. Amar, Robert D. Shamburek, Alan T. Remaley
Lipoprotein Metabolism Section, Cardiovascular and Pulmonary Branch, National Heart,
Lung and Blood Institute, NIH, Bethesda, MD, USA

Contents

Abstract

Lecithin:cholesterol acyltransferase (LCAT) is a key enzyme in lipoprotein metabolism that enables the maturation of high-density lipoprotein (HDL) particles. LCAT esterifies free cholesterol on the surface of HDL, forming cholesteryl esters which then partition into the lipoprotein core, resulting in the formation of mature spherical HDL particles. In this review, we explore the biochemistry of LCAT and its role in reverse cholesterol transport and lipid metabolism. We also describe fish eye disease and familial LCAT deficiency, two genetic disorders of LCAT. Given its role in facilitating the flux of cholesterol from peripheral tissue to the liver, we also examine the role of LCAT in the pathogenesis of atherosclerosis in animal models and human studies. Finally, we describe recent efforts in developing LCAT-based therapeutics.

1. INTRODUCTION

Lecithin:cholesterol acyltransferase (LCAT) has been an enzyme of great interest in lipoprotein metabolism for many decades. In 1935, Sperry first postulated the presence of a serum enzyme that modifies cholesterol when he noted a decrease in the level of free cholesterol after incubation of human serum at 37°C [1]. Glomset later showed, in 1962, the existence of a serum enzyme that mediates the transfer of a fatty acid from the sn-2 position of phosphatidylcholine (lecithin) to cholesterol to form cholesteryl ester and lysolecithin [2]. The enzyme was later named "lecithin:cholesterol acyltransferase (LCAT)". In 1966, Glomset determined that the esterification of cholesterol occurred to a greater extent on high-density lipoproteins (HDL) than on low-density lipoproteins (LDL) [3]. Around the same time, Norum and Gjone reported on a case of three sisters with only trace levels of serum cholesteryl ester and low HDL cholesterol (HDL-C) due to a deficiency of LCAT [4]. Glomset then synthesized all the discoveries at the time to write his landmark review in 1968, describing how LCAT participates in the removal of excess cellular cholesterol in peripheral cells by the reverse cholesterol transport (RCT) pathway [5].

In this chapter, we will review various aspects of LCAT, including its biochemistry, its role in HDL metabolism and atherosclerosis, and the diseases associated with LCAT deficiency. In addition, we will discuss the latest efforts in translating this knowledge into new therapies and diagnostic tests for cardiovascular disease.

2. LCAT BIOCHEMISTRY
2.1. Reactions Catalyzed by LCAT

The overall reaction for LCAT, which results in the conversion of phosphatidylcholine and cholesterol to cholesteryl ester and lysophosphatidyl choline, is shown in Figure 7.1. LCAT mediates this reaction in two steps. First, it preferentially cleaves fatty acids from the sn-2 position of phospholipids by a phospholipase A-2 activity. LCAT also has an acyltransferase activity

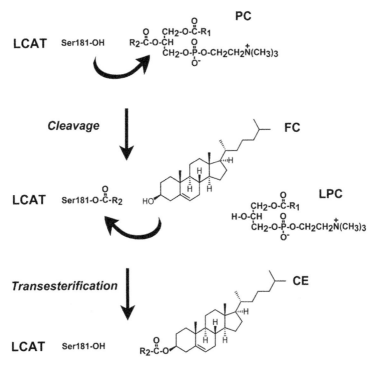

Figure 7.1 *Diagram of LCAT Reaction.* Step 1: Cleavage—Phospholipase cleavage of fatty acid from sn2 position of phosphatidylcholine. Step 2: Transesterification—Transfer of fatty acid bound to Ser181 of LCAT to the hydroxyl group of cholesterol or other sterol acceptor.

and can transesterify the cleaved fatty acid onto the hydroxyl group of the A-ring of cholesterol to produce cholesteryl ester. It performs both of these steps in the absence of an external energy source, such as ATP. Phosphatidylcholine is the preferred phospholipid substrate for LCAT, but it can also act on other phospholipids, such as phosphatidylethanolamine [6]. LCAT also has a preference for unsaturated fatty acids in the sn-2 position, and the preferred length of the fatty acid can vary depending on the species [6]. Cholesterol is the preferred sterol substrate for LCAT, but LCAT can also esterify other closely related sterols [7].

In the absence of sufficient sterols to act as an acyl acceptor, only a reaction with the phospholipase activity of LCAT can occur. In this case, the bound fatty acid is released from the enzyme by hydrolysis to yield a lysophospholipid and free fatty acid. The presence of serum albumin as an acceptor of the fatty acid markedly stimulates this reaction. If there are sufficiently high levels of lysophosphatidylcholine present on a lipoprotein particle, LCAT can also mediate the reverse reaction, causing the reacylation of lysophosphatidylcholine.

In the reaction shown in Figure 7.1, LCAT acts mostly on HDL particles. LCAT is activated by apoA-I, the main protein component on HDL, and then utilizes the lipids on HDL, namely cholesterol and phospholipids, as its substrates. Newly formed HDL and is particularly efficient in promoting the LCAT reaction (Section 3 and Figure 7.3). Esterification of cholesterol converts discoidal HDL into a spherical-shaped HDL. Cholesteryl esters are much more hydrophobic than free cholesterol, causing the esterified cholesterol to partition from the surface of a discoidal HDL into the core, leading to the transformation of discoidal HDL into spherical HDL.

LCAT can also be found on LDL particles, but it is not as active on LDL as on HDL. The affinity of human LCAT for LDL is 2–4-fold lower than for HDL [8]. The apparent V_{max}/K_m of LCAT on LDL has been reported to be only 6.5% of that of HDL. (More mature spherical forms of HDL (1.3% for HDL_2, 16% for HDL_3) are also relatively poor substrates compared to discoidal HDL [8]). Nonetheless, a substantial fraction, sometimes as much as 30%, of the overall cholesterol esterification is thought to occur directly on LDL. Activity on LDL is referred to as β-LCAT activity, whereas activity on HDL is termed α-LCAT activity. Some mutations in LCAT are thought to preferentially affect esterification on HDL, and these mutations may lead to preferential loss of α-LCAT activity and partial retention of βLCAT activity. In addition to the cholesteryl esters directly produced on apoB-containing lipoproteins by LCAT, cholesteryl esters can also be transferred to LDL and other apoB-containing lipoproteins from HDL by the action of CETP. Either way, LCAT is the primary source of cholesteryl esters on apoB-containing lipoproteins. The remainder of cholesteryl esters present on lipoproteins are derived from the liver and intestine when VLDL and chylomicrons are secreted and are produced by one of the two the intracellular enzymes for Acyl-CoA:cholesterol acyltransferase (ACAT) [9].

2.2. Effect of Apolipoproteins on LCAT Activity

As discussed earlier, apoA-I activates LCAT and natural, as well as engineered, mutations in apoA-I have been reported to interfere with this activation [10]. ApoA-I helix 6 (amino acids 143–164) and, to a lesser extent, helix 7 (165–187) are important for LCAT activation. Three arginine residues in apoA-I (R149, R153, and R160) have been suggested to be involved in LCAT binding via multiple hydrogen bonds between the arginine guanidinium group and the polar lipid groups on HDL [11–12]. A natural mutation V156E in this region is reported to completely abolish LCAT activation [13]. Further underscoring the importance of this region, deletion of helix 6 not only blocks esterification but acts as a dominant negative mutation for LCAT activation in vivo [14]. Another natural mutation called apoA-I Milano (R173C) in helix 7 is also an inefficient activator of LCAT [15]. Besides interacting with LCAT, apoA-I may also influence LCAT activity by altering the presentation of its phosphatidyl-choline substrate to LCAT [16]. Other regions of apoA-I important for LCAT interaction and binding have been studied [10,13,17].

Besides apoA-I, the other exchangeable apolipoproteins also appear to have some ability to activate LCAT, but probably less efficiently. For example, although apoE on rHDL particles do not activate LCAT as well as apoA-I [18], apoE may be the primary activator of LCAT on apoB-containing particles [19] and on lipoprotein particles in cerebrospinal fluid [20]. In fact, apoE has been described as a potential disease modifier of familial LCAT deficiency [21]. Other apolipoproteins (such as apoC-I [22]) and apoA-IV [23] may also activate LCAT, whereas apoA-II may inhibit LCAT, perhaps in part by competing for apoA-I [22].

2.3. Effect of Lipids on LCAT Activity

While the presence of the lipid substrates for LCAT, stimulate its activity, the presence of some other lipids can inhibit the LCAT reaction. Enrichment with sphingomyelin, for example, has been shown to inhibit LCAT activity, possibly by affecting interfacial binding of LCAT or by competition with phosphatidylcholine substrate [24,25].

Hydroperoxides of phosphatidylcholine, generated during mild oxidation of lipoprotein lipids, can also inhibit LCAT [26]. Although oxidation inhibits transfer of long-chain acyl groups to cholesterol, it stimulates LCAT's ability to transfer short-chain acyl groups from short-chain polar phosphatidylcholine produced during oxidation to lysophosphatidylcholine.

C18 *trans* unsaturated fatty acids have also been found to inhibit LCAT [27]. Phosphatidylcholines containing 18:1*trans* or 18:2*trans* fatty acids were less effective at stimulating LCAT compared to similar phospholipids containing *cis* isomers. Moreover, the position occupied by the *trans* unsaturated fatty acids in phosphatidylcholine can also affect LCAT activation. 18:1*trans* fatty acid was more inhibitory when present at the *sn*-2 position of phosphatidylcholine versus the *sn*-1 position, when paired with a 16:0 fatty acid in the sn-1 position [27].

Key:

^ = aa insertion.
+ = Multiple aa's inserted.
X = Termination codon
x = Frameshift leading to early termination
Engraved: FLD
Outline: FED
Plain text: Not classified as FLD or FED
*Stars: Catalytic triad (Ser 181, Asp 345, His 377)
Hashtag (plain): N-glycosylation
@ Hashtag (outline): O-glycosylation

Figure 7.2 *LCAT Amino Acid Sequence with Genetic Mutations.* Numbering system is the "old" numbering system (+1 is first amino acid (aa) after signal peptide cleavage). Add 24 aa's to the "old" to get the "new" numbering system, in which +1 is first aa of signal

2.4. Structure of LCAT

The gene for LCAT on chromosome 16 encodes for a protein containing 440 amino acids, with an estimated protein molecular weight of 49.5 kDa [13]. After cleavage of a 24-amino acid-long signal peptide, it has a predicted protein molecular weight of approximately 47 kDa, but the fully mature protein in plasma has an approximate molecular weight of 67 kDa, due to N-linked glycosylation at Asn 20, 84, 272 and 384 and O-linked glycosylation at Thr 407 and Ser 409 [13].

There has been no detailed analysis of its structure, but a model for LCAT has been proposed based on secondary structure analysis and homology to related proteins [28]. According to this model, LCAT forms a globular shaped protein, containing seven β strands connected by four α-helices separated by loops [28]. It contains six cysteines, four of which (Cys50-Cys 74 and Cys 313-Cys356) form disulfide bonds. Cys-31 and Cys-184 remain as free sulfhydryls and are believed to be located near the active site and account for the sensitivity of LCAT to inhibition by 5,5-dithiobis (2-nitrobenzoic acid) (DTNB) [29].

The catalytic triad of LCAT includes Ser181-Asp345-His377 residues and is shielded by a hydrophobic lid (residues 50–74) that may alter the accessibility of the active site to its lipid substrates [28]. By mutational analysis, Ser181 was determined to be the active nucleophile involved in the catalysis and participates in the formation of acyl-LCAT intermediate [23,29]. A phospholipid binding region of the protein has been reported to be located near the active site in residues spanning from 154 to 171 [30]. Residues 50–74 are also reported to be part of the interfacial binding domain of LCAT [31].

The full amino acid sequence of the LCAT protein and the position of natural mutations, resulting in familial LCAT deficiency (FLD) or fish-eye disease (FED), are shown in Figure 7.2. Interestingly, many of the natural mutations in LCAT fall outside of the regions known to be important for the catalytic function of the enzyme and are presumably located in regions important in maintaining its overall structure.

2.5. Regulation of LCAT Gene Expression

The liver is the primary site of LCAT synthesis, but other tissues, such as neural tissue and testes, make a small amount of LCAT [32]. In 1991, a minimal LCAT promoter (71 bp) was described containing a TATA box, an LFAI

---

peptide, before cleavage. Cys50-Cys74 (italics, underlined): Disulfide spanning lid. Deleting DFF (56-58) in vitro eliminates all activity (a and b). Amino acids 154-171 (smaller font, italics, bold, no underline) can form an amphipathic helix with a strong affinity for phospholipids, and may participate in interaction of LCAT with lipid surfaces.

motif, and two Sp1 binding sites [33]. The activity of the promoter was entirely dependent on the Sp1 sites [33]. Sp1 was confirmed as an activator of the LCAT promoter, and Sp3 functions as a dose-dependent repressor of Sp1-mediated LCAT activation [34]. Feister and colleagues [35] identified an IL-6 responsive element at -1514 to -1508 bp and showed that STAT3 mediated the effect of IL-6 on the LCAT gene promoter. TGF-β decreased both the LCAT activity in medium from treated HepG2 cells and the LCAT mRNA level by posttranscriptional mechanisms [36]. Finally, docosahexaenoic acid, arachidonic acid, and eicosapentaenoic acid all decrease LCAT mRNA and protein in HepG2 cells, relative to palmitic acid [37]. Besides these studies, there has been limited investigation on the regulation of LCAT synthesis.

2.6. Biochemistry of Lipoprotein X (LpX)

Lipoprotein X (LpX) is an abnormal cholesterol- and phospholipid-rich lipoprotein particle that is poor in neutral lipids (cholesteryl esters and triglycerides). LpX forms in patients with a genetic defect in LCAT [38]. Unlike normal lipoproteins, which have a micellar-like arrangement of a single layer of phospholipids surrounding a hydrophobic core of cholesteryl ester and triglycerides, LpX has a vesicular structure. Because of the high ratio of amphipathic surface lipids (unesterified cholesterol and phospholipids) to neutral core lipids, LpX forms a bilayer of phospholipids or even a multilamellar phospholipid arrangement, which results in its "onion-like" appearance on electron microscopy. LpX particles are heterogeneous in size (30–100 nm) and can have a density between LDL and VLDL. The composition of LpX is also quite heterogeneous, but in general it contains by mass ~65% phospholipid, mostly phosphatidylcholine, and ~25% free (unesterified) cholesterol. Albumin is a major protein component and is mostly located inside the aqueous core of LpX. It also contains relatively small amounts of exchangeable apolipoproteins, such as apoA-I, apoE, and apoC, presumably bound to its phospholipid surface, but LpX does not contain apoB.

LpX may also appear in patients with cholestasis from various causes, most likely because of the regurgitation of bile salt micelles rich in phospholipid into the plasma compartment [39]. After entering the plasma compartment and dilution of the bile salts below their critical micelle concentration, the bile salt micelles spontaneously rearrange to form LpX particles, with a vesicular structure. The genesis of LpX in patients with FLD is not known, but the low level of plasma cholesteryl esters most likely contributes to its formation in this disorder. It has not been definitively established, but it has been proposed, as will be discussed later, that the presence of LpX particles in FLD is a major causative factor in the development of renal disease in these patients [40].

2.7. LCAT Assays

Numerous assays have been developed for measuring LCAT activity or mass [41], but there is no standard procedure, which can possibly contribute to differences observed between clinical studies on LCAT. A commonly used procedure involves thin-layer chromatography and the measurement of the amount of radiolabeled cholesteryl ester produced, using reconstituted HDL containing radiolabeled free cholesterol as the substrate [42]. Recently, a higher throughput LC-MS procedure has been described for measuring LCAT, which uses a stable isotope of cholesterol [43]. It is important to note, however, as was mentioned by Gillett and Owen [41], that all existing methods of measurement of LCAT activity in plasma are inherently sensitive to the amount of lipoproteins, especially HDL, in the test plasma. HDL cholesterol present in the plasma sample will compete with the exogenous labeled substrate and decrease the measured LCAT activity. In order to minimize this problem, several different volumes of plasma should be tested, but ideally the smallest amount of plasma needed for detection should be used [41,44]. The extent of cholesterol esterification should also be kept below 10–15% to minimize product inhibition and maintain first-order kinetics.

A variation of the LCAT activity assay is the cholesterol esterification rate (CER) assay [45]. In the CER assay, a small amount of radiolabeled cholesterol is directly added to plasma or LDL-depleted plasma and the percentage that is esterified over time is measured, usually after separation from free cholesterol by thin-layer chromatography. This assay is not only dependent upon the amount of LCAT present but also on the type of endogenous lipoprotein particles present in a sample because no exogenous HDL is added. It has been shown that samples enriched in preβ-HDL, because it is a better substrate for LCAT, show increased cholesterol esterification with this assay [45].

3. THE ROLE OF LCAT ON REVERSE CHOLESTEROL TRANSPORT

A diagram of the reverse cholesterol transport (RCT) pathway and how LCAT participates in this process is shown in Figure 7.3. The pathway begins with the formation of HDL when apoA-I interacts with the ABCA1

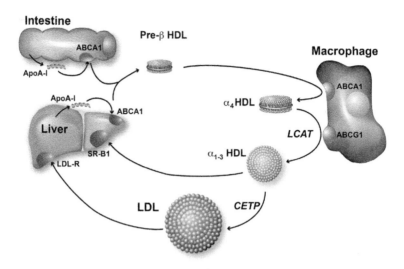

Figure 7.3 *Reverse Cholesterol Pathway.* Pathway begins with the generation of pre-β HDL by liver and intestine by interaction of apoA-I with the ABCA1 transporter. After efflux of additional cholesterol from peripheral cells discoidal (pre-β and α_4) HDL is converted to spherical HDL (α_{1-3}) after esterification by LCAT. Cholesterol is then either returned directly to the liver after interaction with hepatic SR-BI receptors or indirectly after transfer to LDL by CETP and uptake by hepatic LDL receptors.

transporter and acquires phospholipids to form nascent discoidal shaped HDL, with preβ migration on electrophoresis. Most of the newly synthesized cholesterol is made by the liver and to a lesser degree in the intestine. Nascent HDL then interacts with ABCA1 in cells in the periphery and acquires additional phospholipid and cholesterol. It can also acquire cholesterol possibly by the ABCG1 transporter, the SR-BI receptor, and by a passive aqueous diffusion process. HDL at this stage is still either a discoidal shaped HDL with preβ migration or with α-4 migration, but after esterification by LCAT, it is transformed into larger spherical shaped HDL particles with α-1–3 migration, as can be observed by nondenaturing 2D-gel electrophoresis [46]. At this point cholesterol has two possible fates: it can return to the liver as free cholesterol primarily on HDL, or it can be returned as cholesteryl ester via hepatic LDL receptors after transfer to LDL by CETP. Although the SR-BI receptor in the liver can selectively remove cholesteryl esters from HDL, this did not seem to be a major pathway in human tracer studies [47].

According to Glomset's original view [48], HDL helps maintain overall cholesterol homeostasis by removing excess cellular cholesterol from peripheral tissues and returning it to the liver for excretion. Recent studies, however, have revealed that cholesterol homeostasis by most peripheral tissues is relatively unaffected by HDL and the RCT pathway [49]. Nevertheless, the efflux of cholesterol from certain cell types, such as macrophages in the vessel wall, appears to be critical in preventing foam cell formation and inflammation, during the process of atherosclerosis [50]. There is also a growing consensus that simple removal of excess cholesterol from macrophages by HDL and its redistribution of cholesterol among other tissues may be sufficient for reducing atherosclerosis rather than the necessity for delivering it to the liver for net fecal excretion [51]. Another major departure from the original RCT model is that there may be a direct pathway between the plasma compartment and the intestine for the direct excretion of cholesterol called transintestinal cholesterol excretion (TICE) [52].

LCAT is thought to help facilitate the RCT pathway largely because, without LCAT, mature spherical shaped HDL cannot be adequately formed. Discoidal HDL, perhaps because of its small size and/or instability, is rapidly catabolized by the kidney [53] and the overall level of HDL-C and apoA-I in the plasma compartment is remarkably reduced without LCAT. Although discoidal HDL can participate in the initial efflux of cholesterol by ABCA1, the alternative pathways for cholesterol efflux (ABCG1, SR-BI, aqueous diffusion) appear to occur more efficiently with more mature spherical HDL [54]. Human tracer studies have also shown that the majority of cholesteryl esters formed by LCAT in the plasma compartment are removed by the liver [47,55]. Because it is more polar, free cholesterol removed from a peripheral cell by HDL can also readily exchange back and be taken up by another peripheral cell. However, once cholesterol is esterified and is trapped in the core of a lipoprotein, it is more likely to lead to net efflux [56] and removal from the body by hepatic excretion [47,55]. It is difficult to demonstrate with most in vitro assays that LCAT increases cholesterol efflux from cells, but if the ratio of cell donors to HDL acceptor particles is increased to a more physiologic level, LCAT has been shown to increase the net cholesterol efflux from cells [56].

Other properties of LCAT that can potentially affect HDL metabolism and atherosclerosis have also been described [57]. For example, LCAT also hydrolyses oxidized phospholipids and may help protect against oxidized LDL, and it can possibly modulate the binding of thrombin and platelet aggregation, enhance insulin sensitivity, and regulate glucocorticoid levels, by providing a source of sterols for steroid synthesis in the adrenal glands [57]. Besides modulating the level of HDL-C, the level of LDL-cholesterol

(LDL-C) is also remarkably affected by LCAT [58,59]. This may be due in part to the direct esterification of cholesterol that can occur on LDL, but is most likely due to the fact that a significant fraction of cholesterol on LDL is derived from HDL after transfer of cholesteryl esters by CETP or by simple exchange of free cholesterol from HDL [47]. Thus, patients with a genetic defect of LCAT not only have low levels of HDL-C but also relatively low levels of LDL-C, which, as described later, may help limit any increased cardiovascular risk from their low HDL.

4. FAMILIAL LCAT DEFICIENCY AND FISH-EYE DISEASE: CLINICAL SIGNS AND SYMPTOMS

Genetic LCAT deficiency is a rare autosomal recessive disorder with an incidence of less than 1 in 200,000, although the frequency is likely to be higher because of misdiagnosis or underdiagnosis. Mutations in the LCAT gene are associated with either partial or complete absence of plasma LCAT activity, leading to FED or FLD, respectively.

FLD was first described in 1967 as a new genetic metabolic disorder by Norum and Gjone [4]. A 33-year-old Norwegian woman and her two sisters presented with corneal opacities, low HDL, proteinuria, normochromic anemia, mild hypoalbuminemia, and hyperlipidemia. Plasma cholesterol, triglyceride, and phospholipid levels were all increased, but cholesteryl ester and lysophosphatidylcholine were decreased. A renal biopsy showed the accumulation of foam cells in the glomerular tufts, despite the patient having normal blood urea nitrogen (BUN) and creatinine.

A 61-year-old Swedish woman was then later described by Carlson and Philipson [60] who presented with low HDL and severe corneal opacities. The plasma cholesterol and cholesteryl ester levels were normal, triglycerides were increased, and there was no anemia or proteinuria. The plasma LCAT activity was normal, and the syndrome at that time, was felt to be distinct from FLD and was named "fish-eye disease". Similar clinical features were found in two older sisters and her father. Subsequent biochemical studies demonstrated that although total LCAT from patients with FED appeared to be relatively preserved, there was a marked reduction in LCAT activity on HDL.

Originally, it was thought the different manifestation of FLD and FED may be due to the residual amount of LCAT activity present either on HDL or LDL particles [61]. It is now more widely believed that these two disorders lie on a continuum, with FLD patients having a more profound decrease in total LCAT activity [62]. Because of the variety of different types of assays and substrates used for measuring LCAT activity, it has not been possible, however, to reliably predict whether a patient with a mutation in LCAT will develop just FED or FLD. Clinically, the absence of proteinuria and other signs of renal damage are often used to distinguish between these two disorders, although the development of renal disease often takes decades and thus the prognosis of a patient with a new LCAT mutation may remain uncertain.

More detailed features of FLD are described later but the main hallmark findings of FLD are extremely low or absent HDL, mild to moderate hypertriglyceridemia, the development of cloudy cornea in the teenage years, followed by early asymptomatic proteinuria [63]. Normochromic anemia often develops over the next decade, but it is typically mild. Proteinuria often steadily progresses to nephrotic syndrome and end-stage renal disease in the fourth or fifth decade of life. The clinical features of FED are also extremely low or absent HDL and development of cloudy corneas in the teenage years, but the absence of any significant proteinuria and renal disease [63]. Heterozygotes for FLD or FED have no outward clinical symptoms and only have a mildly reduced HDL-C (20–30 mg/dL, 0.52–0.78 mmole/L) and may be at an increased risk for cardiovascular disease.

4.1. Lipoprotein Findings

The main clinical laboratory finding of FLD is an increased percentage of unesterified cholesterol in plasma (> 75%) and the absence of significant levels of HDL-C (< 10 mg/dL, < 0.26 mmol/L). It is important to note that some homogenous direct assays for HDL-C will erroneously measure higher levels of cholesterol possibly due to cross reaction with LpX, but the near absence of HDL can be confirmed by lipoprotein electrophoresis. ApoA-I levels are also typically reduced (< 30 mg/dL) but are not as profoundly decreased as in Tangier disease or in patients with apoA-I mutations, because apoA-I in FLD and FED can often also be found on apoB-containing lipoproteins. LDL-C and apoB are variably reduced in both FLD and FED. Triglyceride levels are also variable and can range from normal to markedly elevated, depending on the degree of proteinuria and renal dysfunction. Unfortunately, there is no routine clinical laboratory method for detecting LpX, but depending on the gel system, it can sometimes be observed by lipoprotein electrophoresis and is thought to form primarily in FLD and not FED patients.

4.2. Ocular Findings

Corneal cholesterol deposits can occur in other disorders of low HDL, but the cornea in LCAT deficiency has the distinct diffuse cloudy appearance of a fish eye as can be seen in Figure 7.4. Despite the cloudy appearance of the cornea, most FLD and FED patients retain relatively normal vision under good lighting conditions. Vision can be impaired in conditions that create glare, such as from bright light or dim light during night driving and some FLD and FED patients are eventually treated with corneal transplants.

Ophthalmological examination of these patients typically shows diffuse haziness of the corneal stroma, with more pronounced opacity near the limbus, often arranged in diffuse, grayish, circular bands [63,64]. Upon slit-lamp examination, small grayish, granular dots are often seen in all layers of the cornea, with a prominent ring-like collection at the periphery leading to an indistinct limbus. Light microscopy often shows preferential localization of vacuoles in the anterior half of the cornea. Electron microscopy has demonstrated that the vacuoles contain electron-dense deposits present in the anterior stroma and in the Bowman layer. Filipin staining indicates the presence of unesterified cholesterol throughout the corneal stroma and Bowman layer but increased staining is absent from the epithelium, Descemet's membrane, and the endothelium [64]. The membranous deposits in the vacuoles of FLD and FED patients probably mostly contain unesterified cholesterol and phospholipids, which suggests that LCAT either directly or because of its effect on increasing HDL levels normally

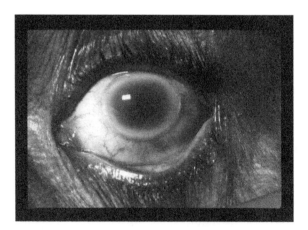

Figure 7.4 Photograph of eye from patient with FLD, showing characteristic clouding of the cornea.

leads to the removal of unesterified cholesterol and phospholipids from the cornea, but this process is inadequate in FLD and FED patients.

4.3. Renal Manifestations

Renal failure is the major cause of morbidity and mortality in FLD. Asymptomatic proteinuria, which usually begins in the teenage years, is often first picked up incidentally many years before the diagnosis of FLD. Albumin is a major component of the proteinuria and can be detected by dipsticks, or from random spot urine samples with increased protein/creatinine ratio and albumin/creatinine ratio, or can be more accurately quantified from a 24-hour urine collection. The urine of these patients usually contains trace amounts of red blood cells and may contain granular and hyaline casts.

Many FLD patients have a prolonged history of proteinuria (1–2 g/24 hours) before their BUN and creatinine levels show a substantial increase. They often go on to develop nephrotic syndrome in the fourth and fifth decade of life, with symptomatic edema and hypertension. The rate of progression of renal disease is unpredictable, with some patients rapidly going from mild proteinuria to a rapid deterioration in their renal function even in their early 20s.

On renal biopsy, focal segmental glomerular sclerosis is often observed in FLD patients. Other common findings include mesangial expansion, a mild increase in mesangial cellularity, and irregular thickening of the glomerular capillary walls, with vacuolization of the glomerular basement membrane due to intramembranous lipid deposits, resulting in a typical "foamy" appearance. Diffuse tubular atrophy with thickening of the tubular basement membranes, along with focal interstitial fibrosis and mononuclear cells infiltrates, can also be found. Immunofluorescence microscopy can reveal bright granular staining for C3 in the capillary loops, mesangium, and arteriolar walls, whereas peripheral capillary loops can stain weakly with antibodies for C1q and fibrinogen [65–67].

Electron microscopy often shows the presence of typical lipid electron-lucent vacuoles and electron-dense lamellar structures in the mesangial matrix, glomerular and tubular basement membrane, and Bowman's capsule [63]. The basement membrane has irregular thickening and fused endothelial foot processes. Capillary walls are abnormal, showing loss of endothelial cells, irregular thickening of the basement membrane, and fused endothelial foot processes. It often has the appearance of "moth-eaten holes" in the mesangial and basement membrane and can contain lamellar osmiophilic deposits.

The pathogenesis of the renal disease in FLD is not completely understood. It has been proposed that it is caused by the presence of lipoprotein

X (LpX) particles and/or possibly large molecular weight LDL (LM-LDL), which also accumulates in FLD [66,68]. The absence of renal disease in FED is presumably prevented by the residual LCAT activity, and the lack of formation of a significant amount of LpX. LpX and LM-LDL appear to become trapped in capillary loops of the glomerulus and induce endothelial damage and vascular injury [66,68]. However, these abnormal lipoprotein particles are not always detected in FLD patients with renal disease [65], and at this time there is not a standard routine diagnostic test for their detection. Infiltrating monocytes can also further contribute to renal damage by taking up large amounts of the deposited lipoprotein-derived lipids, which results in foam cell formation and the stimulation of a wide variety of cytokines and growth factors, causing fibrosis and glomerulosclerosis [69]. The presence of C3 and of electron-dense deposits indistinguishable from those present in immune complex diseases [66,67] has also led to the suggestion that a component of the renal damage in FLD may be immune complex and complement mediated.

4.4. Hematologic Findings

Most FLD patients have a mild chronic normochromic anemia with hemoglobin ranging from 10–12 g/dL, during the second to fourth decade of life. Peripheral smears typically show *poikilocytosis and target cells*. Red cell membranes have increased accumulations of unesterified cholesterol and phosphatidylcholine. These membrane abnormalities can lead to a mild hemolytic anemia or shortened red cell half-life. Severe proteinuria or the worsening of the BUN and creatinine results in chronic anemia (hemoglobin 8–10 mg/dL). This may contribute to diminished compensatory erythropoiesis, sometimes requiring transfusions or erythropoiesis-stimulating agents. Bone marrow aspiration can show the presence of sea-blue histiocytes and other foam-like cells can also be found in bone marrow, spleen, and kidney, presumably due to the excess intracellular storage of lipid from the cellular uptake of LpX and the slow rate of removal due to low HDL.

5. ANIMAL MODELS EXPLORING THE FUNCTIONAL ROLE OF LCAT

Clinical findings in FLD and FED clearly indicate that LCAT plays role a in HDL metabolism and in renal and ocular disease. The role of LCAT in the pathogenesis of atherosclerosis, however, is not as clear despite its purported role in the removal of cholesterol by the RCT pathway. As a consequence of this uncertainty, the role of LCAT in atherosclerosis has been tested in at least four different animal species [6,70,71], as described in Table 7.1.

Table 7.1 Animal Models on the Role of LCAT in Atherosclerosis

Species	Model	Key Findings	References
Mice	Transgenic, human LCAT gene	Hyperalphalipoproteinemia, appearance of second HDL peak, rich in apoE. LCAT transgene exacerbated diet-induced atherosclerosis. CETP corrected dysfunctional HDL.	72–74
Mice	Transgenic, human LCAT gene	Increased HDL, efficient apoA-I, and LCAT interaction.	93
Mice	Transgenic, human LCAT gene	Increased HDL, decreased VLDL, LDL. LCAT transgene did not protect against diet-induced atherosclerosis.	94, 95
Hamster	Adenovirus with human LCAT gene	Increased the level and size of HDL, increased bile cholesterol excretion.	96
Mice	AA virus with LCAT gene	LCAT has minimal effects on macrophage reverse cholesterol transport in vivo.	78
Rabbits	Transgenic	Hyperalphalipoproteinemia; decreased level of apoB-containing lipoproteins, probably via LDL receptor; protection against diet-induced atherosclerosis.	58, 79, 80, 97
Rabbits	Adenovirus with LCAT gene	Increased HDL, protected heterozygous LDL–receptor deficient rabbits against diet-induced atherosclerosis.	98
Squirrel monkey	Adenovirus with LCAT	Increased HDL, decreased level of apoB-containing lipoproteins.	59
Mice	LCAT-KO	Markedly reduced total and HDL cholesterol; susceptibility to glomerulopathy. Diet-induced atherosclerosis was decreased. Adrenal glucocorticoid output was significantly diminished.	82–85, 92
Mice	LCAT-KO	High-fat diet without cholate induced atherosclerosis.	86
Mice	LCAT-KO and SREBP1a transgenic	LpX particles were accumulated in plasma of the mice developing a spontaneous glomerulopathy.	88
Mice	Double ABCA1 and LCAT-KO	Deletion of the two major factors in RCT was not accompanied by severe block of cholesterol transport.	99
Mice	Double LCAT-KO and LDLR-KO	LCAT-KO protects from diet-induced insulin resistance and obesity.	100

One of the first animal models tested for the effect of LCAT on athero-sclerosis was transgenic mice [72], which expressed very high levels of LCAT, but a limitation of the model is that they do not express CETP. Most likely because of the lack of CETP and the inability to transfer cholesteryl esters formed on HDL to apoB-containing lipoproteins, increased LCAT expression in transgenic mice markedly increased the level of HDL-C, leading to the accumulation of an abnormally large species of HDL rich in apoE. Further analysis showed that LCAT transgenic mice were not protected against diet-induced atherosclerosis, and, in fact, had more prominent atherosclerosis than control mice. This is probably due to the dysfunctional HDL that accumulated in these mice, which were shown to be defective in the delivery of cholesterol to the liver via the SR-BI receptor [73,74]. Similar findings were found in other LCAT transgenic mouse models of diet-induced atherosclerosis [75,76].

In experiments with adenoviral-mediated overexpression of LCAT, large cholesteryl ester-rich HDL particles were also observed in the plasma mice, which were poor in promoting cholesterol efflux from extrahepatic tissues [77]. LCAT overexpression in mice with adeno-associated virus also did not increase the flux of cholesterol from radiolabeled macrophages implanted into the peritoneum [78], suggesting that LCAT is not rate limit-ing in this process. The co-expression of CETP in mice along with increased LCAT was able to correct, to some degree, the formation of dysfunctional HDL and decreased the development of diet-induced atherosclerosis com-pared to LCAT transgenic mice without CETP [74].

The beneficial interaction between CETP and LCAT was especially observed in experiments in LCAT transgenic rabbits, which possess endog-enous CETP [79, 80]. In this case, after 17 weeks of an atherogenic diet, the total surface of aortic lesions in LCAT transgenic rabbits was 7-fold smaller than in control New Zealand White rabbits. It is important to point out that in rabbits the LCAT transgene not only increased the level of HDL-C, but also lowered the concentration of pro-atherogenic apoB-containing lipo-proteins [58]. This finding was completely dependent on the presence of LDL receptors [81], suggesting that the delivery of cholesteryl esters pro-duced by LCAT to the liver after the transfer by CETP to apoB-containing lipoproteins is a critical step in the RCT pathway. The effect of LCAT in lowering the level of pro-atherogenic apoB-containing lipoproteins was also confirmed in experiments with adenoviral delivery of LCAT gene to squirrel monkeys, which also express endogenous CETP [59]. Increased LCAT activity was associated with a 2-fold increase in HDL-C and a 37% reduction in LDL-C due to accelerated catabolism of LDL.

Several different LCAT knockout (KO) mouse models on a wide variety of diets have also been tested for atherosclerosis, as well as the effect of the lack of LCAT in diabetes and renal disease (Table 7.1). In one of the first developed LCAT-KO mice, deficiency of LCAT was accompanied by a markedly reduced plasma level of total cholesterol, cholesteryl esters, HDL cholesterol, and apoA-I [82,83], whereas nascent, or preβ, HDL was elevated [82]. The ratio of unesterified cholesterol/cholesteryl esters was also markedly increased in LCAT-KO mice in comparison with control mice, confirming the complete inhibition of the esterification of cholesterol in plasma [82,83]. LCAT-KO males also had elevated levels of triglycerides, similar to what has been observed in some FLD and FED patients [82].

Despite their low levels of HDL, LCAT-deficient mice had significantly reduced atherosclerosis when fed the Paigen high-fat, high-cholesterol (HFHC) diet containing cholate [84]. In LDL receptor-KO mice, as well as in CETP transgenic mice, kept on HFHC diet, and in apoE-KO mice on normal chow diet, LCAT deficiency appeared to be atheroprotective [84]. Similar results were obtained by Ng and colleagues in experiments on apoE-KO mice [85]. In all these cases, LCAT deficiency was associated with a significant decrease in the cholesterol content of pro-atherogenic apoB-containing lipoproteins, which may have overridden the presumably deleterious effect from low HDL from LCAT deficiency. It is important to note that when LCAT-KO/apoE-KO mice were placed on a high-fat diet without cholate, the mice had an elevated level of apoB lipoproteins and increased incidence of atherosclerosis [86]. Taken together, it appears that the anti-atherogenic effect of LCAT deficiency closely correlates with its ability to lower plasma levels of apoB lipoproteins, possibly through upregulation of the LDL receptor [84].

LCAT-deficient mice have also been useful in understanding the mechanisms of progressive loss of renal function observed in FLD patients [63]. As previously discussed, FLD patients often have LpX particles in their plasma [87] and may be involved in the pathogenesis of renal injury [63]. LCAT-KO mice kept on the HFHC diet for 16 weeks accumulated LpX [84]. In this study, renal lesions were only detected in the group of mice that accumulated LpX, which supports the hypothesis of a causal role of LpX in renal disease. Another interesting model, supporting the role of LpX in development of renal insufficiency, was described by Zhu and colleagues [88]. They crossed LCAT-KO mice with SREBP1a transgenic mice, which constitutively overexpressed the amino terminal segment of the SREBP1a variant, resulting in a dramatic increase in hepatic lipogenesis and overproduction of VLDL [89].

Plasma of LCAT-KO/SREBP1a transgenic mice was enriched in LpX particles and when kept on regular chow diet spontaneously developed renal abnormalities similar to those seen in FLD patients [88].

Interestingly, LCAT was recently found in various animal models to also have other effects besides those on lipid metabolism. LCAT deficiency was able to protect mice, especially females, from high-fat, high-sucrose diet-induced obesity and insulin resistance [90,91]. At least three insulin-sensitive tissues were affected by LCAT deficiency, namely, the liver, adipose tissue, and skeletal muscle [91]. An important role of LCAT in the generation of HDL as a source of cholesterol for steroid synthesis was also described [83,92]. Adrenals from LCAT-KO mice demonstrated a marked 40–50% lower glucocorticoid response to adrenocorticotropic hormone exposure, endotoxemia, or fasting [92].

Overall, the results from various animal models of LCAT indicate that there is a complex interaction between LCAT and atherosclerosis, which depends on the diet, and can be modulated by other proteins in the RCT pathway, such as CETP and the LDL receptor. An important outcome of this work is that it appears that the relationship between LCAT and the level of pro-atherogenic lipoproteins appears to have a stronger effect on atherosclerosis than the effect of LCAT on HDL levels.

6. LCAT AND ITS ROLE IN ATHEROSCLEROSIS IN HUMAN STUDIES

Like the situation in the animal models, the role of LCAT in the pathogenesis of atherosclerosis in human studies is also mixed despite the epidemiologic evidence for a protective effect of HDL on cardiovascular disease [101]. Even though overall HDL levels are decreased in LCAT deficiency, preß-HDL particles accumulate, which are particularly good at effluxing cholesterol by the ABCA1 transporter. It has been proposed that cholesterol efflux by ABCA1 may be more important in protecting against development of atherosclerosis and could explain why most FLD and FED patients do not exhibit overt premature cardiovascular disease [63,102]. In addition, the low levels of LDL-C in FLD and FED patients may be below the threshold for the initial development of atherosclerosis, thus making their low HDL levels and decreased cholesterol flux by the RCT pathway less relevant. In Table 7.2, we classified the various clinical trials, imaging studies, biomarker studies, and genomic studies into either a pro-atherogenic, anti-atherogenic, or mixed group, depending on the evidence for the role of LCAT in atherosclerosis.

Table 7.2 Clinical Trials on the Role of LCAT in Atherosclerosis.

Study	Year	Study Cohort	Key Findings
Pro-atherogenic			
Wells et al. [103]	1986	100 male hospitalized patients with 0–3 vessel disease	LCAT activity increased with (1–3 vessel) coronary artery disease compared to 0-vessel involvement or healthy control.
Calabresi et al. [104]	2009	40 carriers of LCAT gene mutations from 13 Italian families and 80 matched controls	LCAT mutation carriers were found to have decreased cIMT with adjustment for other factors such as smoking, BMI, etc.
Dullart et al. [105]	2010	PREVEND cohort: 116 men with CVD compared to 111 male controls	Subjects with CVD found to have greater LCAT activity, even at a given HDL-C level.
Haase et al. [106]	2012	Copenhagen City Heart Study ($n = 10,281$) Copenhagen General Population Study ($n = 50,523$)	Genetically decreased HDL-C due to LCAT polymorphism does not correlate to increased risk of MI.
Anti-atherogenic			
Hovig et al. [109]	1973	90 patients with angiogram established single-, double-, or triple-vessel disease compared to 30 healthy controls.	Patients with coronary artery disease found to have decreased LCAT activity and lower HDL levels.
Ayyobi et al. [114]	2004	Canadian family with FLD: 2 homozygotes and 8 heterozygotes for LCAT mutations	Ultrasound-guided cIMT greater in 6 out of 8 heterozygotes compared to controls. 4 heterozygotes found to have plaques.
Hovingh [115]	2005	Five Dutch families with FED: 9 homozygous and 47 heterozygous subjects with LCAT mutations, 58 family controls	Heterozygotes found to have increased ultrasound-guided cIMT, decreased HDL-C, increased triglycerides, and C-reactive protein.
Scarpioni et al. [110]	2008	Case study on 1 FLD subject	Severe vascular disease reported in a FLD subject with triple-vessel CAD, angina at rest, and PAD.

Continued

Table 7.2 Clinical Trials on the Role of LCAT in Atherosclerosis.—cont'd

Study	Year	Study Cohort	Key Findings
Sethi [112]	2010	95 people with IHD, 110 people without IHD matched by age, gender, and HDL-C	Individuals with IHD had lower LCAT activity in low and high HDL-C groups. IHD was also associated with high preβ1–HDL concentration.
Duivenvoorden et al. [116]	2011	40 LCAT mutation carriers and age-matched controls	MRI and ultrasound-guided cIMT increased in LCAT mutation carriers.
van den Bogaard [117]	2012	45 LCAT mutation carriers and age-matched controls	LCAT mutation carriers had greater carotid-femoral PWV that correlated positively with ultrasound- and MRI-guided cIMT.
Mixed/Gender Specific			
Holleboom et al. [118]	2010	EPIC-NORFOLK cohort: 933 men and women who prospectively acquired CAD compared to 1852 age matched controls	Plasma LCAT levels have no correlation to whole cohort CAD risk. Highest quartile of LCAT concentration correlates to CAD in women.
Calabresi et al. [119]	2011	IMPROVE cohort: 540 individuals with at least three cardiovascular risk factors	Plasma LCAT concentrations have no correlation to whole cohort CAD risk. Increasing concentration of LCAT correlates to CAD in women.

cIMT: Carotid intima media thickness; CVD: Cardiovascular disease; CAD: Coronary artery disease; PWV: Pulse wave velocity; PAD: Peripheral arterial disease; IHD: Ischemic heart disease

6.1. Pro-Atherogenic Studies

In an early study in 1986, a cohort of male hospitalized patients with cardiovascular disease was classified from zero- to three-vessel disease [103]. Patients with coronary artery disease (one- to three-vessel disease) were found to have greater levels of LCAT activity compared to healthy controls or patients with zero-vessel disease, thus implicating the atherogenic effect of LCAT. Increased LCAT activity was also correlated to decreased unesterified cholesterol and increased lysolecithin. Interestingly, a cohort of patients with myocardial infarction in the same study also had decreased LCAT activity [103].

In a more recent study, 40 LCAT mutation carriers from 13 different Italian families were evaluated for the presence of atherosclerosis by carotid ultrasound by Calabresis and colleagues [104]. Published in 2009, the results showed decreased average and maximum values of carotid intima-media thickness (cIMT), an index of atherosclerosis, compared to 80 controls matched by age and gender. Average cIMT was 0.58 mm in carriers compared to 0.65 mm in controls ($p = 0.0003$). Maximum cIMT was also lower in carriers with 0.91 mm compared to 1.12 mm in controls ($p = 0.0027$). The decrease in IMT in carriers remained statistically significant when adjusted for age, sex, BMI, smoking status, hypertension, family history of cardiovascular disease, ultrasound device, and total cholesterol, HDL-C, and triglyceride levels, thus suggesting that LCAT in fact decreases protection against atherosclerosis.

In the PREVEND cohort published in 2010, 116 men who developed cardiovascular disease (CVD) were included in a prospective case-control study. Compared to 111 male controls, subjects with CVD were found to have 5% greater LCAT activity ($p = 0.027$). Similarly, for a given HDL level, higher LCAT activity was associated with slightly higher risk of CVD. These associations led the investigators to conclude that higher LCAT activity does not necessarily correlate to atherosclerotic protection and may even weaken the cardioprotective effects of HDL cholesterol [105].

A recent genome-wide association study (GWAS) examining the effect of genetic polymorphisms in LCAT also found that low HDL-C was not associated with increased risk for cardiovascular disease [106], similar to what has recently been described for other genetic factors that modulate HDL-C levels [107,108]. In this large study, 10,281 individuals from the Copenhagen City Heart Study and 50,523 individuals from the Copenhagen General Population Study were analyzed for expression of a single nucleotide polymorphism of LCAT in subjects who developed a myocardial infarction (MI) [106]. LCAT S208T variant identified in individuals with low HDL-C levels in both cohorts surprisingly did not correlate with incidences of MI in the whole population. This study thus led the investigators to conclude that while decreased levels of plasma HDL-C increases risk of MI, genetically induced decrease of HDL-C due to LCAT mutation does not increase risk of ischemic heart disease.

6.2. Anti-Atherogenic Studies

Shortly after the first description of FLD, autopsy studies reported that FLD patients have increased lipid deposition in the renal veins and arteries, with additional deposition in splenic cells leading to splenomegaly [109]. Although premature atherosclerosis is not common in FLD and FED patients, there are several reported cases of clinically apparent cardiovascular disease. For example, in 2008, Scarpioni and colleagues [110] published a case of severe vascular disease in a patient with FLD. The patient had angina at rest with three-vessel coronary artery disease and lower limb peripheral arterial obstruction, with necrosis of two digits of the left foot. The patient died at age 42 after a right femoral-axillo artery bypass and a thigh amputation [110].

In 1994, a cohort of 90 patients diagnosed to have single-, double-, or triple-vessel disease by angiogram were found to have decreased LCAT activity compared to 30 healthy control subjects without any clinical coronary artery disease. Low LCAT levels were also accompanied by low HDL levels in these patients [111]. A more recent study from a subset of patients from the Copenhagen heart study has shown that patients with both high and low HDL-C but with cardiovascular disease have low LCAT activity compared to an HDL-C matched control group without cardiovascular disease [112]. Interestingly, a low level of LCAT was associated with high preβ HDL-C, which has been shown to be a positive risk maker for CVD [113].

In contrast to the carotid ultrasound study by Calabresi and colleagues [104], several other similar studies have shown an inverse correlation between LCAT activity and cIMT. These studies mainly show that decreases in HDL-C due to mutation or absence in LCAT lead to increased cIMT, thus indicating increased risk for cardiovascular disease. In 2004, after a 25-year follow up of a Canadian family with FLD, increased cIMT was noted in six of the eight heterozygotes compared to controls [114]. Four of the heterozygotes even had distinct plaques. Two homozygotes had minimal increase in cIMT and normal endothelial function. In a study published in 2005, members from five Dutch families with FED were analyzed for cIMT and serum lipids and biomarkers [115]. Nine homozygotes for LCAT mutation were found to have slightly increased cIMT (0.73 mm) compared to 47 heterozygotes (0.623 mm). Due to the small study sample, these values did not reach statistical significance. Heterozygotes interestingly were found to be at a statistically higher risk than homozygotes. Compared to 58 family controls, heterozygotes were found to have a 36% decrease in HDL-cholesterol, 23% increase in triglyceride, and 2.2-fold increase in C-reactive protein on average. Carotid IMT was also increased at 0.623 mm compared to 0.591 mm ($p = 0.0015$) in family controls. It was hypothesized that very low levels of LDL-C in homozygote FLD patients, the primary modulator of atherosclerosis, most likely counteracts the harmful effects of low HDL-C. However, in heterozygote patients with higher LDL levels, the protection against low HDL-C level is lost.

Carotid IMT measurement by ultrasound has been criticized for its two-dimensional analysis of a three-dimensional phenomenon. Therefore, magnetic resonance imaging (MRI) has recently been espoused by investigators as a more reliable tool for measuring IMT. In a study of 40 carriers of LCAT mutation compared to age-matched controls, carriers were found to have greater normalized wall index (0.34 ± 0.07 vs. 0.31 ± 0.04, $p = 0.002$), greater mean wall area (17.3 ± 8.5 mm^2 vs. 14.2 ± 4.1 mm^2, $p = 0.01$), and greater total wall volume ($1,039 \pm 508$ mm^3 vs. 851 ± 247 mm^3, $p = 0.01$ vs. controls) of carotid arteries [116]. Ultrasound-guided carotid IMT also demonstrated an increase in carriers by 12.5% (0.72 ± 0.33 mm vs. 0.64 ± 0.15 mm, $p = 0.14$). All LCAT mutation carriers with clinical cardiovascular disease were excluded from this study [116].

In a study published in 2012 [117], 45 carriers of LCAT mutations and their age-matched controls were evaluated for cholesterol levels, carotid-femoral pulse wave velocity (PWV), and carotid IMT using ultrasound and MRI. After excluding those with prior CVD, mutation carriers were found to have significantly higher PWV, a marker for arterial stiffness, at 7.7 ± 2.0 and 6.9 ± 1.6 m/s, $p < 0.05$. PWV was also positively correlated to cIMT by ultrasound (R 0.50, $p < 0.001$ for carriers and R 0.36, $p < 0.04$ for controls) and 3.0T MRI (R 0.54, $p < 0.001$ for carriers and R 0.58, $p < 0.001$ for controls). As expected, LCAT mutation carriers also had lower HDL-C (32 ± 12 vs. 59 ± 16 mg/dL in controls; $p < 0.0001$) and higher triglyceride levels (median 116 [IQR 80–170] vs. 71 [IQR 53–89] mg/dL in controls; $p < 0.001$) [117].

6.3. Mixed/Gender-Specific Results

In contrast to the previous two types of studies showing either a pro- or anti-atherogenic effect of LCAT, many studies have failed to show conclusively either a positive or negative association of LCAT with cardiovascular disease. In the 2010 EPIC-NORFOLK study [118], mixed findings were reported between males and females in relation to LCAT activity and atherogenic risk factors. Nine hundred and thirty-three men and women prospectively enrolled into the study, who developed CAD, were compared to 1,852 matched healthy controls. The highest quartile of plasma LCAT concentration was associated with an increased risk of CAD in women (OR 1.35, CI 0.87–2.09) and a decreased risk of CAD in men (OR 0.71, CI 0.51–0.91), with adjustment for Framingham risk score. Mixed-gender analysis showed no correlation between level of LCAT expression and atherogenic risk [118]. In the multicenter, longitudinal IMPROVE study published in 2011 [119], 540 individuals with at least three cardiovascular risk factors were enrolled for cIMT visualization and serum analysis. Similar to the findings above, no correlation between plasma LCAT concentration and cIMT were seen in the whole cohort of 247 women and 293 men. However, in gender-specific analysis, cIMT increased in women with increasing quartile of LCAT protein level. The association was not significant, however, when adjusted for age, HDL-C, and triglycerides [119].

In light of these clinical studies, one can conclude that there is an absence of clear and compelling evidence for or against the atherogenic properties of LCAT. In fact, the effect of LCAT on atherosclerosis is probably relatively modest and is likely affected by other dietary and genetic factors as has been shown in the animal studies. It is also important to remember a limitation of many of these clinical studies is that correlation does not always equal causation. It may be, for example, that increased LCAT activity observed in patients with cardiovascular disease may nevertheless still be a beneficial compensatory factor. Also, a wide variety of assays have been used for either measuring the activity or protein mass of LCAT, which can possibly account for some of the divergent results. The jury is still out on whether FLD or FED patients have increased atherosclerosis, but the preponderance of the evidence to date and the results of the latest imaging studies do seem to indicate that they probably do have some increased risk though may be partially protected from their low LDL-C.

7. LCAT-BASED THERAPIES

Because of the uncertainties of the role of LCAT in the pathogenesis of atherosclerosis [70], there have been limited efforts in using LCAT as a target for drug development. Early work, however, on a small-molecule activator of LCAT [120,121] and recombinant LCAT itself [122] as a possible therapy have been described.

The activation of LCAT by a small molecule, particularly one that is orally available, could potentially be valuable in enhancing the reverse cholesterol transport pathway and thus could be widely used for the treatment of patients with pre-existing atherosclerotic disease. A small, sulfhydryl-reactive compound has been identified by Amgen from a high throughput screen to activate LCAT [120,121]. The molecule, called compound A, contains a pyrazine-2-carbonitrile ring and a thiadiazole ring linked by sulfur. It is believed to interact with one of the two sulfhydryl groups near the active site of LCAT and that doses of approximately 10 μM can increase LCAT activity by more than 5-fold in plasma from multiple species, including man [43]. The toxicity of compound A and its specificity for reacting with only LCAT, however, have not been described. Mice treated intraperitoneally with 20 mg/kg of compound A show more than a 5-fold increase in LCAT activity, and plasma HDL-C levels peaked at about 2 times the baseline at 3 hours after treatment [43]. Shortly after treatment, non–HDL-C decreased but then later increased by 24 hours, presumably from the transfer of the esterified cholesterol from HDL to apoB-containing lipoproteins. Interestingly, hamsters treated for 2 weeks with compound A not only had an increase in HDL-C but also a 2-fold increase in the content of bile salts in the gall bladder, suggesting that compound A may increase the RCT pathway [43]. The ability of compound A to promote cholesterol efflux from macrophages and to reduce atherosclerosis in animal models has not been reported and no human clinical trials on small molecule compounds that activate LCAT have been started.

Recombinant LCAT (rLCAT) has been developed by AlphaCore Pharma for two possible indications, namely for the treatment of acute coronary syndrome and for the treatment of FLD. The treatment for acute coronary syndrome would be for the rapid stabilization of patients for reducing the likelihood of a future cardiovascular event similar to what has been tested in several clinical trials with acute intravenous infusion of reconstituted HDL [123]. In the clinical trials with HDL therapy, a weekly intravenous infusion of HDL for 4–5 weeks has been shown to rapidly reduce plaque size to

levels comparable to what has been observed after 2 years of statin treatment
[123]. In mice expressing human apoA-I, infusion of human rLCAT was
found to rapidly raise HDL-C levels by approximately 2-fold, and elevated
levels of HDL-C persisted for at least 3 days [122]. Like HDL therapy, the
intravenous infusion of rLCAT may systemically mobilize cholesterol from
atherosclerotic plaques and thus rapidly stabilize ACS patients. In prelimi-
nary studies, treatment of rabbits on a high-fat diet with rLCAT has been
shown to increase fecal excretion of sterols and to reduce atherosclerotic
plaques [124]. A Phase I clinical trial of stable cardiovascular patients treated
with rLCAT has recently been completed [125], but the results have not yet
been published.

The main goal for treating FLD patients with rLCAT is to prevent the
development of renal disease, although the other problems associated with
FLD, such as anemia and corneal opacities, could also be improved with
therapy. Currently, FLD patients are primarily treated symptomatically,
although some patients have been treated with renal or corneal transplants.
In a preclinical animal study, human rLCAT was used to treat LCAT-KO
mice by either intravenous, intramuscular, or subcutaneous routes, and all
routes of treatment were found to rapidly normalize their lipoprotein profile [122]. After rLCAT treatment, HDL-C increased to normal levels, and
the abnormal apoB-containing lipoproteins and LpX were markedly
decreased. The addition of rLCAT to serum from a patient with FLD also
showed rapid normalization of their lipoprotein profile, suggesting that by
generating cholesteryl esters rLCAT can remodel preformed abnormal
lipoprotein particles that accumulate in FLD and are believed to cause renal
disease [122]. The changes observed in the lipoprotein profile from LCAT-
KO mice treated with rLCAT persisted for several days, suggesting that
weekly treatment of FLD patients with rLCAT may possibly be sufficient to
prevent the development of renal disease. Whether rLCAT treatment can
also be used to partly reverse pre-existing renal disease and some of the
other complications of FLD is not known and will have to be tested in
clinical trials of FLD patients, which are now underway.

8. CONCLUSION

It has been more than 50 years since the discovery of LCAT by Glomset
[2]. Although we have come a long way in our understanding of this
important enzyme in HDL metabolism, there are still many mysteries left
unsolved. The early work in the development of LCAT-based therapies is
promising and will hopefully stimulate further efforts in this area. Although

the epidemiologic data and clinical trials are inconclusive as to whether LCAT may be beneficial in preventing atherosclerosis, results from these types of studies are not necessarily informative or predictive on the effect of acutely raising LCAT activity with a small molecule or with rLCAT and may vary depending on the patient. New therapies for modulating LCAT will have to be tested in appropriate animal models and ultimately clinical trials.

REFERENCES

[1] Sperry WM. Cholesterol esterase in blood. J Biol Chem 1935;111(2):467–78.
[2] Glomset JA. The mechanism of the plasma cholesterol esterification reaction: plasma fatty acid transferase. Biochim Biophys Acta 1962;65:128–35.
[3] Glomset JA, Janssen ET, Kennedy R, Dobbins J. Role of plasma lecithin:cholesterol acyltransferase in the metabolism of high density lipoproteins. J Lipid Res 1966;7(5): 638–48.
[4] Norum KR, Gjone E. Familial serum-cholesterol esterification failure. A new inborn error of metabolism. Biochim Biophys Acta 1967;144(3):698–700.
[5] Glomset JA. The plasma lecithins:cholesterol acyltransferase reaction. J Lipid Res 1968;9(2):155–67.
[6] Rousset X, Vaisman B, Amar M, Sethi AA, Remaley AT. Lecithin: cholesterol acyltransferase–from biochemistry to role in cardiovascular disease. Curr Opin Endocrinol Diabetes Obes 2009;16(2):163–71.
[7] Rogers MA, Liu J, Kushnir MM, Bryleva E, Rockwood AL, Meikle AW, et al. Cellular pregnenolone esterification by acyl-CoA:cholesterol acyltransferase. J Biol Chem 2012;287(21):17483–92.
[8] Kosek AB, Durbin D, Jonas A. Binding affinity and reactivity of lecithin cholesterol acyltransferase with native lipoproteins. Biochem Biophys Res Commun 1999;258(3): 548–51.
[9] Chang TY, Li BL, Chang CC, Urano Y. Acyl-coenzyme A:cholesterol acyltransferases. Am J Physiol Endocrinol Metab 2009;297(1):E1–9.
[10] Sorci-Thomas MG, Bhat S, Thomas MJ. Activation of lecithin:cholesterol acyltransferase by HDL ApoA-I central helices. Clin Lipidol 2009;4(1):113–24.
[11] Cho KH, Durbin DM, Jonas A. Role of individual amino acids of apolipoprotein A-I in the activation of lecithin:cholesterol acyltransferase and in HDL rearrangements. J Lipid Res 2001;42(3):379–89.
[12] Roosbeek S, Vanloo B, Duverger N, Caster H, Breyne J, De Beun I, et al. Three arginine residues in apolipoprotein A-I are critical for activation of lecithin:cholesterol acyltransferase. J Lipid Res 2001;42(1):31–40.
[13] Jonas A. Lecithin cholesterol acyltransferase. Biochim Biophys Acta 2000;1529(1-3): 245–56.
[14] Sorci-Thomas MG, Thomas M, Curtiss L, Landrum M. Single repeat deletion in ApoA-I blocks cholesterol esterification and results in rapid catabolism of delta6 and wild-type ApoA-I in transgenic mice. J Biol Chem 2000;275(16):12156–63.
[15] Sorci-Thomas MG, Zabalawi M, Bharadwaj MS, Wilhelm AJ, Owen JS, Asztalos BF, et al. Dysfunctional HDL containing L159R ApoA-I leads to exacerbation of atherosclerosis in hyperlipidemic mice. Biochim Biophys Acta 2012;1821(3):502–12.
[16] Segrest JP, Jones MK, Catte A, Thirumuruganandham SP. Validation of previous computer models and MD simulations of discoidal HDL by a recent crystal structure of apoA-I. J Lipid Res 2012;53(9):1851–63.

[17] Jones MK, Catte A, Li L, Segrest JP. Dynamics of activation of lecithin:cholesterol acyltransferase by apolipoprotein A-I. Biochemistry 2009;48(47):11196–210.

[18] Rye KA, Bright R, Psaltis M, Barter PJ. Regulation of reconstituted high density lipoprotein structure and remodeling by apolipoprotein E. J Lipid Res 2006;47(5):1025–36.

[19] Zhao Y, Thorngate FE, Weisgraber KH, Williams DL, Parks JS. Apolipoprotein E is the major physiological activator of lecithin-cholesterol acyltransferase (LCAT) on apolipoprotein B lipoproteins. Biochemistry 2005;44(3):1013–25.

[20] Hirsch-Reinshagen V, Donkin J, Stukas S, Chan J, Wilkinson A, Fan J, et al. LCAT synthesized by primary astrocytes esterifies cholesterol on glia-derived lipoproteins. J Lipid Res 2009;50(5):885–93.

[21] Baass A, Wassef H, Tremblay M, Bernier L, Dufour R, Davignon J. Characterization of a new LCAT mutation causing familial LCAT deficiency (FLD) and the role of APOE as a modifier gene of the FLD phenotype. Atherosclerosis 2009;207(2):452–7.

[22] Soutar AK, Garner CW, Baker HN, Sparrow JT, Jackson RL, Gotto AM, et al. Effect of the human plasma apolipoproteins and phosphatidylcholine acyl donor on the activity of lecithin: cholesterol acyltransferase. Biochemistry 1975;14(14):3057–64.

[23] Emmanuel F, Steinmetz A, Rosseneu M, Brasseur R, Gosselet N, Attenot F, et al. Identification of specific amphipathic alpha-helical sequence of human apolipoprotein A-IV involved in lecithin:cholesterol acyltransferase activation. J Biol Chem 1994;269(47):29883–90.

[24] Subbaiah PV, Liu M. Role of sphingomyelin in the regulation of cholesterol esterification in the plasma lipoproteins. Inhibition of lecithin-cholesterol acyltransferase reaction. J Biol Chem 1993;268(27):20156–63.

[25] Subbaiah PV, Jiang XC, Belikova NA, Aizezi B, Huang ZH, Reardon CA. Regulation of plasma cholesterol esterification by sphingomyelin: effect of physiological variations of plasma sphingomyelin on lecithin-cholesterol acyltransferase activity. Biochim Biophys Acta 2012;1821(6):908–13.

[26] Subbaiah PV, Liu M. Disparate effects of oxidation on plasma acyltransferase activities: inhibition of cholesterol esterification but stimulation of transesterification of oxidized phospholipids. Biochim Biophys Acta 1996;1301(1-2):115–26.

[27] Subbaiah PV, Subramanian VS, Liu M. Trans unsaturated fatty acids inhibit lecithin: cholesterol acyltransferase and alter its positional specificity. J Lipid Res 1998;39(7): 1438–47.

[28] Peelman F, Vinaimont N, Verhee A, Vanloo B, Verschelde JL, Labeur C, et al. A proposed architecture for lecithin cholesterol acyl transferase (LCAT): identification of the catalytic triad and molecular modeling. Protein Sci 1998;7(3):587–99.

[29] Francone OL, Fielding CJ. Effects of site-directed mutagenesis at residues cysteine-31 and cysteine-184 on lecithin-cholesterol acyltransferase activity. Proc Natl Acad Sci USA 1991;88(5):1716–20.

[30] Peelman F, Goethals M, Vanloo B, Labeur C, Brasseur R, Vandekerckhove J, et al. Structural and functional properties of the 154-171 wild-type and variant peptides of human lecithin-cholesterol acyltransferase. Eur J Biochem 1997;249(3):708–15.

[31] Adimoolam S, Jonas A. Identification of a domain of lecithin-cholesterol acyltransferase that is involved in interfacial recognition. Biochem Biophys Res Commun 1997; 232(3):783–7.

[32] Warden CH, Langner CA, Gordon JI, Taylor BA, McLean JW, Lusis AJ. Tissue-specific expression, developmental regulation, and chromosomal mapping of the lecithin: cholesterol acyltransferase gene. Evidence for expression in brain and testes as well as liver. J Biol Chem 1989;264(36):21573–81.

[33] Meroni G, Malgaretti N, Pontoglio M, Ottolenghi S, Taramelli R. Functional analysis of the human lecithin cholesterol acyl transferase gene promoter. Biochem Biophys Res Commun 1991;180(3):1469–75.

[34] Hoppe KL, Francone OL. Binding and functional effects of transcription factors Sp1 and Sp3 on the proximal human lecithin:cholesterol acyltransferase promoter. J Lipid Res 1998;39(5):969–77.

[35] Feister HA, Auerbach BJ, Cole LA, Krause BR, Karathanasis SK. Identification of an IL-6 response element in the human LCAT promoter. J Lipid Res 2002;43(6):960–70.

[36] Skretting G, Gjernes E, Prydz H. Regulation of lecithin:cholesterol acyltransferase by TGF-beta and interleukin-6. Biochim Biophys Acta 1995;1255(3):267–72.

[37] Kuang YL, Paulson KE, Lichtenstein AH, Lamon-Fava S. Regulation of the expression of key genes involved in HDL metabolism by unsaturated fatty acids. Br J Nutr 2012;108(8):1351–9.

[38] Narayanan S. Biochemistry and clinical relevance of lipoprotein X. Ann Clin Lab Sci 1984;14(5):371–4.

[39] Narayanan S, Lipoprotein- X. CRC Crit Rev Clin Lab Sci 1979;11(1):31–51.

[40] Lynn EG, Siow YL, Frohlich J, Cheung GT, Karmin O. Lipoprotein-X stimulates monocyte chemoattractant protein-1 expression in mesangial cells via nuclear factor-kappa B. Kidney Int 2001;60(2):520–32.

[41] Gillett MPT, Owen JS. Cholesterol esterifying enzymes—lecithin:cholesterol acyl-transferase (LCAT) and acylcoenzyme A:cholesterol acyltransferase (ACAT). In: Converse CA, Skinner ER, editors. Lipoprotein analysis: a practical approach. Oxford; New York: IRL Press;Oxford University Press; 1992. p. 187–201.

[42] Vaisman BL, Remaley AT. Measurement of lecithin-cholesterol acyltransferase activity with the use of a peptide-proteoliposome substrate. Methods Mol Biol 2013;1027:343–52.

[43] Chen Z, Wang SP, Krsmanovic ML, Castro-Perez J, Gagen K, Mendoza V, et al. Small molecule activation of lecithin cholesterol acyltransferase modulates lipoprotein metabolism in mice and hamsters. Metabolism 2012;61(4):470–81.

[44] Parks JS, Gebre AK, Furbee JW. Lecithin-cholesterol acyltransferase. Assay of cholesterol esterification and phospholipase A2 activities. Methods Mol Biol 1999;109:123–31.

[45] Dobiasova M, Frohlich J, Sedova M, Cheung MC, Brown BG. Cholesterol esterification and atherogenic index of plasma correlate with lipoprotein size and findings on coronary angiography. J Lipid Res 2011;52(3):566–71.

[46] Asztalos BF, Schaefer EJ. High-density lipoprotein subpopulations in pathologic conditions. Am J Cardiol 2003;91(7A):12E–7E.

[47] Schwartz CC, VandenBroek JM, Cooper PS. Lipoprotein cholesteryl ester production, transfer, and output in vivo in humans. J Lipid Res 2004;45(9):1594–607.

[48] Glomset JA. The plasma lecithins:cholesterol acyltransferase reaction. J Lipid Res [Review] 1968;9(2):155–67.

[49] Osono Y, Woollett LA, Marotti KR, Melchior GW, Dietschy JM. Centripetal cholesterol flux from extrahepatic organs to the liver is independent of the concentration of high density lipoprotein-cholesterol in plasma. Proc Natl Acad Sci USA 1996;93(9):4114–9.

[50] Liu Y, Tang C. Regulation of ABCA1 functions by signaling pathways. Biochim Biophys Acta 2012;1821(3):522–9.

[51] Rosenson RS, Brewer Jr HB, Davidson WS, Fayad ZA, Fuster V, Goldstein J, et al. Cholesterol efflux and atheroprotection: advancing the concept of reverse cholesterol transport. Circulation 2012;125(15):1905–19.

[52] van der Velde AE, Vrins CL, van den Oever K, Seemann I, Oude Elferink RP, van Eck M, et al. Regulation of direct transintestinal cholesterol excretion in mice. Am J Physiol Gastrointest Liver Physiol 2008;295(1); G203–G2G8.

[53] Horowitz BS, Goldberg IJ, Merab J, Vanni TM, Ramakrishnan R, Ginsberg HN. Increased plasma and renal clearance of an exchangeable pool of apolipoprotein A-I in subjects with low levels of high density lipoprotein cholesterol. J Clin Invest 1993;91(4):1743–52.

[54] Tall AR. Cholesterol efflux pathways and other potential mechanisms involved in the athero-protective effect of high density lipoproteins. J Intern Med 2008;263(3):256–73.

[55] Turner S, Voogt J, Davidson M, Glass A, Killion S, Decaris J, et al. Measurement of reverse cholesterol transport pathways in humans: in vivo rates of free cholesterol efflux, esterification. excretion. J Am Heart Assoc 2012;1(4):e001826.

[56] Czarnecka H, Yokoyama S. Regulation of cellular cholesterol efflux by lecithin:cholesterol acyltransferase reaction through nonspecific lipid exchange. J Biol Chem 1996; 271(4):2023–8.

[57] Kunnen S, Van Eck M. Lecithin:cholesterol acyltransferase: old friend or foe in athero-sclerosis? J Lipid Res [Review] 2012;53(9):1783–99.

[58] Brousseau ME, Santamarina-Fojo S, Vaisman BL, Applebaum-Bowden D, Berard AM, Talley GD, Brewer Jr HB, Hoeg JM. Overexpression of human lecithin:cholesterol acyltransferase in cholesterol-fed rabbits: LDL metabolism and HDL metabolism are affected in a gene dose-dependent manner. J Lipid Res 1997;38(12):2537–47.

[59] Amar MJA, Shamburek RD, Vaisman B, Knapper CL, Foger B, Hoyt RF, et al. Adeno-viral expression of human lecithin-cholesterol acyltransferase in nonhuman primates leads to an antiatherogenic lipoprotein phenotype by increasing high-density lipoprotein and lowering low-density lipoprotein. Metab-Clin Exp [Article] 2009;58(4):568–75.

[60] Carlson LA, Philipson B. Fish-eye disease. A new familial condition with massive corneal opacities and dyslipoproteinaemia. Lancet 1979;2(8149):922–4.

[61] Carlson LA, Holmquist L. Evidence for deficiency of high density lipoprotein lecithin: cholesterol acyltransferase activity (alpha-LCAT) in fish eye disease. Acta Med Scand 1985;218(2):189–96.

[62] Lacko AG, Pritchard PH. International Symposium on Reverse Cholesterol Transport. Report on a meeting. J Lipid Res 1990;31(12):2295–9.

[63] Santamarina-Fojo S, Hoeg JM, Assmann G, Brewer HB. Lecithin cholesterol acyl-transferase deficiency and fish eye disease. In: Scriver CR, Beaudet AL, Sly WS, Valle D, editors. The Metabolic and Molecular Bases of Inherited Disease. 8th ed. New York: McGraw-Hill; 2001. p. 2817–33.

[64] Cogan DG, Kruth HS, Datilis MB, Martin N. Corneal opacity in LCAT disease. Cornea 1992;11(6):595–9.

[65] Borysiewicz LK, Soutar AK, Evans DJ, Thompson GR, Rees AJ. Renal failure in familial lecithin: cholesterol acyltransferase deficiency. Q J Med 1982;51(204): 411–26.

[66] Imbasciati E, Paties C, Scarpioni L, Mihatsch MJ. Renal lesions in familial lecithin-cholesterol acyltransferase deficiency. Ultrastructural heterogeneity of glomerular changes. Am J Nephrol 1986;6(1):66–70.

[67] Lager DJ, Rosenberg BF, Shapiro H, Bernstein J. Lecithin cholesterol acyltransferase deficiency: ultrastructural examination of sequential renal biopsies. Mod Pathol 1991;4(3):331–5.

[68] Gjone E, Blomhoff JP, Skarbovik AJ. Possible association between an abnormal low density lipoprotein and nephropathy in lecithin: cholesterol acyltransferase deficiency. Clin Chim Acta 1974;54(1):11–8.

[69] Munshi R, Johnson A, Siew ED, Ikizler TA, Ware LB, Wurfel MM, et al. MCP-1 gene activation marks acute kidney injury. J Am Soc Nephrol 2011;22(1):165–75.

[70] Rousset X, Shamburek R, Vaisman B, Amar M, Remaley AT. Lecithin cholesterol acyltransferase: an anti- or pro-atherogenic factor? Curr Atheroscler Rep 2011; 13(3):249–56.

[71] Kunnen S, Van Eck M. Lecithin: cholesterol acyltransferase: old friend or foe in atherosclerosis? J Lipid Res [Review] 2012;53(9):1783–99.

[72] Vaisman BL, Klein HG, Rouis M, Berard AM, Kindt MR, Talley GD, et al. Overexpres-sion of human lecithin cholesterol acyltransferase leads to hyperalphalipoproteinemia in transgenic mice. J Biol Chem 1995;270(20):12269–75.

[73] Berard AM, Foger B, Remaley A, Shamburek R, Vaisman BL, Talley G, et al. High plasma HDL concentrations associated with enhanced atherosclerosis in transgenic mice overexpressing lecithin-cholesteryl acyltransferase. Nat Med 1997;3(7):744–9.

[74] Foger B, Chase M, Amar MJ, Vaisman BL, Shamburek RD, Paigen B, et al. Cholesteryl ester transfer protein corrects dysfunctional high density lipoproteins and reduces aortic atherosclerosis in lecithin cholesterol acyltransferase transgenic mice. J Biol Chem 1999;274(52):36912–20.

[75] Mehlum A, Muri M, Hagve TA, Solberg LA, Prydz H. Mice overexpressing human lecithin:cholesterol acyltransferase are not protected against diet-induced atherosclerosis. APMIS 1997;105(11):861–8.

[76] Berti JA, de Faria EC, Oliveira HC. Atherosclerosis in aged mice over-expressing the reverse cholesterol transport genes. Braz J Med Biol Res 2005;38(3):391–8.

[77] Alam K, Meidell RS, Spady DK. Effect of up-regulating individual steps in the reverse cholesterol transport pathway on reverse cholesterol transport in normolipidemic mice. J Biol Chem 2001;276(19):15641–9.

[78] Tanigawa H, Billheimer JT, Tohyama J, Fuki IV, Ng DS, Rothblat GH, et al. Lecithin: cholesterol acyltransferase expression has minimal effects on macrophage reverse cholesterol transport in vivo. Circulation 2009;120(2):160–9.

[79] Hoeg JM, Vaisman BL, Demosky Jr SJ, Meyn SM, Talley GD, Hoyt Jr RF, et al. Lecithin:cholesterol acyltransferase overexpression generates hyperalpha-lipoproteinemia and a nonatherogenic lipoprotein pattern in transgenic rabbits. J Biol Chem 1996;271(8): 4396–402.

[80] Hoeg JM, Santamarina-Fojo S, Berard AM, Cornhill JF, Herderick EE, Feldman SH, et al. Overexpression of lecithin:cholesterol acyltransferase in transgenic rabbits prevents diet-induced atherosclerosis. Proc Natl Acad Sci USA 1996;93(21):11448–53.

[81] Brousseau ME, Kauffman RD, Herderick EE, Demosky Jr SJ, Evans W, Marcovina S, et al. LCAT modulates atherogenic plasma lipoproteins and the extent of atherosclerosis only in the presence of normal LDL receptors in transgenic rabbits. Arterioscler Thromb Vasc Biol 2000;20(2):450–8.

[82] Sakai N, Vaisman BL, Koch CA, Hoyt Jr RF, Meyn SM, Talley GD, et al. Targeted disruption of the mouse lecithin:cholesterol acyltransferase (LCAT) gene. Generation of a new animal model for human LCAT deficiency. J Biol Chem 1997;272(11):7506–10.

[83] Ng DS, Francone OL, Forte TM, Zhang J, Haghpassand M, Rubin EM. Disruption of the murine lecithin:cholesterol acyltransferase gene causes impairment of adrenal lipid delivery and up-regulation of scavenger receptor class B type I. J Biol Chem 1997;272(25):15777–81.

[84] Lambert G, Sakai N, Vaisman BL, Neufeld EB, Marteyn B, Chan CC, et al. Analysis of glomerulosclerosis and atherosclerosis in lecithin cholesterol acyltransferase-deficient mice. J Biol Chem 2001;276(18):15090–8.

[85] Ng DS, Maguire GF, Wylie J, Ravandi A, Xuan W, Ahmed Z, et al. Oxidative stress is markedly elevated in lecithin:cholesterol acyltransferase-deficient mice and is paradoxically reversed in the apolipoprotein E knockout background in association with a reduction in atherosclerosis. J Biol Chem 2002;277(14):11715–20.

[86] Furbee Jr JW, Sawyer JK, Parks JS. Lecithin:cholesterol acyltransferase deficiency increases atherosclerosis in the low density lipoprotein receptor and apolipoprotein E knockout mice. J Biol Chem 2002;277(5):3511–9.

[87] Forte T, Nichols A, Glomset J, Norum K. The ultrastructure of plasma lipoproteins in lecithin:cholesterol acyltransferase deficiency. Scand J Clin Lab Invest Suppl 1974; 137:121–32.

[88] Zhu X, Herzenberg AM, Eskandarian M, Maguire GF, Scholey JW, Connelly PW, et al. A novel in vivo lecithin-cholesterol acyltransferase (LCAT)-deficient mouse expressing predominantly LpX is associated with spontaneous glomerulopathy. Am J Pathol 2004;165(4):1269–78.

[89] Shimano H, Horton JD, Hammer RE, Shimomura I, Brown MS, Goldstein JL. Overproduction of cholesterol and fatty acids causes massive liver enlargement in transgenic mice expressing truncated SREBP-1a. J Clin Invest 1996;98(7):1575–84.

[90] Li L, Naples M, Song H, Yuan R, Ye F, Shafi S, et al. LCAT-null mice develop improved hepatic insulin sensitivity through altered regulation of transcription factors and suppressors of cytokine signaling. Am J Physiol Endocrinol Metab 2007;293(2):E587–94.

[91] Ng DS. Lecithin cholesterol acyltransferase deficiency protects from diet-induced insulin resistance and obesity-novel insights from mouse models. Vitam Horm 2013;91:259–70.

[92] Hoekstra M, Korporaal SJ, van der Sluis RJ, Hirsch-Reinshagen V, Bochem AE, Wellington CL, et al. LCAT deficiency in mice is associated with a diminished adrenal glucocorticoid function. J Lipid Res 2013;54(2):358–64.

[93] Francone OL, Gong EL, Ng DS, Fielding CJ, Rubin EM. Expression of human lecithin-cholesterol acyltransferase in transgenic mice. Effect of human apolipoprotein AI and human apolipoprotein AII on plasma lipoprotein cholesterol metabolism. J Clin Invest 1995;96(3):1440–8.

[94] Mehlum A, Staels B, Duverger N, Tailleux A, Castro G, Fievet C, et al. Tissue-specific expression of the human gene for lecithin: cholesterol acyltransferase in transgenic mice alters blood lipids, lipoproteins and lipases towards a less atherogenic profile. Eur J Biochem 1995;230(2):567–75.

[95] Mehlum A, Gjernes E, Solberg LA, Hagve TA, Prydz H. Overexpression of human lecithin:cholesterol acyltransferase in mice offers no protection against diet-induced atherosclerosis. APMIS 2000;108(5):336–42.

[96] Zhang AH, Gao S, Fan JL, Huang W, Zhao TQ, Liu G. Increased plasma HDL cholesterol levels and biliary cholesterol excretion in hamster by LCAT overexpression. FEBS Lett 2004;570(1-3):25–9.

[97] Brousseau ME, Wang J, Demosky Jr SJ, Vaisman BL, Talley GD, Santamarina-Fojo S, et al. Correction of hypoalphalipoproteinemia in LDL receptor-deficient rabbits by lecithin:cholesterol acyltransferase. J Lipid Res 1998;39(8):1558–67.

[98] Van Craeyveld E, Lievens J, Jacobs F, Feng Y, Snoeys J, De Geest B. Apolipoprotein A-I and lecithin:cholesterol acyltransferase transfer induce cholesterol unloading in complex atherosclerotic lesions. Gene Ther 2009;16(6):757–65.

[99] Hossain MA, Tsujita M, Akita N, Kobayashi F, Yokoyama S. Cholesterol homeostasis in ABCA1/LCAT double-deficient mouse. Biochim Biophys Acta Mol Cell Biol Lipids [Article] 2009;1791(12):1197–205.

[100] Li L, Hossain MA, Sadat S, Hager L, Liu L, Tam L, et al. Lecithin cholesterol acyltransferase null mice are protected from diet-induced obesity and insulin resistance in a gender-specific manner through multiple pathways. J Biol Chem 2011;286(20):17809–20.

[101] Castelli WP, Garrison RJ, Wilson PW, Abbott RD, Kalousdian S, Kannel WB. Incidence of coronary heart disease and lipoprotein cholesterol levels. The Framingham Study. JAMA 1986;256(20):2835–8.

[102] Calabresi L, Simonelli S, Gomaraschi M, Franceschini G. Genetic lecithin:cholesterol acyltransferase deficiency and cardiovascular disease. Atherosclerosis 2012;222(2): 299–306.

[103] Wells IC, Peitzmeier G, Vincent JK. Lecithin: cholesterol acyltransferase and lysolecithin in coronary atherosclerosis. Exp Mol Pathol 1986;45(3):303–10.

[104] Calabresi L, Baldassarre D, Castelnuovo S, Conca P, Bocchi L, Candini C, et al. Functional lecithin: cholesterol acyltransferase is not required for efficient atheroprotection in humans. Circulation 2009;120(7):628–35.

[105] Dullaart RP, Perton F, van der Klauw MM, Hillege HL, Sluiter WJ. High plasma lecithin:cholesterol acyltransferase activity does not predict low incidence of cardiovascular events: possible attenuation of cardioprotection associated with high HDL cholesterol. Atherosclerosis 2010;208(2):537–42.

[106] Haase CL, Tybjaerg-Hansen A, Qayyum AA, Schou J, Nordestgaard BG, Frikke-Schmidt RLCAT. HDL cholesterol and ischemic cardiovascular disease: a Mendelian randomization study of HDL cholesterol in 54,500 individuals. J Clin Endocrinol Metab 2012;97(2):E248–56.

[107] Frikke-Schmidt R, Nordestgaard BG, Stene MC, Sethi AA, Remaley AT, Schnohr P, et al. Association of loss-of-function mutations in the ABCA1 gene with high-density lipoprotein cholesterol levels and risk of ischemic heart disease. JAMA 2008;299(21):2524–32.

[108] Boes E, Coassin S, Kollerits B, Heid IM, Kronenberg F. Genetic-epidemiological evidence on genes associated with HDL cholesterol levels: a systematic in-depth review. Exp Gerontol 2009;44(3):136–60.

[109] Hovig T, Gjone E. Familial plasma lecithin: cholesterol acyltransferase (LCAT) deficiency. Ultrastructural aspects of a new syndrome with particular reference to lesions in the kidneys and the spleen. Acta Pathol Microbiol Scand A 1973;81(5):681–97.

[110] Scarpioni R, Paties C, Bergonzi G. Dramatic atherosclerotic vascular burden in a patient with familial lecithin-cholesterol acyltransferase (LCAT) deficiency. Nephrol Dial Transplant 2008;23(3):1074; author reply -5.

[111] Solajic-Bozicevic N, Stavljenic-Rukavina A, Sesto M. Lecithin-cholesterol acryltransferase activity in patients with coronary artery disease examined by coronary angiography. Clin Investig 1994;72(12):951–6.

[112] Sethi AA, Sampson M, Warnick R, Muniz N, Vaisman B, Nordestgaard BG, et al. High pre-beta1 HDL concentrations and low lecithin: cholesterol acyltransferase activities are strong positive risk markers for ischemic heart disease and independent of HDL-cholesterol. Clin Chem 2010;56(7):1128–37.

[113] Kane JP, Malloy MJ. Prebeta-1 HDL and coronary heart disease. Curr Opin Lipidol 2012;23(4):367–71.

[114] Ayyobi AF, McGladdery SH, Chan S. John Mancini GB, Hill JS, Frohlich JJ. Lecithin: cholesterol acyltransferase (LCAT) deficiency and risk of vascular disease: 25 year follow-up. Atherosclerosis 2004;177(2):361–6.

[115] Hovingh GK, Hutten BA, Holleboom AG, Petersen W, Rol P, Stalenhoef A, et al. Compromised LCAT function is associated with increased atherosclerosis. Circulation 2005;112(6):879–84.

[116] Duivenvoorden R, Holleboom AG, van den Bogaard B, Nederveen AJ, de Groot E, Hutten BA, et al. Carriers of lecithin cholesterol acyltransferase gene mutations have accelerated atherogenesis as assessed by carotid 3.0-T magnetic resonance imaging [corrected]. J Am Coll Cardiol 2011;58(24):2481–7.

[117] van den Bogaard B, Holleboom AG, Duivenvoorden R, Hutten BA, Kastelein JJ, Hovingh GK, et al. Patients with low HDL-cholesterol caused by mutations in LCAT have increased arterial stiffness. Atherosclerosis 2012;225(2):481–5.

[118] Holleboom AG, Kuivenhoven JA, Vergeer M, Hovingh GK, van Miert JN, Wareham NJ, et al. Plasma levels of lecithin:cholesterol acyltransferase and risk of future coronary artery disease in apparently healthy men and women: a prospective case-control analysis nested in the EPIC-Norfolk population study. J Lipid Res 2010;51(2):416–21.

[119] Calabresi L, Baldassarre D, Simonelli S, Gomaraschi M, Amato M, Castelnuovo S, et al. Plasma lecithin:cholesterol acyltransferase and carotid intima-media thickness in European individuals at high cardiovascular risk. J Lipid Res 2011;52(8):1569–74.

[120] Frank Kayser ML, Bei Shan, Jian Zhang, Mingyue Zhou, inventor Amgen Inc., assignee. Methods for Treating Atherosclerosis. United States patent US 2008/0096900 A1. http://www.freepatentsonline.com/y2008/0096900.html. Accessed 04/24/2008.

[121] Zhou MFP, Zhang J. Novel small molecule LCAT activators raise HDL levels in rodent models. Arterioscler Thromb Vasc Biol 2008;28:E65–6.

[122] Rousset X, Vaisman B, Auerbach B, Krause BR, Homan R, Stonik J, et al. Effect of recombinant human lecithin cholesterol acyltransferase infusion on lipoprotein metabolism in mice. J Pharmacol Exp Ther 2010;335(1):140–8.

[123] Remaley AT, Amar M, Sviridov D. HDL-replacement therapy: mechanism of action, types of agents and potential clinical indications. Expert Rev Cardiovasc Ther 2008;6(9):1203–15.

[124] Zhou MY, Sawyer J, Kelley K, Fordstrom P, Chan J, Tonn G, et al. Lecithin cholesterol acyltransferase promotes reverse cholesterol transport and attenuates atherosclerosis progression in New Zealand white rabbits. Circulation 2009;120(18); S1175–S117S.

[125] Alphacore Pharma L, National Heart, Lung, and Blood Institute, NIH. Effect of ACP-501 on safety, tolerability, pharmacokinetics and pharmacodynamics in subjects with coronary artery disease. ClinicalTrialsgov Identifier: NCT01554800. Available at: http://clinicaltrials.gov/ct2/show/NCT01554800. Accessed 7/29/13.

CHAPTER 8

Cholesteryl Ester Transfer Protein Inhibitors: A Hope Remains

Akihiro Inazu

Department of Clinical Laboratory Science, School of Health Sciences, Kanazawa University, Kanazawa, Japan

Contents

Abstract

Naturally CETP-deficient animals and genetic cholesteryl ester transfer protein (CETP) deficiency caused by TaqIB polymorphism in human are relatively resistant to atherosclerosis including coronary artery disease (CAD). CETP inhibitors were developed for new therapeutic measures against atherosclerotic vascular disease through increasing high-density lipoprotein cholesterol (HDL-C) and decreasing low-density lipoprotein cholesterol (LDL-C). Although a clinical trial with torcetrapib was terminated due to hypertension-related side effects, two other compounds, anacetrapib and evacetrapib, are under Phase III clinical trials. Although additional failure of dalcetrapib suggested that the hypertension-related adverse effect is not only the cause of failure of torcetrapib, but also the validity of the CETP inhibitor itself is questionable, this chapter summarizes that a hope remains in CETP inhibitors as a potential agent to reduce residual CAD risk in some clinical settings. Rationale for the CETP inhibitor development is discussed from clinical and experimental insights of lipoprotein phenotype, functional activity on LDL and HDL, and role of CETP activity in relation to inflammation. Structure and function relationship between the N-terminal hydrophobic tunnel of CETP and cholesteryl ester (CE)/triglycerides (TG) with or without a CETP inhibitor is discussed. CETP antibody may have a differential potential on directional selectivity of neutral lipid transfer in plasma lipoproteins.

1. INTRODUCTION

In the previous edition of this text in 2010, I had written a chapter on plasma cholesteryl ester transfer protein (CETP) in relation to human pathophysiology of genetic CETP deficiency [1]. Since then, more reports have been published to fill a gap in the CETP research area. Furthermore, more progress has been made in the development of CETP inhibitors. This chapter includes recent knowledge on the CETP structure and function relationship, and emerging evidence of CETP inhibitors in the last 5 years, along with my opinion.

2. HDL-TG AS A KEY COMPONENT DETERMINING NEUTRAL LIPID TRANSFER

CETP is a 74-kD glycoprotein consisting of 476 amino acids and N-glycosylation. Its crystal structure reveals a banana-shaped molecule with N- and C-terminal β-barrel domains, a central β-sheet, and an ~60 Å-long hydrophobic central cavity. A long tunnel has a space for hydrophobic 2 molecules of cholesteryl ester (CE) or triacylglycerol (TG) and plugged by an amphiphilic phosphatidylcholine (PC) at each end. C-terminal amino acids of 433, 443, 457, and 459 appear to be close to the tunnel neck [2]. By an optimized negative-staining electron microscope protocol, CETP

C-terminal is observed to be more globular and N-terminal is more tapered at the end [3]. HDL and CETP form a binary complex, which could be seen as a tadpole-shape: CETP protruding from spherical HDL surface. The banana-shaped CETP has a concave surface protruding at an approximately 45-degree angle from the HDL surface. Since PC-binding pores are located at central β-sheet of CETP, the pore is close to the HDL surface, which is composed of phospholipid (PL) layers. Thus, CETP bridges HDL to LDL or VLDL to form ternary complex. Based on the asymmetric structure of CETP, N-terminal CETP prefers to bind to HDL and C-terminal end binds to VLDL or LDL. Furthermore, recent studies suggest that the distal portion flexibility of N-terminal β-barrel domain is considerably greater in solution than in crystal and it remains hydrophobic in solution [4].

It is believed that CETP mediates hetero-exchange of TG and CE by moving between VLDL and HDL like a shuttle (Figure 8.1). If a shuttle model is correct, CETP adopts a conformation to enable binding to a large lipoprotein like VLDL in addition to HDL. However, in a recent model proposed by Charles and Kane [5], based on a recent experiment data, it appears to depend on the ternary complex between CETP-HDL and VLDL. The process includes sensing, penetration, and docking of CETP. CETP penetrates ~50 Å into HDL with the N-terminal β-barrel domain,

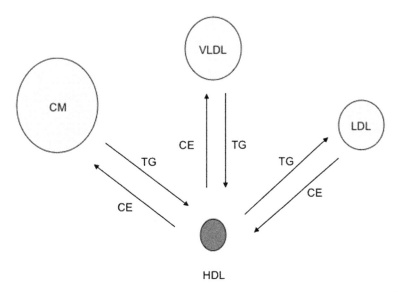

Figure 8.1 *HDL as a Playmaker for the Neutral Lipid Exchange.* HDL-TG is a source of TG in LDL by CETP-mediated lipid transfer. In addition, VLDL could convert into LDL after lipolysis of the core TG.

while penetrating LDL or VLDL only 20–25 Å through its distal C-terminal β-barrel domain since the outer PL shell of lipoproteins is 18–27 Å thick. CETP, phospholipid transfer protein (PLTP), and lipopolysaccharide (LPS)-binding protein (LBP) belong to the tubular lipid-binding (TULIP) domain superfamily [6].

3. HDL METABOLISM IN CETP DEFICIENCY

Lipoprotein phenotype in CETP deficiency has been well investigated; as have high HDL and low LDL, but fewer consistent findings were found in TG metabolism. Concentration of preβ1-HDL is inconsistent between homo- and heterozygotes with CETP deficiency [7]. Mild reduction of CETP activity found in heterozygotes had low levels of preβ1-HDL. However, complete CETP deficiency had oppositely higher levels of preβ1-HDL, suggesting that reduction of CETP activity is not linearly correlated with preβ1-HDL levels. Preβ1-HDL is believed to be an efficient acceptor for ATP-binding cassette transporter A1 (ABCA1)-mediated cholesterol efflux activity. Preβ1-HDL is converted to spherical HDL via lecithin:cholesterol acyltransferase (LCAT)-mediated cholesterol esterification. Therefore, either increased ABCA1 or PL transfer activity or decreased LCAT activity would be associated with increased levels of preβ1-HDL. A cause of increased preβ1-HDL found in homozygous CETP deficiency is currently unknown, but it is likely associated with decreased LCAT activity rather than accelerated lipolysis of VLDL/chylomicron (CM) [8,9].

As shown in a kinetic study [10], fractional catabolic rate (FCR) of apoA-I is decreased in CETP deficiency, however, cholesterol/CE clearance rate from HDL is not established in CETP deficiency. Scavenger receptor class B type I (SR-BI) receptor mediates selective uptake of HDL lipids in the liver, as SR-BI deficiency is a cause of increased HDL levels in humans [11].

RCT from peripheral tissues appeared to be pro-atherogenic if FC/CE in HDL would transfer to the VLDL-intermediate density lipoprotein (IDL)-low-density lipoprotein (LDL) pathway via CETP activity, and FC/CE could be reutilized in VLDL formed in the liver after lipoprotein uptake of VLDL-IDL-LDL by the liver receptors, such as LDL receptors. In contrast, the SR-BI pathway selectively promoted cholesterol secretion from plasma HDL into bile [12], thereby it is anti-atherogenic (Figure 8.2). The relationship between RCT and LDL receptor activity is discussed later in the section on statin in perspective of CETP inhibitor.

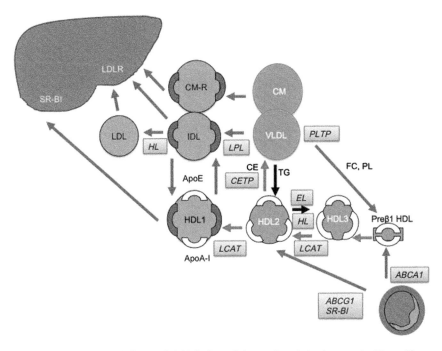

Figure 8.2 *HDL as a Cholesterol Vehicle from Atherosclerotic Lesions to the Liver.* There are two pathways; one is a direct pathway mediated by apoE-rich HDL and SR-BI receptor in the liver, and the other is an indirect pathway mediated by IDL-LDL or chylomicron remnants and LDLR in the liver.

4. LDL METABOLISM IN CETP DEFICIENCY

LDL-C consists largely of CE, which is esterified by LCAT in HDL. CETP-mediated CE transferred from HDL to VLDL is a major determinant of LDL-C, because LDL-C tends to be lower in complete CETP deficiency. However, a reciprocal transfer of TG from VLDL to HDL is diminished in CETP deficiency, consistent with findings of TG-rich VLDL and TG-poor HDL in CETP deficiency [13]. The CETP-deficient homozygotes had polydisperse LDL subclasses from IDL-like particles to small and dense LDL on a native polyacrylamide gel, suggesting that complete CETP deficiency would inhibit interconversion of lipids between LDL subclasses [14,15]. In contrast, partial CETP deficiency increased LDL size. Heteroexchange of CE and TG between VLDL and LDL via HDL leads to formation of TG-rich LDL, which consequently becomes small and dense LDL after lipolysis of the core TG. Thus, decreased formation of TG-rich LDL is expected in heterozygous CETP deficiency, resulting in increases in relatively larger LDL subclasses.

5. VLDL METABOLISM IN CETP DEFICIENCY

Postprandial lipemia is the most common risk of cardiovascular disease. Although predominant lipoproteins appearing in the postprandial periods have been discussed, they appear to be VLDL remnants [16]. In postprandial periods challenged by oral fat load, plasma TG response was diminished in the hetero- and homozygous CETP deficiency [17]. It is possible that CETP deficiency induced LDL receptor expression in the liver, which may accelerate TG-rich lipoprotein clearance.

However, the relationship between CETP activity and the magnitude of postprandial lipemia is controversial, as it appears to be dependent on the metabolic context of subjects. In women who have lower hepatic lipase activity than men, it has been shown that low CETP activity would deteriorate postprandial TG response [18], since CETP mediates the transfer of CE from HDL particles to VLDL in exchange for TG. When VLDL increases postprandially, VLDL act as CE acceptors for CETP activity, thereby an increased net rate of TG transfer from VLDL to HDL/LDL is expected. As a result, TG-rich HDL and LDL are avidly bound to hepatic lipase that effectively hydrolyzed TG, and they become HDL3 and small dense LDL, respectively. Thus, in this step, CETP helps to lower TG in plasma via enhancing lipolysis of TG in LDL and HDL fractions.

6. ANTIOXIDANT ACTIVITY IN CETP DEFICIENCY

Serum paraoxonase 1 (PON1) activity was increased in two cases with homozygous CETP deficiency but PON1 activity/apoA-I level ratio are comparable to controls [19]. The oxidized LDL (oxLDL) levels were positively correlated with apoB, PLTP activity, but negatively with CETP activity in the general population [20]. Since CETP enhances the ability of HDL to inhibit LDL oxidation in vitro, low CETP activity state may be susceptible to oxidative stress [21]. However, CETP-deficient subjects did not reveal elevated levels of oxLDL and 8-isoprostane, nor decreased levels of paraoxonase activity [22].

Plasma PLTP concentration was increased in CETP deficiency by +57%, but PLTP activity was not increased [23]. The role of increased PLTP mass is currently unclear in CETP deficiency.

7. DIFFERENCE OF LIPOPROTEIN PHENOTYPE BETWEEN HOMOZYGOTES AND HETEROZYGOTES WITH CETP DEFICIENCY

CETP may have dual aspects of its atherogenicity (Table 8.1). On one hand of pro-atherogenicity, CETP would increase CE contents in VLDL-IDL; after LPL and hepatic lipase-mediated lipolysis, VLDL-IDL becomes CE-rich LDL. If VLDL is concurrently increased in combined hyperlipidemia (also called mixed hyperlipidemia) or in postprandial periods, heteroexchange of CE and TG between VLDL and LDL results in formation of small, dense LDL via HDL-mediated lipid transfer. Similarly, TG–rich HDL produced by CETP-mediated TG transfer enhances catabolism of HDL. On the other hand, CETP-mediated CE net-transfer from HDL to VLDL-IDL-LDL is beneficial as long as hepatic LDL receptor activity is not saturated. CETP may help lipoprotein conversion among HDL subclasses via recycling from large HDL to small HDL, including preβ1–HDL formation.

CETP deficiency increased HDL-C and decreased LDL-C levels in adults. Thus, CETP inhibition may delay cholesterol clearance as plasma HDL-C levels increase. However, heterozygous CETP deficiency in a fetus showed decreased LDL-C without HDL-C changes [24], which may be associated with concurrently decreased LCAT activity.

Table 8.1 Summary of Anti- and Pro-Atherogenic Aspects in CETP Deficiency

Parameter	Homozygote	Heterozygote	Author, Year, Ref
HDL-C (mg/dL)	164	66	Inazu, 1990 [25]
ApoE-rich HDL1	Very high	High	Koizumi, 1985 [26], Inazu, 2008 [17]
Preβ1–HDL	Increased	Decreased	Asztalos, 2004 [7]
Cholesterol esterification rate	Very low	Low	Oliveira, 1997 [8]
ABCG1/SR-BI-mediated chol efflux	Very high	High	Matsuura, 2006 [27], Miwa, 2009 [28]
LDL-C (mg/dL)	77	111	Inazu, 1990 [25]
LDL size	Polydispersed	Large	Yamashita, 1988 [14], Brown, 1989 [15], Wang, 2002 [29]
Lp(a)	Decreased	Not reported	unpublished*

*Inazu et al. (1993)

8. SOURCE OF CETP AND CHOLESTERYL ESTER TRANSFER (CET) DETERMINANT

CETP mRNA is abundant in tissues of liver, spleen, and adipose. Net CE transfer rate is determined not only by plasma CETP activity but also VLDL levels, which is a potent CE acceptor for the CETP-mediated lipid transfer. Some studies suggest that increased CET is a stronger risk factor than plasma CETP mass or activity. CET is associated with PAF-AH activity, which is also known as lipoprotein-associated phospholipase A2 [30]. Since free fatty acid (FFA) generated by phospholipase activity by PAF-AH may increase binding of CETP to LDL, CET is accelerated.

8.1. Cardiovascular Disease risk in CETP Deficiency and Single-Nucleotide Polymorphisms (SNPs) in the CETP Gene

The low CETP genotype of TaqIB2 was associated with decreased prevalence of coronary disease as well as increased HDL-C levels and large LDL size in men of the Framingham Heart Study [31]. However, a more recent study measuring plasma CETP activity suggested that lower plasma CETP activity was unexpectedly associated with greater cardiovascular risk (myocardial infarction, stroke, or heart failure) with relative risk 1.4 [32]. The reason for this apparent discrepancy is unknown. As such, it is important to define whether the cause of the decrease in plasma CETP activity is genetic or environmental, since acute phase reaction and inflammation would decrease CETP expression. It is possible that lower CETP levels may be just a surrogate marker for inflammation rather than the genetic effect in the latter study, as discussed later.

Anti-atherogenic effect of genetically lower CETP levels caused by the TaqI B2 allele has been reproduced in a recent meta-analysis [33]. In heterozygous CETP deficiency, the Honolulu Heart Study showed that heterozygotes are anti-atherogenic at least when they have increased HDL-C > 60 mg/dL [34]. Homozygous CETP deficiency has been mainly found in Japan. Only very few cases are found in the other populations, including European descendants. Thus, epidemiological studies that have been made in a relatively small number in Japan have suggested mixed results in the coronary artery disease (CAD) risk [35,36].

Longevity is expected in some CETP-deficient heterozygous subjects [25,37]. A promising effect on longevity has been reported in Ashkenazi Jewish population, where increased homozygosity of I405V was found in offspring of individuals with exceptional longevity [38]. Furthermore, the homozygosity of I405V was associated with slower memory decline and lower incident dementia [39].

Meta-analysis of CETP SNPs associated with low CETP activity and high HDL-cholesterol levels are anti-atherogenic such as TaqIB (rs708272) and -629C>A (rs1800775) [33,40]. In prospective cohort studies, the Copenhagen City Heart Study and the Women's Genome Health Study showed that genetically low CETP activity is anti-atherogenic in men and women [41,42].

However, some recent studies reported inconsistent results. Hiura and colleagues [43] reported that the minor allele of rs3764261, located in the CETP promoter, is associated with elevated HDL levels and unexpectedly increased myocardial infarction (MI) risk in a Japanese population. Similarly, SNPs located between intron 8 and exon 9, which were associated with exon 9 skipping, manifested an increased MI risk in men [44].

Opposite trends that TaqIB2 is associated with an increased vascular risk were found in statin-treated cardiovascular patients and very high-risk populations who had increased HDL and C-reactive protein (CRP) levels and low PAF-AH activity [45,46]. These opposite results need to be fully investigated to learn whether or not that finding is related with reverse causality.

8.2. CETP in Relation to Inflammation and Adiposity

In a prospective observational study of patients with stable CAD in Germany (KAROLA study), low CETP was associated with increased risk for death with an adjusted hazard ratio of 1.84 [47]. In the Ludwigshafen Risk and Cardiovascular Health (LURIC) study, CETP levels are lower in smokers, diabetics, and unstable CAD. CETP showed a negative correlation with CRP and IL-6 and a positive correlation with homocysteine and adiponectin. Low CETP is associated with increased hazard ratios for death after multivariate adjustment [48]. During experimental endotoxemia, decreased activity of CETP and LCAT was found; in contrast, PLTP activity was increased in human subjects [49].

The common allele of TaqIB (i.e., TaqIB1) increased CETP activity and decreased HDL levels, and was associated with insulin resistance and metabolic syndrome [50]. Hyperalphalipoproteinemic subjects tend to have decreased CRP levels [51]. However, there is no evidence that CETP deficiency had low CRP levels.

9. DRUG DESIGNS IN CLINICAL TRIALS

CETP inhibitors are shown in Figure 8.3.

9.1. Torcetrapib (Pfizer)

Torcetrapib is a compound of tetrahydroquinoline that binds to CETP, forming CETP-HDL complex in plasma. It is a noncompetitive inhibitor that could bind to CETP reversibly. The IC_{50} values for CETP activity are 17–79 nM. The binding site on CETP for torcetrapib is in the lipid-binding pocket near N-terminal CETP [52]. Favorable lipoprotein profiles of increased HDL2 and large LDL are shown with increasing HDL (+106%) and decreasing LDL (−17%) with 240 mg torcetrapib in humans. In December 2006, the Phase III trial of the ILLUMINATE Study, a combination study with atorvastatin, was terminated because of excess death (hazard ratio 1.58) and major cardiovascular events (hazard ratio 1.25) in the combination arm [53].

The principle cause of death appeared to be related to hypertension-related vascular events. Later, torcetrapib is found to be associated with high aldosterone levels, which are associated with increased aldosterone synthase

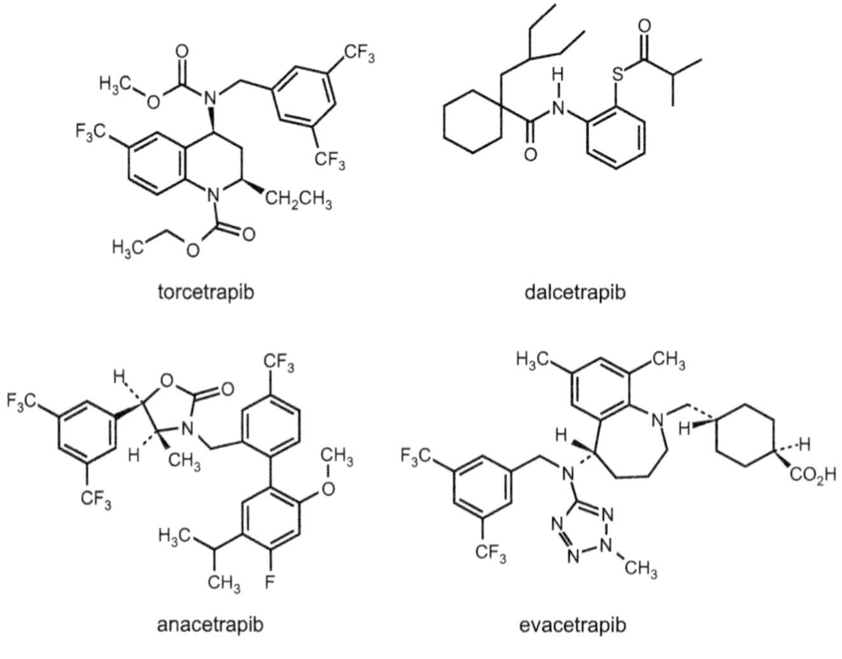

Figure 8.3 CETP Inhibitors.

(CYP11B2) [54] and endothelial dysfunction [55]. Blood pressure was increased in spontaneously hypertensive rats treated with torcetrapib, but not Wistar-Kyoto rats [56]. Torcetrapib induced a sustained impairment of endothelial function; decreased eNOS mRNA, protein, and nitric oxide (NO) release; and stimulated vascular ROS and endothelin-1 production in addition to aldosterone. Since rats and mice are deficient in CETP activity, these hormonal changes related in artery tonus are independent of CETP activity.

Furthermore, imaging trials of coronary and carotid arteries were negative, but a post hoc analysis of the IVUS study of the ILLUSTRATE Study indicated that the only highest HDL group (HDL-C > 87 mg/dL) showed the regression of coronary atherosclerosis [57]. Beneficial effects of increased HDL2 are associated with higher cholesterol efflux via SR-BI or ABCG1 pathways [58]. Moreover, torcetrapib attenuates the atherogenicity of postprandial TG-rich lipoproteins in type IIB hyperlipidemia [59].

Despite the failure in the clinical trials, torcetrapib is considered to induce RCT efficacy. Increased RCT from peripheral macrophages to feces is considered to be anti-atherogenic. A selective uptake of CE from HDL in the liver was increased by 1.7-fold in the treatment of torcetrapib in CETP-transgenic mice [58]. In hamsters, a naturally CETP-expressed species, it was shown that increased cholesterol excretion in the feces was found during CETP inhibition by torcetrapib alone [60].

9.2. Dalcetrapib (RO4607381, Roche; JTT-705, JT)

This compound, formerly named JTT-705, is structurally different from fluorine-containing structures of torcetrapib, anacetrapib, and evacetrapib, because it has an ortho-thio-anilide core and it requires Cys-13 of CETP molecule to form a covalent disulfide bond, thereby dalcetrapib irreversibly binds to CETP [61]. The SH group of Cys-13 resides at the bottom of the lipid-binding pocket of CETP [62]. Inhibiting CETP activity is relatively mild (IC_{50} 0.4-10 μM), accordingly it would increase HDL-C levels modestly.

Indeed, dalcetrapib is not associated with increased aldosterone and high blood pressure [63]. Moreover, no clinically relevant changes in lymph nodes, or other safety parameters, were found in phase I and phase II trials [64]. Clinical outcome study is expected in dal-OUTCOMES using 600 mg dalcetrapib in patients ($N = 15,600$), which was initiated in 2008. Thus, one might expect that dalcetrapib is more effective because it is a weak CETP inhibitor maintaining interconversions of HDL subclasses. Moreover, dal-VESSEL is focused on modulation of vascular function such as endo-thelial function by CETP inhibition. The dal-PLAQUE has been initiated to assess the impact of dalcetrapib on atherosclerotic plaque development using PET-CT and MRI [65].

However, unfortunately, in May 2012, the dal-OUTCOMES Phase III trials were terminated because of a lack of clinically meaningful efficacy, which is recommended by the independent Data and Safety Monitoring Board. All the studies in the dal-HEART program were terminated. Although it is unclear why the benefits of increasing HDL levels were not seen in that study, it may be explained by potential adverse effects as fol-lows: the median CRP level was 0.2 mg/L higher and mean systolic blood pressure was 0.6 mmHg higher with dalcetrapib as compared with a placebo in patients with a recent acute coronary syndrome [66]. It is reasonable to state that dalcetrapib is a weak inhibitor without an effect on reducing LDL-C and TG, which may be one reason for the failure of the dal-OUTCOMES.

9.3. Anacetrapib (MK-0859, Merck)

Anacetrapib has a triad of trifluoromethyl-benzene derivative like torcetra-pib but it has a distinct biaryl moiety. Although this compound effectively elevates HDL-C along with lowering LDL-C as well as torcetrapib, anace-trapib is not associated with increased aldosterone and high blood pressure [67]. It is a noncompetitive inhibitor binding to CETP reversibly. The IC_{50} values for CETP activity are 10–17 nM [68].

The Determining the Efficacy and Tolerability of CETP Inhibition with anacetrapib (DEFINE) Study was published in 2010 [69]. Patients with CAD or at high risk who were taking a statin were included in a random-ized, double-blinded, placebo–controlled trial to receive 100 mg of anacetra-pib or placebo. LDL-C was decreased from 81 mg/dL to 45 mg/dL (−40%) and HDL-C was increased from 41 mg/dL to 101 mg/dL (+138%) as com-pared with a placebo with acceptable side effects. In addition to its LDL-C lowering effect, anacetrapib decreased plasma Lp(a) levels by −50% [70].

In a detailed analysis of lipoprotein subfraction by density gradient ultra-centrifugation in healthy individuals treated with anacetrapib 20 mg, medium and small LDL levels were decreased, whereas very small and dense LDL levels were increased, which is compatible with LDL subclasses found in severe CETP deficiency, but not in partial deficiency [71].

Large reduction in LDL-C needs to be confirmed by another method than the Friedewald formula (calculated LDL-C $= TC - HDL-C - TG/5$), since the method would underestimate LDL-C, because VLDL-C, which is estimated as TG/5, is lower in CETP deficiency and patients treated with anacetrapib than controls [1,72]. However, the direct HDL-C method would underestimate HDL-C levels in plasmas with apoE-rich HDL found in CETP deficiency, therefore the direct HDL-C method would overestimate calculated LDL-C through the formula [1].

As anti-atherogenicity of HDL, HDL after treatment with niacin or anacetrapib exhibits potent ability to suppress macrophage toll-like receptor 4-mediated inflammatory responses. The increased HDL fraction is rich in apoE and LCAT, but not in PAF-AH activity [73].

In 2011, the REVEAL trial (Randomized Evaluation of the Effects of Anacetrapib Through Lipid Modification) was started with a daily dose of 100 mg anacetrapib in patients with CAD with statin therapy [74]. This study will recruit 30,000 CAD patients 50 years of age or older. Their LDL-C levels will be controlled with atorvastatin with total cholesterol < 155 mg/dL, then patients will be randomized to have anacetrapib or not. It is expected that the REVEAL trial will provide valuable results by January 2017.

9.4. Evacetrapib (LY2484595, Eli Lilly)

A novel benzazepine compound is a potent, selective CETP inhibitor [75]. It contains a quinoline core like torcetrapib and the 3,5-bis-trifluoromethylbenzyl group but also a methyl tetrazole and cyclohexane carboxylic acid side chain. Evacetrapib inhibited human recombinant CETP (5.5 nM IC50) and CETP activity in human plasma (36 nM IC50) as well as torcetrapib and anacetrapib. Importantly, evacetrapib did not induce aldosterone and cortisol biosynthesis in a human adrenal cortical carcinoma cell line.

Evacetrapib was evaluated in patients with high LDL levels as monotherapy or in combination with statins [76]. Evacetrapib at 100 mg/day increased HDL-C (+54, ~+129%) and decreased LDL-C (−14, ~−36%) in the monotherapy through decreasing CETP activity (−50, ~−89%) but increasing CETP mass (+64, ~+137%).

10. STRUCTURAL DIFFERENCE OF CETP INHIBITORS IN THE CAVITY OF CETP

All compounds appeared to be related to increased plasma CETP mass—up to a 3-fold increase, which is contrast with antisense therapy. The reason for the increase in mass is not fully understood, although it may be related to decreased clearance of CETP in plasma. The CETP-CETP inhibitor complex is increased with HDL as seen in the electromicroscope.

The CETP inhibitors are buried deeply within the CETP protein, shifting the bound CE in the N-terminal pocket of the long hydrophobic tunnel and displacing the PL from the pocket. The lipids in the C-terminal pocket of the hydrophobic tunnel remain unchanged. Polar residues of Gln-199, Ser-230, and His-232 are found in the inhibitor-binding site. For example, torcetrapib occupies a volume of ~12 Å × 12 Å × 7 Å within the N-terminal pocket of the CETP tunnel [77]. Thus, torcetrapib binding physically interferes with PL binding and forces CE into a position that is presumably unfavorable for lipid transfer by blocking the narrow passage. The trifluoromethyl group of the torcetrapib projects deeply into the N-terminal pocket, sub-pocket formed by Ile-11, Cys-13, Ile-215, and the aromatic faces of His-232 and Phe-263. The binding site of dalcetrapib, Cys-13, is located in between the side chains of His-232 and Phe-263 as indicated in the model by Liu and colleagues [77]. Thus, dalcetrapib binding to the CETP is time dependent in the disulfide bond formation to Cys-13. However, other compounds with trifluoromethyl group are competitive in CETP binding.

11. DIFFERENTIAL EFFECTS AMONG CETP INHIBITORS

11.1. Differences of Levels of Preβ1-HDL and HDL2, and Cholesterol Efflux Capacity

Like PLTP, CETP itself is a conversion factor of HDL subclasses. CETP increased the size of HDL from HDL3 to HDL2, giving formation of smaller HDL particles of ~8 nm. In vitro levels of preβ1-HDL levels are varied after incubation with dalcetrapib or torcetrapib/anacetrapib. Torcetrapib/anacetrapib decreased preβ1-HDL levels in the concentration-dependent manner, but dalcetrapib did not decrease them [78]. A similar finding was found when neutralizing antibody TP1 was incubated in human plasma [79]: complete inhibition of CETP activity would retard preβ1-HDL formation. However, ex vivo analysis of plasmas of CETP-deficient humans resulted in opposite data (Table 8.1), partial inhibition would result in low levels of preβ1-HDL, but complete inhibition would increase in the preβ1-HDL levels.

Torcetrapib increased plasma larger HDL2 particles, which are increased postprandially up to 8 hours and act as active cholesterol acceptor via SR-BI and ABCG1-dependent cholesterol efflux pathway [80].

Cholesterol efflux was also increased to HDL from anacetrapib-treated hamsters via both the ABCA1 and ABCG1/SR-BI pathways. Indeed, anacetrapib induced HDL-C levels rich in cholesteryl linoleate (18:2), which is compatible with findings in CETP deficiency [81].

Khera and colleagues [82] reported that cholesterol efflux capacity was negatively associated with CAD risk independently of HDL-C levels. The capacity was determined ex vivo that radiolabeled J774 macrophage cells were incubated with apoB-depleted serum from patients for 4 hours, reflecting HDL capacity for cholesterol efflux activity mediated by ABCA1, ABCG1, and SR-BI pathways as well as aqueous diffusion. Thus, among several HDL functions, acceptor capacity for cholesterol efflux was likely enhanced in patients with CETP inhibitors.

11.2. Differences of in vivo Macrophage-Derived Reverse Cholesterol Transport (RCT) Among CETP Inhibitors

In a method involving administrating a radioactive cholesterol-labeled macrophage in the peritoneum, Tanigawa and colleagues [83] have measured direct RCT activity from peripheral macrophages to liver, bile, and feces. In LDL receptor-KO mice, CETP cDNA adeno-associated virus mediated transfection promotes cholesterol to the liver, but not to bile and feces. In contrast, in SR-BI-KO mice, CETP cDNA transfection increased cholesterol loss in the feces, indicating that overall RCT induced by CETP is not via SR-BI, but through LDL receptor in the liver in mice.

In 0.3% of cholesterol diet-induced combined hyperlipidemia of hamsters, an increase of aortic cholesterol content is correlated with higher cholesterol/TG ratio in the liver as well as increased plasma levels of non-HDL-cholesterol (3.8-fold) and increased CETP activity (+40%). In the gene expression of the cholesterol-fed hamster, mRNA levels of ABCA1 and ABCG5 increased, but the levels of LDL receptor and SR-BI decreased in the liver. In vivo, cholesterol efflux activity from macrophages to plasma and to bile/feces was decreased despite increased HDL-C levels (+90%) in hamsters [84], suggesting that HDL levels do not directly reflect efficacy of macrophage-derived RCT.

Using a hamster macrophage, RCT of radiolabeled cholesterol from the macrophages is maintained in the experiments with dalcetrapib, but it is diminished in studies with torcetrapib and anacetrapib [78]. The apparent difference of the induced cholesterol efflux activity appeared to be correlated with levels of preβ1-HDL; namely, dalcetrapib would maintain the levels, but strong inhibitors such as torcetrapib and anacetrapib, decreased them. Thus, dalcetrapib may have a unique lipoprotein profile such as preserved levels of preβ1-HDL, but it would be interesting to know whether or not it is due to a weaker inhibitor or a compound-specific effect.

Torcetrapib increased HDL-C, accelerating a secretion of cholesterol and bile acids in feces in hamsters but not in humans [85]. Several regulations in lipid homeostasis in hamsters are different from those in humans. In the liver of hamsters, dietary cholesterol feedings increased hepatic expression levels of ABCG5/G8 and PCSK9, but decreased CYP7A, with increasing bile cholesterol secretion. Therefore, decreasing both expression of LDLR and bile acid formation deteriorated magnitude of dyslipidemia in hamsters [86,87]. Although dyslipidemic hamsters are statin-resistant, the LDL-lowering drug berberine upregulates RCT with torcetrapib [88].

Th effect of anacetrapib on macrophage-to-feces RCT in hamster models is conflicting [89]. Although anacetrapib failed to show induced RCT in normolipidemic hamsters in a previous study [78], a recent study showed that administration to dyslipidemic hamsters resulted in improved RCT under the condition of strongly inhibited CETP activity by −94% [89].

11.3. Effects on Paraoxonase, PAF-AH (Lp-PLA2), and Anti-Inflammatory Activity

Serum CRP reduction was not reported in any compound, although anti-oxidative enzymes were substantially changed (Table 8.2). In an ex-vivo study, anti-inflammatory properties of HDL were maintained in hamsters treated by anacetrapib as well as in controls [90].

11.4. Vascular Effects

Flow-mediated dilation (FMD) of the brachial artery was increased by 41% in patients with low HDL-C (< 46 mg/dL) treated with dalcetrapib 600 mg, but that effect was not seen in patients with higher HDL-C in the

Table 8.2 Changes of PON1 and PAF-AH Activity in CETP Inhibitors

Drug	Subjects	PON1	PAF-AH	Author, Year, Ref
Dalcetrapib	Low HDL	Increased (+41%)	nd	Bisoendial, 2005 [91]
Dalcetrapib	CHD	nd	Increased (+17%)	Lüsher, 2012 [92]
Anacetrapib	Dyslipidemia	nd	No change	Yvan-Charvet, 2010 [73]

nd, not determined

baseline [93]. However, in the dal–VESSEL randomized clinical trial, FMD was not changed during the treatment with dalcetrapib 600 mg [92].

12. PERSPECTIVE OF CETP INHIBITOR

12.1. Glucose Tolerance, Diabetes Incidence During CETPi

High HDL syndrome is often associated with low prevalence of diabetes mellitus [36]. In vitro studies suggested that HDL may offer an antidiabetic effect by an increased pancreas beta-cell insulin secretion through mediation by ABCA1 and ABCG1 transporters [94]. Moreover, HDL may activate AMP-activated protein kinase in skeletal muscle [95], thereby accelerating glucose uptake.

Plasma CETP activity is increased in obesity or metabolic syndrome, but it is decreased in Type 2 diabetes [96]. This may be related to out-of-regulation of SREBP1 and 2 in skeletal muscles and adipose tissues of Type 2 diabetes [97]. However, CETP gene TaqI B2 allele is protective in diabetes, suggesting genetically low CETP activity is beneficial in macroangiopathy development of coronary disease, arteriosclerosis obliterans, and cerebral vascular disease [98].

Thus, it would be interesting to know whether or not impact on cardiovascular events by torcetrapib are stronger in diabetic patients involved in the ILLUMINATE trial. Conversely, Barter and colleagues [99] recently reported in the analysis of the ILLUMINATE trial that torcetrapib decreased HOMA-IR in the torcetrapib/atorvastatin arm as compared to the atorvastatin arm, which is associated with increased insulin sensitivity. Similarly, torcetrapib induced decrease in HOMA-IR in obese insulin-resistant CETP-apoB100 transgenic mice [100].

12.2. CETPi in Relation to Combination Therapy with HMG-CoA Reductase Inhibitor (statin) or Other Drugs

Statins per se would decrease plasma CETP levels modestly [101], but on-stain CETP is inversely related to coronary outcomes in a large clinical trial–based cohort [102]. However, Barter and colleagues [103] negated the idea of adverse interaction between atorvastatin and torcetrapib based on findings of the ILLUMINATE trial. Indeed, higher doses of atorvastatin appeared to protect against the harmful effect caused by torcetrapib. A recent study suggested that low CETP phenotype linked with genotype of TaqI B2 may predict increased mortality in statin-treated men in contrast with the fact that the genotype is associated with lower coronary risk in a meta-analysis [33,45]. Thus, the role of low CETP activity is conflicting in the statin-treated population. It should be investigated in a prospective manner.

Plasma CET is not only associated with CETP activity, but also other modulators: VLDL mass and FFA contents of lipoproteins. Thus, either a fibrate or a PAF-AH inhibitor may be good a candidate for combination therapy with the CETP inhibitor since fibrate will decrease VLDL levels and PAF-AH inhibitors decrease CET by decreasing LDL-FFA levels.

12.3. CETPi in Relation to apoE-rich HDL Levels

Reverse cholesterol transport is enhanced by increase in apoE-rich HDL levels. Xanthohumol, a prenylated chalcone derived from natural products, is a CETP inhibitor. The compound was shown to prevent atherosclerosis in CETP-transgenic mice. Importantly, other factors such as LCAT, apoE, SR-BI, and LDLR, which are upregulated in the liver, accelerate RCT along with increased apoE-rich HDL levels [104].

12.4. Infectious Disease Risk in CETPi

Torcetrapib-related excess death appeared to be related to noncardiovascular events such as malignancy and/or infection. Low CETP activity may be associated with high mortality as suggested by a recent prospective study in hospitalized patients [105]. In that study, each 1 mg/dL increase in HDL decreased the odds of severe sepsis by 3% during hospitalization, suggesting a role of HDL as an LPS scavenger. Similarly, recombinant HDL decreased LPS-induced inflammatory response in patients with liver cirrhosis [106]. Thus, increased HDL would protect from infection.

However, the reduction of plasma CETP was associated with mortality in hospitalized patients [105]. It may reflect severe infection reducing CETP expression in hematopoietic cells. Furthermore, in vitro studies it is unlikely that torcetrapib has a direct effect on LBP and bactericidal/permeability increasing protein (BPI) function, nor an inhibitory effect on the interaction with LPS [107].

12.5. Potential of CETPi Against C-Terminal Polypeptide

Vaccine-induced antibodies were tested earlier in rabbits [108]. The epitope consisted of C-terminal CETP (461–476) and the peptide of tetanus toxin, therefore CETP inhibition was expected in the ternary complex of HDL-CETP-VLDL or HDL-CETP-LDL. The approach results in a decrease in CETP activity by −24%, increasing HDL-C levels by +42% with reduced aortic atherosclerosis in cholesterol-fed rabbits. The approach was tested in human clinical trials, but the phase II failed to meet the primary endpoint of increasing HDL-C levels [109]. Thus, low concentrations of anti-human CETP antibody need an efficient adjuvant formulation. This approach would be interesting because the antibody inhibits CETP activity through C-terminal CETP, which is involved in the interaction with lipid transfer acceptors such as VLDL or LDL, but not in interconversion among HDL subclasses.

13. CONCLUSION

Anti-atherogenicity of low CETP activity appears to be dependent on the cause, whether it is genetic or environmental. Also, it is unclear how much lower CETP activity would be beneficial in human atherosclerosis. Since CETP inhibitors such as anacetrapib and evacetrapib have been tested in Phase III trials, it is expected that those trials will provide results on vascular endpoints by 2017. As the structure and function relationship between the hydrophobic tunnel of CETP and CE/TG and PL has been disclosed, different inhibitors targeting the other domains are promising. Also, CETP antibody therapy awaits further investigation.

REFERENCES

[1] Inazu A. Plasma cholesteryl ester transfer protein (CETP) in relation to human pathophysiology. In: Komoda T, editor. The HDL handbook. Burlington: MA: Academic Press; 2010. p. 35–60.
[2] Qiu X, Mistry A, Ammirati MJ, Chrunyk BA, Clark RW, Cong Y, et al. Crystal structure of cholesteryl ester transfer protein reveals a long tunnel and four bound lipid molecules. Nat Struct Mol Biol 2007;14:106–13.
[3] Zhang L, Yan F, Zhang S, Lei D, Charles MA, Cavigiolio G, et al. Structural basis of transfer between lipoproteins by cholesteryl ester transfer protein. Nature Chem Biol 2012;8:342–9.
[4] Lei D, Zhang X, Jiang S, Cai Z, Rames MJ, Zhang L, et al. Structural features of cholesteryl ester transfer protein: a molecular dynamics simulation study. Proteins 2013;81:415–25.
[5] Charles MA, Kane JP. New molecular insights into CETP structure and function: a review. J Lipid Res 2012;53:1451–8.
[6] Kopec KO, Alva V, Lupas AN. Bioinfomatics of the TULIP domain superfamily. Biochem Soc Trans 2011;39:1033–8.
[7] Asztalos B, Horvath KV, Kajinami K, Nartsupha C, Cox CE, Batista M, et al. Apolipoprotein composition of HDL in cholesteryl ester transfer protein deficiency. J Lipid Res 2004;45:448–55.
[8] Oliveira HC, Ma L, Milne R, Marcovina SM, Inazu A, Mabuchi H, et al. Cholesteryl ester transfer protein activity enhances plasma cholesteryl ester formation. Studies in CETP transgenic mice and human genetic CETP deficiency. Arterioscler Thromb Vasc Biol 1997;17:1045–52.

[9] Miyazaki O, Fukamachi I, Mori A, Hashimoto H, Kawashiri MA, Nohara A, et al. Formation of prebeta1-HDL during lipolysis of triglyceride-rich lipoprotein. Biochem Biophys Res Commun 2009;379:55–9.

[10] Ikewaki K, Rader DJ, Sakamoto T, Nishiwaki M, Wakimoto N, Schaefer JR, et al. Delayed catabolism of high density lipoprotein apolipoproteins A-I and A-II in human cholesteryl ester transfer protein deficiency. J Clin Invest 1993;92:1650–8.

[11] Vergeer M, Korporaal SJ, Franssen R, Meurs I, Out R, Hovingh GK, et al. Genetic variant of the scavenger receptor BI in humans. N Engl J Med 2011;364:136–45.

[12] Robins SJ, Fasulo JM. High density lipoproteins, but not other lipoproteins, provide a vehicle for sterol transport to bile. J Clin Invest 1997;99:380–4.

[13] Koizumi J, Inazu A, Yagi K, Koizumi I, Uno Y, Kajinami K, et al. Serum lipoprotein lipid concentration and composition in homozygous and heterozygous patients with cholesteryl ester transfer protein deficiency. Atherosclerosis 1991;90:189–96.

[14] Yamashita S, Matsuzawa Y, Okazaki M, Kako H, Yasugi T, Akioka H, et al. Small poydisperse low density lipoproteins in familial hyperalphalipoproteinemia with complete deficiency of cholesteryl ester transfer activity. Atherosclerosis 1988;70:7–12.

[15] Brown ML, Inazu A, Hesler CB, Agellon LB, Mann C, Whitlock ME, et al. Molecular basis of lipid transfer protein deficiency in a family with increased high-density lipoproteins. Nature 1989;342:448–51.

[16] Nakajima K, Nakano T, Tokita Y, Nagamine T, Inazu A, Kobayashi J, et al. Postprandial lipoprotein metabolism:VLDL and chylomicrons. Clin Chim Acta 2011;412:1306–18.

[17] Inazu A, Nakajima K, Nakano T, Niimi M, Kawashiri M, Nohara A, et al. Decreased post-prandial triglyceride response and diminished remnant lipoprotein formation in cholesteryl ester transfer protein (CETP) deficiency. Atherosclerosis 2008;196:953–7.

[18] Parra ES, Urban A, Panzoldo NB, Nakamura RT, Oliveira R, de Faria EC. A reduction of CETP activity, not an increase, is associated with modestly impaired postprandial lipemia and increased HDL-cholesterol in adult asymptomatic women. Lipids Health Dis 2011;10:87.

[19] Noto H, Kawamura M, Hashimoto Y, Satoh H, Hara M, Iso-o N, et al. Mudulation of HDL metabolism by probucol in complete cholesteryl ester transfer protein deficiency. Atherosclerosis 2003;171:131–6.

[20] Ferreira PFC, Zago VHS, D'Alexandri FL, Panzoldo NB, Gidlund MA, Nakamura RT, et al. Oxidized low-density lipoproteins and their antibodies: relationships with the reverse cholesterol transport and carotid atherosclerosis in adults without cardiovascular diseases. Clin Chim Acta 2012;413:1472–8.

[21] Hine D, Mackness B, Mackness M. Cholesteryl-ester transfer protein enhances the ability of high-density lipoprotein to inhibit low-density lipoprotein oxidation. IUBMB Life 2011;63:772–4.

[22] Chantepie S, Bochem AE, Chapman MJ, Hovingh GK, Kontush A. High-density lipoprotein (HDL) particle subpopulations in heterozygous cholesteryl ester transfer protein (CETP) deficiency: maintenance of antioxidative activity. PLoS One 2012;7: e49336.

[23] Oka T, Yamashita S, Kujiraoka T, Ito M, Nagano M, Sagehashi Y, et al. Distribution of human plasma PLTP mass and activity in hypo- and hyperalphalipoproteinemia. J Lipid Res 2002;43:1236–43.

[24] Nagasaka H, Chiba H, Kikuta H, Akita H, Takahashi Y, Yanai H, et al. Unique character and metabolism of high density lipoprotein (HDL) in fetus. Atherosclerosis 2002;161:215–23.

[25] Inazu A, Brown ML, Hesler CB, Agellon LB, Koizumi J, Takata K, et al. Increased high-density lipoprotein levels caused by a common cholesteryl-ester transfer protein gene mutation. N Engl J Med 1990;323:1234–8.

[26] Koizumi J, Mabuchi H, Yoshimura A, Michishita I, Takeda M, Itoh H, et al. Deficiency of serum cholesteryl-ester transfer activity in patients with familial hyperalphalipo-proteinaemia. Atherosclerosis 1985;58:175–86.

[27] Matsuura F, Wang N, Chen W, Jiang XC, Tall AR. HDL from CETP-deficient sub-jects shows enhanced ability to promote cholesterol efflux from macrophages in an apoE- and ABCG1-dependent pathway. J Clin Invest 2006;116:1435–42.

[28] Miwa K, Inazu A, Kawashiri M, Nohara A, Higashikata T, Kobayashi J, et al. Choles-terol efflux from J774 macrophages and Fu5AH hepatoma cells to serum is preserved in CETP deficient patients. Clin Chim Acta 2009;402:19–24.

[29] Wang J, Qiang H, Chen D, Zhang C, Zhuang Y. CETP gene mutation (D442G) increases low-density lipoprotein particle size in patients with coronary heart disease. Clin Chim Acta 2002;322:85–90.

[30] Dullaart RPF, Constantinides A, Perton FG, van Leeuwen JJJ, van Pelt JL, de Vries R, et al. Plasma cholesteryl ester transfer, but not cholesterol esterification, is related to lipoprotein-associated phospholipase A2: possible contribution to an atherogenic lipoprotein profile. J Clin Endocrin Metab 2011;96:1077–84.

[31] Ordovas JM, Cupples LA, Corella D, Otvos JD, Osgood D, Martinez A, et al. Association of cholesteryl ester transfer protein-TaqIB polymorphism with variations in lipoprotein subclasses and coronary heart disease risk: the Framingham study. Arterioscler Thromb Vasc Biol 2000;20:1323–9.

[32] Vasan RS, Pencina MJ, Robins SJ, Zachariah JP, Kaur G, D'Agostino RB, et al. Association of circulating cholesteryl ester transfer protein activity with incidence of cardiovascular disease in the Community. Circulation 2009;120:2414–20.

[33] Thompson A, Di Angelantonio E, Sarwar N, Erqou S, Saleheen D, Dullaart RP, et al. Association of cholesteryl ester transfer protein genotypes with CETP mass and activ-ity, lipid levels, and coronary risk. JAMA 2008;299:2777–88.

[34] Curb JD, Abbott RD, Rodriguez BL, Masaki K, Chen R, Sharp DS, et al. A prospec-tive study of HDL-C and cholesteryl ester transfer protein gene mutations and the risk of coronary heart disease in the elderly. J Lipid Res 2004;45:948–53.

[35] Hirano K, Yamashita S, Nakajima N, Arai T, Maruyama T, Yoshida Y, et al. Genetic cholesteryl ester transfer protein deficiency is extremely frequent in the Omagari area of Japan. Marked hyperalphalipoproteinemia caused by CETP gene mutation is not associated with longevity. Arterioscler Thromb Vasc Biol 1997;17:1053–9.

[36] Moriyama Y, Okamura T, Inazu A, Doi M, Iso H, Mouri Y, et al. A low prevalence of coronary heart disease among subjects with increased high-density lipoprotein cho-lesterol levels, including those with plasma cholesteryl ester transfer protein deficiency. Prev Med 1998;27:659–67.

[37] Koropatnick TA, Kimbell J, Chen R, Grove JS, Donlon TA, Masaki KH, et al. A pro-spective study of high-density lipoprotein cholesterol, cholesteryl ester transfer pro-tein gene variants, and healthy aging in very old Japanese-American men. J Gerontol A Biol Sci Med Sci 2008;63A:1235–40.

[38] Barzilai N, Alzmon G, Schechter C, Schaefer EJ, Cupples AL, Lipton R, et al. Unique lipoprotein phenotype and genotype associated with exceptional longevity. JAMA 2003;290:2030–40.

[39] Sanders AE, Wang C, Katz M, Derby CA, Barzilai N, Ozelius L, et al. Association of a functional polymorphism in the cholesteryl ester transfer protein (CETP) gene with memory decline and incidence of dementia. JAMA 2010;303:150–8.

[40] Boekholdt SM, Sacks FM, Jukema JW, Shepherd J, Freeman DJ, McMahon AD, et al. Cholesteryl ester transfer protein TaqIB variant, high-density lipoprotein cholesterol levels, cardiovascular risk, and efficacy of pravastatin treatment. Individual patient meta-analysis of 13677 subjects. Circulation 2005;111:278–87.

[41] Johannsen TH, Frikke-Schmidt R, Schou J, Nordestgaard BG, Tybjaerg-Hansen A. Genetic inhibition of CETP, ischemic vascular disease and mortality, and possible adverse effects. J Am Coll Cardiol 2012;60:2041–8.

[42] Ridker PM, Pare G, Parker AN, Zee RYL, Miletich JP. Chasman DI. Polymorphism in the CETP gene region, HDL cholesterol, and risk of future myocardial infarction. Circ Cardiovasc Genet 2009;2:26–33.

[43] Hiura Y, Shen C-S, Kokubo Y, Okamura T, Morisaki T, Tomoike H, et al. Identification of genetic markers associated with high-density lipoprotein-cholesterol by genome-wide screening in a Japanese population. The Suita Study 2009;73:1119–26.

[44] Papp AC, Pinsonneault JK, Wang D, Newman LC, Gong Y, Johnson JA, et al. Cholesteryl ester transfer protein (CETP) polymorphisms affect mRNA splicing, HDL levels, and sex-dependent cardiovascular risk. PLoS One 2012;7: e31930.

[45] Regieli JJ, Jukema JW, Grobbee DE, Kastelein JJ, Kuivenhoven JA, Zwinderman AH, et al. CETP genotype predicts increased mortality in statin-treated men with proven cardiovascular disease: an adverse pharmacogenetic interaction. Eur Heart J 2008;29:2792–9.

[46] Corsetti JP, Ryan D, Rainwater DL, Moss AJ, Zareba W, Sparks CE. Cholesteryl ester transfer protein polymorphism (TaqIB) associates with risk in postinfarction patients with high C-reactive protein and high-density lipoprotein cholesterol levels. Arterioscler Thromb Vasc Biol 2010;30:1657–64.

[47] Duwensee K, Breiling LP, Tancevski I, Rothenbacher D, Demetz E, Patsch JR, et al. Cholesteryl ester transfer protein in patients with coronary heart disease. Eur J Clin Invest 2010;40:616–22.

[48] Ritsch A, Scharnagl H, Eller P, Tancevski I, Duwensee K, Demetz E, et al. Cholesteryl ester transfer protein and mortality in patients undergoing coronary angiography. The Ludwigshafen Risk and Cardiovascular Health Study. Circulation 2010;121:366–74.

[49] Levels JHM, Pajkrt D, Shultz M, Hoek FJ, van Tol A, Meijers JCM, et al. Alterations in lipoprotein homeostasis during human experimental endotoxemia and clinical sepsis. Biochim Biophys Acta 2007;1771:1429–38.

[50] Sandhofer A, Tatarczyk T, Laimer M, Ritsch A, Kaser S, Paulweber B, et al. The Taq1B-variant in the cholesteryl ester-transfer protein gene and the risk of metabolic syndrome. Obesity 2008;16:919–22.

[51] Pirro M, Siepi D, Lupattelli G, Roscini AR, Schillaci G, Gemelli F, et al. Plasma C-reactive protein in subjects with hypo/hyperalphalipoproteinemias. Metabolism 2003;52:432–6.

[52] Cunningham D, Lin W, Hoth LR, Danley DE, Ruggeri RB, Geoghegan KF, et al. Biophysical and biochemical approach to locating an inhibitor binding site on cholesteryl ester transfer protein. Bioconjug Chem 2008;19:1604–13.

[53] Barter PJ, Caulfield M, Eriksson M, Grundy SM, Kastelein JJ, Komajda M MILLUMINATE Investigators, et al. Effects of torcetrapib in patients at high risk for coronary events. N Engl J Med 2007;357:2109–22.

[54] Clerc RG, Stauffer A, Weibel F, Hainaut E, Perez A, Hoflack J-C, et al. Mechanisms underlying off-target effects of the cholesteryl ester transfer protein inhibitor torcetrapib involve L-type calcium channels. J Hypertens 2010;28:1676–86.

[55] Connelly MA, Parry TJ, Giardino EC, Huang Z, Cheung WM, Chen C, et al. Torcetrapib produces endothelial dysfunction independent of cholesteryl ester transfer protein inhibition. J Cardiovasc Pharmacol 2010;55:459–68.

[56] Simic B, Hermann M, Shaw SG, Bigler L, Stalder U, Dörries C, et al. Torcetrapib impairs endothelial function in hypertension. Eur Heart J 2012;33:1615–24.

[57] Nicholls SJ, Tuzcu EM, Brennan DM, Tardif JC, Nissen SE. Cholesteryl ester transfer protein inhibition, high-density lipoprotein raising, and progression of coronary atherosclerosis: insights from ILLUSTRATE (Investigation of Lipid Level Management Using Coronary Ultrasound to Assess Reduction of Atherosclerosis by CETP Inhibition and HDL Elevation). Circulation 2008;118:2506–14.

[58] Catalano G, Julia Z, Frisdal E, Vedie B, Fournier N, Le Goff W, et al. Torcetrapib differentially modulates the biological activities of HDL2 and HDL3 particles in the reverse cholesterol transport pathway. Arterioscler Thromb Vasc Biol 2009;29:268–75.

[59] Guerin M, Le Goff W, Duchene E, Julia Z, Nguyen T, Thuren T, et al. Inhibition of CETP by torcetrapib attenuates the atherogenicity of postprandial TG-rich lipoproteins in type IIB hyperlipidemia. Arterioscler Thromb Vasc Biol 2008;28:148–54.

[60] Tchoua U, D'Souza W, Mukhamedova N, Blum D, Niesor E, Mizrahi J, et al. The effect of cholesteryl ester transfer protein overexpression and inhibition on reverse cholesterol transport. Cardiovasc Res 2008;77:732–9.

[61] Okamoto H, Yonemori F, Wakitani K, Minowa T, Maeda K, Shinkai H. A cholesteryl ester transfer protein inhibitor attenuates atherosclerosis in rabbits. Nature 2000;406:203–7.

[62] Davidson MH. Update of CETP inhibition. J Clin Lipidol 2010;4:394–8.

[63] Stein EA, Roth EM, Rhyne JM, Burgess T, Kallend D, Robinson JG. Safety and tolerability of dalcetrapib (RO4607381/JTT-705): results from a 48-week trial. Eur Heart J 2010;31:480–8.

[64] Stalenhoef AF, Davidson MH, Robinson JG, Burgess T, Duttlinger-Maddux R, Kallend D, et al. Efficacy and safety of dalcetrapib in type 2 diabetes mellitus and/or metabolic syndrome patients, at high cardiovascular disease risk. Diabetes Obes Metab 2012;14:30–9.

[65] Fayad ZA, Mani V, Woodward M, Kallend D, Bansilal S, Pozza J, et al. Rationale and design of dal-PLAQUE: a study assessing efficacy and safety of dalcetrapib on progression or regression of atherosclerosis using magnetic resonance imaging and 18F-fluorodeoxyglucose positron emission tomography/computed tomography. Am Heart J 2011;162:214–21.

[66] Schwartz GG, Olsson AG, Abt M, Ballantyne CM, Barter PJ, Brumm J dal-OUTCOMES Investigators, et al. Effects of dalcetrapib in patients with a recent acute coronary syndrome. N Engl J Med 2012;367:2089–99.

[67] Krishna R, Anderson MS, Bergman AJ, Jin B, Fallon M, Cote J, et al. Effect of the cholesteryl ester transfer protein inhibitor, anacetrapib, on lipoproteins in patients with dyslipidaemia and on 24-h ambulatory blood pressure in healthy individuals: two double-blind, randomised placebo-controlled phase I studies. Lancet 2007;370: 1907–14.

[68] Ranalletta M, Bierilo KK, Chen Y, Milot D, Chen Q, Tung E, et al. Biochemical characterization of cholesteryl ester transfer protein inhibitors. J Lipid Res 2010;51:2739–52.

[69] Cannon CP, Shah S, Dansky HM, Davidson M, Brinton EA, Gotto AM, et al. Determining the Efficacy and Tolerability Investigators, et al. Safety of anacetrapib in patients with or at high risk for coronary heart disease. N Engl J Med 2010;363: 2406–15.

[70] Bloomfield D, Carlson GL, Sapre A, Tribble D, McKenney JM, Littlejohn 3rd TW, et al. Efficacy and safety of the cholesteryl ester transfer protein inhibitor anacetrapib as monotherapy and coadministered with atorvastatin in dyslipidemic patients. Am Heart J 2009;157:352–60.

[71] Krauss RM, Wojnooski K, Orr J, Geaney JC, Pinto CA, Liu Y, et al. Changes in lipoprotein subfraction concentration and composition in healthy individuals treated with the CETP inhibitor anacetrapib. J Lipid Res 2012;53:540–7.

[72] Davidson M, Liu SX, Barter P, Brinton EA, Cannon CP, Gotto AM, et al. Measurement of LDL-C after treatment with the CETP inhibitor anacetrapib. J Lipid Res 2013;54:467–72.

[73] Yvan-Charvet L, Kling J, Pagler T, Li H, Hubbard B, Fisher T, et al. Cholesterol efflux potential and antiinflammatory properties of high-density lipoprotein after treatment with niacin or anacetrapib. Arterioscler Thromb Vasc Biol 2010;30:1430–8.

[74] Gutstein DE, Krishna R, Johns D, Surks HK, Dansky HM, Shah S, et al. Anacetrapib, a novel CETP inhibitor: Pursuing a new approach to cardiovascular risk reduction. Clini Pharmacol Ther 2012;91:109–22.

[75] Cao G, Beyer TP, Zhang Y, Schmidt RJ, Chen YQ, Cockerham SL, et al. Evacetrapib is a novel, potent, and selective inhibitor of cholesteryl ester transfer protein that elevates HDL cholesterol without inducing aldosterone or increasing blood pressure. J Lipid Res 2011;52:2169–76.

[76] Nicholls SJ, Brewer HB, Kastelein JJP, Krueger KA, Wang M-D, Shao M, et al. Effects of the CETP inhibitor evacetrapib administered as monotherapy or in combination with statins on HDL and LDL cholesterol. JAMA 2011;306:2099–109.

[77] Liu S, Mistry A, Reynolds JM, Lloyd DB, Grifor MC, Perry DA, et al. Crystal structures of cholesteryl ester transfer protein in complex with inhibitors. J Biol Chem 2012;287:37321–9.

[78] Niesor EJ, Magg C, Ogawa N, Okamoto H, von der Mark E, Matile H, et al. Modulating cholesteryl ester transfer protein activity maintains efficient preβ1-HDL formation and increases reverse cholesterol transport. J Lipid Res 2010;51:3443–54.

[79] Lagrost L, Gambert P, Dangremont V, Athias A, Lallemant C. Role of cholesteryl ester transfer protein (CETP) in the HDL conversion process as evidenced by using anti-CETP monoclonal antibodies. J Lipid Res 1990;31:1569–75.

[80] Bellanger N, Julia Z, Villard EF, Khoury PE, Duchene E, Chapman MJ, et al. Functionality of postprandial larger HDL2 particles is enhanced following CETP inhibition therapy. Atherosclerosis 2012;221:160–8.

[81] Bisgaier CL, Siebenkas MV, Brown ML, Inazu A, Koizumi J, Mabuchi H, et al. Familial cholesteryl ester transfer protein deficiency is associated with triglyceride-rich low density lipoproteins containing cholesteryl esters of probable intracellular origin. J Lipid Res 1991;32:21–33.

[82] Khera AV, Cuchel M, de la Llera-Moya M, Rodrigues A, Burke MF, Jafri K, et al. Cholesterol efflux capacity, high-density lipoprotein function, and atherosclerosis. N Engl J Med 2011;364:127–35.

[83] Tanigawa H, Billheimer JT, Tohyama J, Zhang Y, Rothblat G, Rader DJ. Expression of cholesteryl ester transfer protein in mice promotes macrophage reverse cholesterol transport. Circulation 2007;116:1267–73.

[84] Tréguier M, Briand F, Boubacar A, André A, Magot T, Nguyen P, et al. Diet-induced dyslipidemia impairs reverse cholesterol transport in hamsters. Eur J Clin Invest 2011;41:921–8.

[85] Brousseau ME, Diffenderfer MR, Millar JS, Nartsupha C, Asztalos BF, Welty FK, et al. Effects of cholesteryl ester transfer protein inhibition on high-density lipoprotein subspecies, apolipoprotein A-I metabolism, and fecal sterol excretion. Arterioscler Thromb Vasc Biol 2005;25:1057–64.

[86] Ness GC, Gertz KR. Hepatic HMG-CoA reductase expression and resistance to dietary cholesterol. Exp Biol Med 2004;229:412–6.

[87] Dong B, Wu M, Li H, Kraemer FB, Adeli K, Seidah NG, et al. Strong induction of PCSK9 gene expression through HNF1alpha and SREBP2: mechanism for the resistance to LDL-cholesterol lowering effect of statins in dyslipidemic hamsters. J Lipid Res 2010;51:1486–95.

[88] Briand F, Thieblemont Q, Muzotte E, Sulpice T. Upregulating reverse cholesterol transport with cholesteryl ester transfer protein inhibition requires combination with the LDL-lowering drug berberine in dyslipidemic hamsters. Arterioscler Thromb Vasc Biol 2013;33:13–23.

[89] Castro-Perez J, Briand F, Gagen K, Wang SP, Chen Y, McLaren DG, et al. Anacetrapib promotes reverse cholesterol transport and bulk cholesterol excretion in Syrian golden hamsters. J Lipid Res 2011;52:1965–73.

[90] Han S, LeVoci L, Fischer P, Wang S-P, Gagen K, Chen Y, et al. Inhibition of cholesteryl ester transfer protein by anacetrapib does not impair the anti-inflammatory properties of high density lipoprotein. Biochim Biophys Acta 2013;1831:825–33.

[91] Bisoendial RJ, Hovingh GK, Harchaoui KE, Levels JHM, Tsimikas S, Pu K, et al. Consequences of cholesteryl ester transfer protein inhibition in patients with familial hypoalphalipoproteinemia. Arterioscler Thromb Vasc Biol 2005;25:e133–4.

[92] Lüsher TF, Taddei S, Kaski JC, Jukema JW, Kallend D, Münzel T, et al. dal-VESSEL Investigators. Vascular effects and safety of dalcetrapib in patients with or at risk of coronary heart disease: the dal-VESSEL randomized clinical trial. Eur Heart J 2012;33:857–65.

[93] Hermann F, Enseleit F, Spieker LE, Periat D, Sudano I, Hermann M, et al. Cholesteryl ester transfer protein inhibition and endothelial function in type II hyperlipidemia. Thromb Res 2009;123:460–5.

[94] Fryirs MA, Barter PJ, Appavoo M, Tuch BE, Tabet F, Heather AK, et al. Effects of high-density lipoproteins in pancreatic beta-cell insulin secretion. Arterioscler Thromb Vasc Biol 2010;30:1642–8.

[95] Drew BG, Duffy SJ, Formosa MF, Natoli AK, Henstridge DC, Penfold SA, et al. High-density lipoprotein modulates glucose metabolism in patients with type 2 diabetes mellitus. Circulation 2009;119:2103–11.

[96] MacLean PS, Vadlamudi S, MacDonald KG, Pories WJ, Barakat HA. Suppression of hepatic cholesteryl ester transfer protein expression in obese humans with the development of type 2 diabetes mellitus. J Clin Endocrinol Metab 2005;90:2250–8.

[97] Ducluzeau PH, Perretti N, Laville M, Andreelli F, Vega N, Riou JP, et al. Regulation by insulin of gene expression in human skeletal muscle and adipose tissue. Evidence for specific defects in type 2 diabetes. Diabetes 2001;50:1134–42.

[98] Kawasaki I, Tahara H, Emoto M, Shoji T, Nishizawa Y. Relationship between TaqIB cholesteryl ester transfer protein gene polymorphism and macrovascular complications in Japanese patients with type 2 diabetes. Diabetes 2002;51:871–4.

[99] Barter PJ, Rye KA, Tardif JC, Waters DD, Boekholdt SM, Breazna A, et al. Effect of torcetrapib on glucose, insulin, and hemoglobin A1c in subjects in the investigation of lipid level management to understand its impact in atherosclerotic events (ILLUMINATE) trial. Circulation 2011;124:555–62.

[100] Briand F, Thieblermont Q, Andre A, Ouguerram K, Sulpice T. CETP inhibitor torcetrapib promotes reverse cholesterol transport in obese insulin-resistant CETP-apoB100 transgenic mice. Clin Trans Sci 2011;4:414–20.

[101] Inazu A, Koizumi J, Kajinami K, Kiyohara T, Chichibu K, Mabuchi H. Opposite effects on serum cholesteryl ester transfer protein levels between long-term treatments with pravastatin and probucol in patients with primary hypercholesterolemia and xanthoma. Atherosclerosis 1999;145:405–13.

[102] Khera AV, Wolfe ML, Cannon CP, Qin J, Rader DJ. On-statin cholesteryl ester transfer protein mass and risk of recurrent coronary events (from the pravastatin or atorvastatin evaluation and infection therapy-thrombolysis in myocardial infarction 22 [PROVE IT-TIMI 22] study). Am J Cardiol 2010;106:451–6.

[103] Barter PJ, Rye K-A, Beltangady MS, Ports WC, Duggan WT, Boekholdt SM, et al. Relationship between atorvastatin dose and the harm caused by torcetrapib. J Lipid Res 2012;53:2436–42.

[104] Hirata H, Yimin Segawa S, Ozaki M, Kobayashi N, Shigyo T, Chiba H. Xanthohumol prevents atherosclerosis by reducing arterial cholesterol content via CETP and apolipoprotein E in CETP-Transgenic mice. PLoS One 2012;7: e49415.

[105] Grion CMC, Cardoso LTQ, Perazolo TF, Garcia AS, Barbosa DS, Morimoto HK, et al. Lipoproteins and CETP levels as risk factors for severe sepsis in hospitalized patients. Eur J Clin Invest 2010;40:330–8.
[106] Galbois A, Thabut D, Tazi KA, Rudler M, Mohammadi MS, Bonnefont-Rousselot D, et al. Ex vivo effects of high-density lipoprotein exposure on the lipopolysaccharide-induced inflammatory response in patients with severe cirrhosis. Hepatology 2009;49:175–84.
[107] Clark RW, Cunningham D, Cong Y, Subashi TA, Tkalcevic GT, Lloyd DB, et al. Assessment of cholesteryl ester transfer protein inhibitors for interaction with proteins involved in the immune response to infection. J Lipid Res 2010;51:967–74.
[108] Rittershaus CW, Millar DP, Thomas LJ, Picard MD, Honas CM, Emmett CD, et al. Vaccine-induced antibodies inhibit CETP activity in vivo and reduce aortic lesions in a rabbit model of atherosclerosis. Arterioscler Thromb Vasc Biol 2000;20; 2016–2012.
[109] Ryan US, Rittershaus CW. Vaccines for the prevention of cardiovascular disease. Vascul Pharmacol 2006;45:253–7.

CHAPTER 9

HDL Apoprotein Mimetic Peptides as Anti-Inflammatory Molecules

Godfrey S. Getz, Catherine A. Reardon
Department of Pathology, The University of Chicago, Chicago, IL, USA

Contents

Abstract

HDL and apoA-I have been shown in epidemiological studies and in animal models to reduce the risk of atherosclerosis. Synthetic peptides modeling the repeating amphipathic α-helices in apoA-I, the major apoprotein of HDL, were originally designed to study the lipid binding and other physical properties of the helices. The interest in apoA-I mimetic peptides was greatly stimulated by the observation that some, but not all, of the peptides significantly reduce atherosclerosis in animal models. Both monohelical and tandem helical peptides have been studied in vitro and in vivo for properties that mimic the anti-atherogenic functional properties of HDL and apoA-I, particularly their ability to promote cholesterol efflux and reduce inflammation and oxidation. Mimetic peptides related to apoE, another apoprotein with repeating amphipathic helices, have also been shown to be anti-atherogenic. Both peptide families likely function via common and unique mechanisms. The understanding of how the structural properties of the various mimetic peptides relate to their mechanism of action will facilitate the development of optimal anti-atherogenic therapeutic mimetic peptides.

Key Concepts

- Several apoA-I and apoE mimetic peptides reduce atherosclerosis in animal models.
- Both families of mimetic peptides are anti-inflammatory and promote cholesterol efflux. However, the anti-atherogenic apoA-I mimetic peptides bind oxidized lipids with high affinity, while apoE mimetic peptides reduce plasma lipid levels. Thus, they likely function via common and distinct mechanisms.
- Recent evidence suggests that the anti-atherogenic apoA-I mimetic peptides may function in the intestine to reduce the oxidative burden.
- Understanding the mechanism of action of the various mimetic peptides and how this relates to their structure will facilitate the development of optimal anti-atherogenic peptides.

1. INTRODUCTION

There has been much epidemiological evidence linking low high-density lipoprotein (HDL) levels with an increased incidence of cardiovascular disease [1]. Despite this wealth of information, the precise role of HDL and its components in atherosclerotic disease in humans remains a subject of continuing investigation. The most robust evidence for the role of HDL in atherosclerosis derives from animal experimentation. In these experimental models, the most manipulable component of HDL is its major apoprotein, apoA-I, which, unlike most other HDL components, is probably present on every HDL particle [2,3]. When this apoprotein is overexpressed in murine and rabbit models of atherosclerosis, the extent of atherosclerotic lesions is reduced [4,5]. On the other hand, when animals are deficient in apoA-I, atherosclerotic disease is usually exacerbated [6]. Based upon the epidemiological and experimental results, a number of clinical studies have been launched in attempts to raise HDL levels, with varying results [7]. This has given rise to questions about how plasma HDL is measured [8]. Almost all clinical studies have been based upon the measurement of HDL cholesterol. A number of potential mechanisms by which HDL and/or apoA-I may exert their anti-atherogenic effects have been identified (Table 9.1). At issue is whether the measurement of HDL cholesterol adequately reflects on the functionality of the HDL. Although apoA-I represents two-thirds of the protein of HDL, HDL particles are heterogeneous in both density and composition. As has been pointed out by Heinecke using shotgun proteomics, as many as 50 different protein species, with a variety of molecular functions, are found in total HDL [2]. Of potential importance with respect to HDL functionality, the protein composition of HDL from healthy individuals and individuals with coronary artery disease are not identical [2].

Table 9.1 Anti-Atherogenic Functions of HDL and/or ApoA-I

Promote Cholesterol Efflux and Reverse Cholesterol Transport

Anti-inflammatory
Antioxidant
Antithrombotic
Anti-apoptotic
Promotes nitric oxide production by endothelial cells

Of the variety of anti-atherogenic functions of HDL and/or apoA-I, three have been extensively studied. This includes the promotion of reverse cholesterol transport (i.e., the removal of lipid from cholesterol-loaded cells and tissues and the delivery of cholesterol to the liver for excretion into the bile and feces), their ability to reduce inflammation, and their ability to reduce oxidative stress. These functions are not necessarily independent of each other. At least part of the anti-inflammatory effect may be related to the removal of cholesterol from lipid rafts and the antioxidant function may be related to their anti-inflammatory function. Several approaches have been developed for the assessment of cholesterol efflux and reverse cholesterol transport by HDL and apoA-I in vitro and in vivo [9–11], though the in vivo assessments are not entirely satisfactory. The development of a monocyte chemotactic assay (MCA) has been very valuable in assessing the anti-inflammatory and antioxidative activity of HDL [12]. The assay is based upon the oxidation of low-density lipoprotein (LDL) by co-cultures of primary endothelial cells obtained from hearts to be replaced by transplantation and smooth muscle cells that resemble an "in vitro" artery. The oxidation of LDL by the endothelial cells elicits the production of monocyte chemotactic activity by the cells. Thus, upon the addition of monocytes to the co-culture, some of these cells transmigrate across the cellular layer. The amount of monocyte transmigration is correlated with the extent of oxidation of the LDL. The oxidation of LDL is probably attributable to the endothelial cell 12-lipoxygenase, as the transfection of endothelial cells with an antisense to the mRNA for this enzyme reduces LDL oxidation and the consequent transmigration of monocytes [13]. The addition and continued presence of normal HDL in the co-culture limits the extent of LDL oxidation and the transmigration of monocytes. At least part of this may be attributable to the presence of paraoxonase 1 (PON1) on the HDL, an enzyme capable of hydrolyzing oxidized lipids [14]. This assay has provided an extremely valuable assessment of HDL "function", independent of its action in promoting reverse cholesterol transport. For illustration, when HDL is obtained from patients with coronary artery disease, the chemotaxis of monocytes is not as notably attenuated as with normal HDL [15]. Thus, the assay can be used to assess the inflammatory action of HDL or the HDL inflammatory index.

While the co-incubation of HDL with LDL in the co-culture system is sufficient to attenuate the oxidation of LDL, lipid-free apoA-I is unable to achieve this outcome unless it is pre-incubated with the culture system and removed before the addition of LDL [12,13]. The fact that to be effective in reducing the promotion of monocyte migration apoA-I must be removed before the addition of LDL indicates that the apoprotein is probably sequestering some product of the co-culture system (but at relatively low affinity) that is ultimately responsible for the oxidation of LDL. The experiment with the 12-lipoxygenase antisense argues that it is likely apoA-I removes an oxidized lipid that promotes the oxidation of LDL. We have described the MCA assay in some detail, as it has been used very successfully to characterize some of the functions of the mimetic peptides.

2. ApoA-I AND THE HISTORY OF ITS DERIVED MIMETIC PEPTIDES

As mentioned earlier, apoA-I is the major apoprotein of essentially all HDL. In humans, it is a 243-amino acid protein, encoded in four exons of its gene. Exon 3 encodes the first 43 amino acids while the fourth exon encodes the remaining 200 amino acids. The fourth exon encoded amino acids are responsible for the majority of the functions of apoA-I, including binding to phospholipid, promoting the efflux of cholesterol via ABCA1 and non-ABCA1 mediated pathways, and the activation of lecithin:cholesterol acyltransferase (LCAT), the plasma enzyme responsible for esterification of cholesterol effluxed to apoA-I and HDL [16]. These 200 amino acids are made up of 10 repeating amphipathic α-helices, eight of which contain 22 amino acids and two contain 11 amino acids. Most of the helices are interrupted at their boundaries by the helix-breaking residue, proline. The majority of the helices are class A amphipathic α-helices with hydrophilic residues on the polar face and hydrophobic residues on the non-polar face. Cationic residues, mostly lysines, are found at the interface between the two faces and anionic residues are at the center of the polar surface of the helix.

This overall structure of apoA-I has formed the basis for the development of the apoA-I mimetic peptides. In order to understand the lipid binding properties of the apoprotein, Segrest and colleagues synthesized peptides corresponding to each of the 22 amino acid amphipathic helices

separately. These eight helices are not equivalent in their phospholipid binding ability [17]. To further understanding of these helices, Segrest developed
a model 18-amino acid peptide [18], which, though representing their average biophysical character, was not identical in sequence to any of the individual helices. This peptide was designated 18A. The modification of the
N-terminus by acetylation and the C-terminus by amidation increases the
peptide helicity, its lipid-binding capacity, and its self-association properties.
Subsequently, it became practice to end block the peptides [19,20]. The 18A
peptide solubilizes phospholipids and activates LCAT. When complexed
with lipids, the lysine side chains interact with the phospholipid acyl chains,
and the $NH3^+$ group extends toward the polar face of the helix. This dispo-
sition of the lysine side chain is known as "snorkeling" [21]. This enhances
the peptide's lipid-binding capacity by allowing for a deeper penetration
into the phospholipid bilayer of the particle. If the distribution of positive
and negative residues is reversed, this is called a reverse 18A peptide. This
change lowers the lipid affinity of the peptide [18]. In addition, the length
of the lysine side chain is important because its replacement with homoami-
noalanine, with a shorter side chain, reduces the peptide's capacity for lipid
binding [22].

The end-blocked 18A peptide is referred to as 2F because it contains
two phenylalanine residues at positions 6 and 18 of the helix (Figure 9.1).
The 2F peptide has been the basis for structural changes that have been
made to enhance the biological activity of the peptide. Variants of the peptide have been fabricated with hydrophobic residues in the helix being
replaced with phenylalanine residues, resulting in peptides with different
number and location of the hydrophobic phenylalanine residues (Figure
9.1). These peptides are designated based on the number of phenylalanines
in the peptide. 4F, which is the peptide most studied, has phenylalanines at
positions 3, 6, 14, and 18.

A number of physical properties of this family of peptides have been
examined, including physical shape (based on modeling), hydrophobicity,
solubilization of phospholipids, intrinsic tryptophan fluorescence, helicity
which is increased in the presence of phospholipid, stabilization, and
quenching of a fluorescent probe 2-(3-(diphenylhexatrienyl)propanoyl)-
1-hexadecanoyl-*sn*-glycero-3-phospholcholine (DPH-PC) [23–25]. Since
these peptides are based on apoA-I structure, biological activities relevant to
the function of apoA-I have also been examined. All of the peptides are
capable of activating LCAT with different efficiencies [23,27]. 4F, 5F, and 6F
are the most anti-inflammatory/antioxidative peptides in the MCA [23].

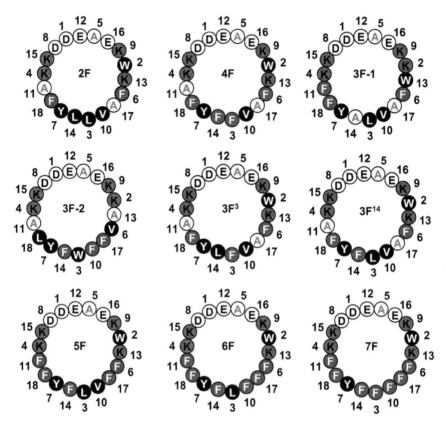

Figure 9.1 *Helical Wheel Representations of the 2F Family of apoA-I Mimetic Peptides.* Hydrophobic residues are black, acidic residues are white with black lettering, basic residues are hatched and the phenylalanine residues (F) are gray.

Unlike apoA-I, these peptides are antioxidative without the need to be removed from the co-culture before the addition of LDL [23,27]. This was one of the first indications that the affinity of at least some of the peptides for oxidation products is significantly higher than that of apoA-I, a suggestion borne out by later experiments that will be discussed as follows.

The positioning of the aromatic phenylalanine residues on the nonpolar face is particularly interesting in relation to the antioxidative properties of the peptide. This is illustrated in the group of 3F variants (3F-1, 3F-2, 3F[3] and 3F[14]), which have their phenylalanines either clustered close to the nonpolar/polar interface or the center of the nonpolar face or are located across the nonpolar face [24]. The proximity of the phenylalanines to the single tryptophan residue also seems to be important. While phenylalanines and tryptophan are mostly positioned at the nonpolar/polar interface, in the case of one of these peptides (3F-2) the tryptophan residue is at the center of the nonpolar face. The biological activity of these 3F peptides with respect to scavenging lipid hydroperoxides from LDL and attenuation of monocyte chemotaxis in the MCA was compared with a number of their physical properties along with that of 4F. The 3F-1 and 3F-2 peptides are as active biologically in these assays as 4F, while the 3F^3 and 3F^{14} variants exhibit very little biological activity [24]. The capacity to solubilize the synthetic phospholipid palmitoyl oleoyl phosphatidylcholine (POPC), the increase in helicity upon binding POPC, and the quenching of DPH-PC fluorescent signal, which is thought to indicate the capacity of the water molecule to penetrate the acyl chain surface, are not correlated with biological activity. On the other hand, tryptophan motional restriction is a property shared by the biologically active 3F-1, 3F-2, and 4F peptides. Thus, a conformational plasticity seems to be correlated with bioactivity.

3. ANTI-ATHEROGENIC EFFECTS OF apoA-I MIMETIC PEPTIDES

The interest in mimetic peptides was greatly stimulated by the observation that 4F, in contrast to 2F [28], significantly reduces atherosclerosis in animal models, including apoE- and LDL receptor (LDLR)-deficient mice [29–34]. 4F is particularly effective in modulating early lesion development but is much less effective on more mature lesions [35]. However, it has been suggested that 4F may synergize with statins in modulating the more advanced lesions [31,33]. The peptide synthesized with natural L-amino acids (L-4F) was effective when administered intraperitoneally, subcutaneously, or intravenously. However, L-4F was not stable when administered orally, presumably due to its susceptibility to proteolysis in the intestinal lumen [29]. This problem was circumvented by fabrication of the peptide with D amino acids. Many of the in vitro physical and biological properties of D-4F and L-4F are similar. In contrast to L-4F, D-4F was anti-atherogenic upon oral administration [29]. In addition, it has been reported that L-4F can be stabilized against proteolysis when fed along with niclosamide [33]. Several other members of the 2F-related peptides have also been shown to attenuate atherosclerosis in animal models (Table 9.2). None of these peptides reduce plasma lipid levels, indicating that other mechanisms account for their anti-atherogenicity.

Table 9.2 Effect of Mimetic Peptides on Atherosclerosis in Animal Models

	Decrease Atherosclerosis	No Effect on Atherosclerosis
ApoA-I-derived mimetics		
	4F	2F
	3F-2	3F[14]
	5F *	4F-Pro-4F
	6F	
	5A	
ApoE-derived mimetics		
	Ac-hE-18A-NH$_2$	Ac-nhE-18A-NH$_2$ (slight increase)
	ATI-5261	
Other apoprotein-derived or general amphipathic peptides		
	mR18L	m18L
	apoJ (113–122) **	
	mSAA2.1 (1–20) †	
	mSAA2.1 (74–103) §	
	KRES [d]	KERS [d]
	FREL [d]	

*Garber DW, Datta G, Chaddha M, Palgunachari MN, Hama SY, Navab M, Fogelman AM, Segrest JP, Anantharamaiah GM. A new synthetic class A amphipathic peptide analogue protects mice from diet-induced atherosclerosis. *J Lipid Res* 2001;42:545-52.

**Navab M, Anantharamaiah GM, Reddy ST, Van Lenten BJ, Wagner AC, Hama S, Hough G, Bachini E, Garber DW, Mishra VK, Palgunachari MN, Fogelman AM. An oral apoJ peptide renders HDL antiinflammatory in mice and monkeys and dramatically reduces atherosclerosis in apolipoprotein E-null mice. *Arterioscler Thromb Vasc Biol* 2005;25:1932-7.

†Tam SP, Ancsin JB, Tan R, Kisilevsky R. Peptides derived from serum amyloid A prevent, and reverse, aortic lipid lesions in apoE−/− mice. *J Lipid Res* 2005;46:2091-101.

§Navab M, Anantharamaiah GM, Reddy ST, Hama S, Hough G, Frank JS, Grijalva VR, Ganesh VK, Mishra VK, Palgunachari MN, Fogelman AM. Oral small peptides render HDL antiinflammatory in mice and monkeys and reduce atherosclerosis in ApoE null mice. *Circ Res* 2005;97:524-32.

As the notable biological effect of the mimetic peptides, especially 4F, is to significantly reduce atherosclerosis, it is not surprising that attention should pass to the cellular mechanisms that may help to account for this anti-atherogenic effect. Currently we do not fully understand the mechanism(s) by which the peptides reduce atherosclerosis. Among the properties of HDL/apoA-I that have been extensively explored with respect to the function of the mimetic peptides are the ability to promote cholesterol efflux (the first step in the process of reverse cholesterol transport) and the attenuation of inflammation and oxidation. In the rest of this chapter we review some of the studies that have been performed in vitro and in vivo to assess the potential anti-atherogenic actions of the peptides.

4. ANTI-INFLAMMATORY PROPERTIES OF apoA-I MIMETIC PEPTIDES WITH CELLS IN CULTURE

4.1. Monohelical Peptides

In evaluating the connection between the potential anti-atherogenic mechanisms and the in vivo influence of the peptides on atherosclerosis, the structure of peptide variants can be quite informative. The ability of 4F to promote cholesterol efflux was noted in several laboratories [23,24,33,36,37], although it is not as effective as lipid-free apoA-I. The efficient promotion of efflux of lipid from cholesterol-loaded cells by lipid-free apoA-I is via its interaction with ABCA1. Its interaction with ABCA1 is believed to promote the translocation of cholesterol from cytoplasmic stores to the plasma membrane followed by the microsolubilization of these lipid-enriched regions of the membrane [38]. The study by Oram and colleagues [36] is of special note as the function of some of the peptides in promoting cholesterol efflux did not correlate with their anti-atherogenic effects in vivo. 2F and 4F were equally effective in stimulating macrophage cholesterol efflux, both in ABCA1-dependent and independent manner. There was no difference in the effect of 2F synthesized from L and D amino acids. 2F, in contrast to 4F, is not anti-atherogenic, suggesting that cholesterol efflux from lesional foam cells may not be a major contributor to the mechanism by which the peptides attenuate atherosclerosis. Unfortunately, we do not yet know whether the bioactive peptides 3F-1 and 3F-2 are distinguished in their promotion of cholesterol efflux from their inactive variants, $3F^3$ and $3F^{14}$.

Oram and colleagues noted additional properties consequent on the interaction of apoA-I and the derived mimetic peptides with ABCA1 on macrophages. ABCA1 is stabilized and JAK2 is auto-phosphorylated, leading to STAT3 activation [36,39]. The activation of the JAK2-STAT3 pathway by apoA-I promotes an anti-inflammatory phenotype in macrophages [39]. The anti-inflammatory function (i.e., STAT3 activation) and cholesterol efflux competence of ABCA1 are independent of one another [39]. This anti-inflammatory effect of the peptides may be an additional mechanism by which the peptides are anti-atherogenic.

4F also influences the function of monocyte-derived macrophages in culture in a fashion similar to apoA-I [40]. Both exhibited a similar efficacy in reducing the expression of CD86, CD11b, and CD11c, surface markers characteristic of activated macrophages, as well as a marked reduction in

CD49d expression, the receptor for VCAM-1. An increase in IL-10 production was noted with both apoA-I and 4F, steering the macrophage toward alternatively activated M2 phenotype. The actions of 4F and apoA-I were not entirely identical since in these experiments 4F increased the expression of fatty acid binding protein 4 by almost 10-fold, an action not shared by apoA-I. Monocyte adhesion to lipopolysaccharide (LPS) stimulated human umbilical vein endothelial cells (HUVECs) was reduced by both apoA-I and 4F, confirming an observation previously made by Gupta and colleagues [41]. The latter study also noted a 4F-mediated reduction in proinflammatory chemokine and cytokine production and adhesion molecule expression by LPS-treated HUVECs, including IL-6, IL-8, IFNγ, TNFα, MCP-1, E- selectin, ICAM 1, and VCAM 1. Taken together these studies suggest that improvement of the inflammatory status of cells in the atherosclerotic lesions may be one mechanism by which the peptides are anti-atherogenic.

4F may also influence chemotaxis of immune cells. In the studies by Smythies and colleagues [40], it was shown that 4F also suppresses the CXCL2 (MIP-1α)-dependent neutrophil chemotaxis, perhaps by decreasing the expression of the CXCR2 chemokine receptor.

Perhaps the most attractive basis for the anti-atherogenic properties of the apoA-I peptides rests on their capacity to bind oxidized lipids with high affinity as demonstrated in vitro. Oxidized lipids are proatherogenic, in part by stimulating inflammatory responses in several different cell types found in the artery wall [42]. Using surface plasmon resonance, Van Lenten and colleagues demonstrated that the 4F peptide has orders of magnitude higher affinity for oxidized lipids than apoA-I [27]. This includes oxidized fatty acids, sterols, and phospholipids. The affinity of 4F for nonoxidized lipids was significantly lower. The high-affinity binding of oxidized lipids by 4F may account for the fact that it does not need to be removed from the MCA assay prior to the addition of LDL to reduce inflammation, unlike apoA-I. This affinity for oxidized lipids is shared with the 3F-2 peptide, which also attenuates atherosclerosis, but not with 3F[14], which is not anti-atherogenic. The affinity for oxidized lipids is also associated with their ability to remove hydroperoxides from LDL and inhibit oxidized phospholipid-induced monocyte chemotaxis [24]. This high-affinity binding of oxidized lipids by the peptides may contribute to the explanation of how these peptides are anti-atherogenic even when present in low concentrations with respect to HDL or apoA-I.

4.2. Tandem apoA-I Peptides

Based upon the repeating pattern of amphipathic helices of apoA-I that is frequently interrupted by proline residues, a pattern that is highly conserved among apoA-I proteins of different species [16], tandem apoA-I mimetic peptides have also be studied. The first tandem symmetrical peptide consisted of two copies of the 18A peptide linked by a proline (37pA). This peptide along with its cousin synthesized from D amino acids is active in cholesterol efflux and in the stabilization of ABCA1 [36,43]. The tandem peptide was more efficient in promoting cholesterol efflux than the monomer [43]. The 37pA tandem peptide also induces neutrophil chemotaxis via the G protein coupled receptor formyl peptide receptor 2, which also binds serum amyloid A (SAA), lipoxin A4, beta amyloid, and HIV envelope protein [44].

Variants of this tandem peptide composed of either two 18A sequences linked by alanine or two 4F sequences linked by proline, alanine, or a seven-amino acid proline-containing interhelical sequence (IHS; derived from the sequence separating helices 4 and 5 of apoA-I) have also been studied [37,43,45]. In the case of the tandem 4F peptides, all are more efficient in promoting cholesterol efflux than is 4F [37]. However, the 4F-Pro-4F tandem peptide did not attenuate atherosclerosis [34], consistent with the lack of an association between the in vitro efflux capacity and anti–atherogenicity of the peptides. The three tandem peptides studied in our laboratories, 4F-Pro-4F, 4F-Ala-4F, and 4F-IHS-4F, are capable of displacing apoA-I but not apoA-II from HDL [45]. The 4F-Pro-4F tandem peptide is better than 4F at binding HDL both in vitro and in vivo. The association of 4F with HDL in vivo is a matter of some controversy.

Sethi and colleagues [46] reported on an asymmetric tandem peptide based on 37pA, in which the second amphipathic helix has alanine residues in place of each of the 5 hydrophobic residues (Figure 9.2). This is designated 5A. The basis for the design of this peptide is that within apoA-I itself, neighboring helices have distinguishable physical properties. The 5A peptide is less hydrophobic than the parent peptide due to the amino acid substitutions in the second helix, but nevertheless is more effective in promoting ABCA1-dependent cholesterol efflux in vitro. The nonspecific efflux by 37pA may be related to its ability to extract lipids from the cell membrane. Both 37pA and 5A are capable of solubilizing lipid, but their effectiveness depends upon the nature of the lipid. 5A solubilizes dimyristoylphosphatidylcholine (DMPC) better than 37pA, but the reverse is the case for a mixture of phospholipids, including some negatively charged

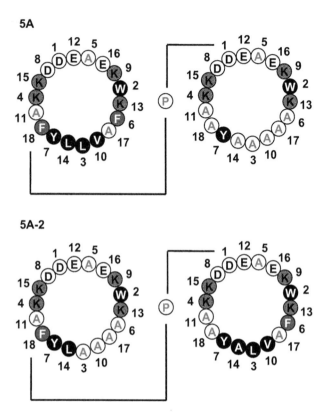

Figure 9.2 *Helical Wheel Representation of 5A Tandem Peptide and Variant.* Hydrophobic residues are black, acidic residues are white with black lettering, basic residues are hatched and the phenylalanine residues (F) are gray.

phospholipids such as are found in cell membranes. 5A complexed with phospholipid has been shown to attenuate TNFα-stimulated endothelial cell expression of ICAM1 and VCAM1 and reduce the production of reactive oxygen species in an ABCA1-dependent process [47]. An interesting variant of 5A involves decreasing the hydrophobicity of the first helix and increasing the hydrophobicity of the second helix, designated 5A-2 (Figure 9.2). The resulting asymmetric peptide has different properties from 5A [46]. For example, 5A-2 in contrast to 5A is not particularly effective in solubilizing phospholipids. The 5A peptide forms an HDL-like particle upon incubation with phospholipids. This peptide–phospholipid complex is almost as effective as the lipid-free 5A peptide in promoting ABCA1 dependent sterol efflux [48]. But this complex, unlike lipid-free 5A, also has the capacity to promote efflux via an ABCG1 process. Increased cholesterol efflux from cells expressing ABCA1 or ABCG1 was also observed using plasma obtained from apoE$^{-/-}$ mice injected with 5A-POPC complexes. The asymmetric peptide 5A, when injected for 13 weeks into 4- or 8-month-old apoE$^{-/-}$ mice maintained on chow, significantly reduced the area of aorta covered with lesions (*en face* analysis of the aorta) [48]. The peptide was injected intraperitoneally three times a week in a complex with either POPC or sphingomyelin and dipalmitoyl phosphatidylcholine. The latter formulation was particularly effective in 4-month-old animals.

More recently, a more detailed structure–function analysis has been conducted using variants of 5A or a tandem peptide containing two canonical type A amphipathic α-helices constructed from only three amino acids—glutamic acid, leucine, and lysine (ELK) [49]. A total of 22 variants were synthesized, varying in hydrophobicity, net charge, and type of α-helix. Their capacity to efflux cholesterol by ABCA1-dependent and independent pathways, to abrogate LDL oxidation, and to attenuate inflammation in monocytes and endothelial cells via measuring the reduction in the expression of CD11b and VCAM1, respectively, was assessed. The optimal structure of the peptide for each of these functions differed. For example, a large hydrophobic face promoted cholesterol efflux, but reduced the anti-inflammatory effect on monocytes. Asymmetry among the helices yielded the most active inhibitors of CD11b expression on monocytes, while increased size of the hydrophobic face was beneficial for inhibition of endothelial cell VCAM1 expression. The optimal peptide concentrations for each of these functional assays were widely different, ranging from 20 μg/mL to 750 μg/mL. For the most part, 5A variants containing cysteine or histidine, amino acids with known antioxidant properties, were relatively inactive in cholesterol efflux, but were active in reducing LDL oxidation. Thus, it appears that different structural features of the mimetic peptides influence these different anti-atherogenic functions. HDL is itself quite versatile in its ability to promote each of these functions. It will be interesting to see which, if any, of these variants of 5A or ELK with different in vitro properties are associated with ability to reduce atherosclerosis.

A series of apoA-I-derived peptides, not strictly mimetics as they involve direct apoA-I-derived sequences, have been tested for lipid efflux properties. It had previously been shown that helix 1 (residues 44–65) and helix 10 (residues 220–241) are the most lipophilic helices [17]. Yet on their own they are inactive in stimulating cholesterol efflux from cells [50]. However, when each is coupled with the 11-amino acid helix, helix 9 (residues 209-219), the resultant 33-amino acid peptides, 1/9 and 9/10, were each as effective as intact apoA-I in promoting cholesterol efflux. They also stabilized ABCA1. Other 33-amino acid tandem peptides derived from apoA-I helices (1/3, 2/9, and 4/9) were not particularly effective in promoting efflux. The precise distribution of charged residues along the 1/9 and 9/10 tandem peptides has an impact on their efflux capacities. Thus, the 10/9 tandem peptide, though having precisely the same amino acid composition, is less effective at effluxing cholesterol than the 9/10 peptide. On the other hand, the 1/9 and 9/1 were more similar in their activities. Peptides based upon the mouse apoA-I sequence were also studied for their ability to promote cholesterol efflux. The sequence of the last helix, helix 10, differs between apoA-I of C57BL/6 and FVB mice. These strains of mice have different susceptibility to the development of atherosclerosis, with the C57BL/6 mice being more susceptible. Amino acids $Q^{225}V^{226}$ in C57BL/6 apoA-I are replaced by $K^{225}A^{226}$ in the protein in FVB mice. The peptides corresponding to the 9/10 helices from these two strains had different cholesterol efflux capabilities, with the C57BL/6 tandem peptide being much more effective than the FVB peptide [51].

5. IN VIVO ANTI-INFLAMMATORY PROPERTIES OF apoA-I-DERIVED MIMETIC PEPTIDES

As mentioned earlier, the most attractive basis for the anti-atherogenic properties of these peptides rests on their capacity to bind oxidized lipids with high affinity as demonstrated in vitro [27]. Several in vivo observations support the antioxidative property of the mimetic peptides. Upon treatment of animals with 4F and related peptides, the levels of oxidized lipids in the plasma and tissues are notably reduced [52,53]. Systemic inflammation is also reduced as measured by the lower plasma levels of SAA [45,54]. Consistent with this is the anti-atherogenic effect of 3F-2, which also binds oxidized lipids, while $3F^{14}$ does not protect against atherosclerosis nor bind oxidized lipids [25]. In addition, we have observed that treatment of apoE$^{-/-}$ mice with 4F increases the plasma titer of a natural antibody to oxidized phospholipid [34] that has been positively associated with reduced atherosclerosis [55].

Endothelial cell dysfunction may contribute to the development of atherosclerosis. 4F exerts a positive influence on vasodilation and vessel wall thickness in hyperlipidemic LDLR$^{-/-}$ and LDLR$^{-/-}$apoA-I$^{-/-}$ mice. In this latter model, D-4F improves endothelial nitric oxide synthase function, though apoA-I containing HDL is required for this action [56].

It is believed that the peptides likely exerted their action by direct effects on lipoproteins in the plasma or on cells in the vessel wall. However, recent studies by Navab and colleagues have offered a new perspective on the action of these peptides [53,54,57]. Peptide concentrations were much higher in the liver and plasma when the same doses of 4F peptide were administered subcutaneously than when it was given orally [57]. On the other hand, the peptide concentration in the intestine and in the feces was comparable regardless of the route of peptide administration. Yet, regardless of the site of administration aortic atherosclerosis, serum SAA levels and serum levels of pro-inflammatory lysophosphatidic acid were equally reduced. This finding is somewhat counterintuitive, as the oral route should result in a higher initial concentration of the peptide being "seen" by the intestine contrasted with the subcutaneous route. This raises two possibilities: either oral D-4F is somehow partially inactivated in the intestinal lumen, or the enterocytes have a saturable binding site for the active peptide, even when presented subcutaneously. This matter requires further attention. Nonetheless, these studies naturally drew attention to the possibility that the intestine may be the major locus of peptide action. The concentration of oxidized fatty acids and derivatives in murine models of atherosclerosis is much higher in the small intestine than in the liver and in other tissues [53].

For therapeutic purposes, the long-term (repeated) oral delivery of peptide presents significant logistic problems. The synthetic production of bioactive peptide, end blocked, and containing D amino acids, would be prohibitively expensive. Fogelman and collaborators have very recently presented a partial resolution of this problem. They found that 6F, when presented orally, is bioactive even when made from L amino acids and not end blocked [58]. The basis for this increased stability and retained activity has yet to be fully explored. End-blocked 6F is more hydrophobic than 4F, more effectively activates LCAT, and is at least as effective in binding oxidized lipids [26,27]. Non-end-blocked L-6F is capable of exerting an anti-atherogenic influence in the LDLR$^{-/-}$ model when incorporated into the high-fat diet at relatively high concentrations. Short-term experiments indicated that 6F enters the enterocytes, but is not detectable in the plasma. If this proves to be the case for long-term treatment with 6F it carries certain implications for the basis of the anti-atherogenic activity of this and probably related members of the mimetic peptide family. Its action is unlikely to be the direct result of the promotion of reverse cholesterol transport, of modulation of the anti-inflammatory properties of HDL, or modulation of the homeostasis of lesional macrophages. 6F treatment results in the reduction in plasma SAA, a reduction in oxidized fatty acids in plasma and enterocytes, and in unsaturated lysophosphatidic acid in the plasma, all of which proved to be correlated with the reduction in lesions. An increase in plasma HDL was also noted, although whether this is based upon a stimulation of the enterocyte contribution to HDL biogenesis remains to be explored. A likely mechanism is that the 6F peptide reduces the enterocyte oxidation pressure, which in turn is translated to a reduction in systemic and vessel wall oxidation pressure [59]. Such a hypothesis could account for many of the observations made in this recent report.

Perhaps the most novel feature of this report is that *Agrobacterium tumefaciens* bacteria-mediated transformation was used to generate transgenic tomato plants expressing the 6F peptide [58]. The consumption of the transgenic plants proved capable of delivering a sufficient level of 6F peptide to reduce atherosclerosis to the animals. Importantly, the raw plants, as freeze-dried tomatoes, could be delivered in the diet without any need for peptide isolation or purification. Though an exciting finding, much further work is required to elucidate all of the properties of 6F, the basis for its stability, and its anti-atherogenic activity.

If the capacity of binding highly bioactive lipids is a/the major basis for the efficacy of the apoA-I mimetic peptides, it is not surprising that they should also improve many other inflammatory pathologies, especially those that depend in part on reactive oxygen species, including oxidized lipids. Indeed, the 4F peptide has been shown to provide benefit in a variety of diseases—influenza, pneumonia, scleroderma, Type 1 and Type 2 diabetes, hepatic fibrosis, hyperlipidemia-induced renal inflammation hyperlipidemia and sickle cell–induced vascular dysfunction, arthritis, and Alzheimer's pathology [60–65]. It has also been shown to decrease airway hyperresponsiveness in a murine model of asthma [66]. The 5A peptide has also been shown to attenuate house dust mite-induced asthma [67], although it has not yet been determined if this peptide binds oxidized lipids. Both 4F and 5F reduce tumor burden in an ovarian tumor model [68], perhaps by reducing tumor cellular oxidative stress [69].

The 4F peptide also reduces LPS-induced endotoxemia [70] and polymicrobial sepsis induced by cecal ligation puncture [71]. This may, in part, be due to the regulation of Toll-like receptors by 4F and leading to reduction in nuclear factor-kappa B (NF-κB) translocation to the nucleus and secretion of inflammatory cytokines by macrophages [72]. This effect is correlated with reductions in cholesterol and caveolin in membrane lipid rafts.

6. HYBRID AND apoE MIMETICS

6.1. Single-Domain Peptides

Among the most abundant HDL apoproteins are apoproteins A-I, A-II, and E. ApoE is a multifunctional protein that associates with VLDL, LDL, and subsets of HDL. It participates in the clearance of these lipoproteins from the plasma, especially the apoB-containing VLDL and LDL, by serving as ligand for the LDL receptor and its family of related receptors. The binding of apoE to proteoglycans is believed to facilitate its ligand function and perhaps also other functions of the protein. Like apoA-I, it is also active in reverse cholesterol transport and has anti-inflammatory and antioxidative activity [73]. Its expression is upregulated in cholesterol-loaded macrophages, and this facilitates the efflux of cholesterol from the cells. These properties are believed to contribute to the anti-atherogenic function of apoE. Within the human population are three major apoE isoforms, apoE2, E3, and E4. ApoE2 and apoE4 differ by a single amino acid from apoE3, the most prevalent isoform. ApoE is composed of repeating amphipathic α-helices. However, unlike apoA-I, the majority of the helices are not separated by proline residues nor are the majority of the helices class A amphipathic helices.

Mimetic peptides derived from the sequence of apoE have been developed. The C-terminal region of apoE (residues 222–299) can duplicate the cholesterol efflux capacity of full-length apoE [74]. The generation of peptides corresponding to the helices in the C-terminal domain indicates that interaction among the neighboring class A and class G amphipathic helices contributes in a major way to the efflux competence of domain. One of the class A helices in this domain (residues 238–266) has been modified by Bielicki and colleagues to yield a 25-amino acid peptide designated AT1-5261 [75]. This peptide differs from its parent sequence by being more hydrophobic, having increased α-helicity and a more acidic polar face. In the lipid-free state ATI-5261 efficiently promotes ABCA1-mediated cholesterol efflux, with a K_m molar efficiency and V_{max} similar to intact apoA-I. Thus, unlike the situation with apoA-I mimetic peptides, this apoE-based peptide was an effective promoter of cholesterol efflux without being composed of tandem repeats. When the peptide was complexed with phospholipid, it was still capable of promoting cholesterol efflux in a partially ABCA1-dependent fashion, though it has a relatively high K_m indicating lower efficiency. The peptide is anti-atherogenic. It decreased aortic lesions in LDLR$^{-/-}$ mice fed a high-fat, high-cholesterol diet for 13 weeks and treated with lipid-free peptide only for the last 6 weeks. It also reduced lesions in the aorta and the aortic sinus of apoE$^{-/-}$ mice fed a high-fat diet for 18 weeks and then shifted to chow diet for 6 weeks during which the animals were treated with lipid-free or phospholipid complexed peptide.

A second apoE derived single-domain peptide has been studied. It is a derivative of residues 138–149 of apoE in which the arginine residues at positions 140 and 145 are replaced with aminoisobutyric acid (Aib), generating the peptide Ac-AS(Aib)LRKL(Aib)KRLL-NH$_2$. The inclusion of Aib enhances the helical structure of the peptide. This peptide, designated COG1410, has been studied in comparison with a control peptide lacking the last two leucine residues (designated peptide 264) [44]. With the intrapulmonary exposure to CXCL1, there is an influx of neutrophils that is attenuated by COG1410, but not by the control peptide 264. This peptide has been studied extensively with respect to its ability to reduce neurological inflammation and sepsis [76,77] but has not yet been studied in atherosclerotic models.

An additional class of amphipathic helical peptides derived from class L (lytic) model peptides has also been studied for their effect on atherosclerosis. These amphipathic helices have a broader hydrophobic face and have positive residues, mostly lysines, in the center of the polar face of the helix [78]. Based on modeling, these peptides are believed to form an inverted wedge. The cationic class L peptide m18L and a further modified peptide in which the lysine (K) residues on the polar face of m18L are substituted with arginine (R) residues (mR18L) have been studied. This K→R substitution reduces the lytic activity of the peptide, despite the fact that the non-polar face is similar in both peptides. Short-term treatment of apoE$^{-/-}$ mice with mR18L resulted in a more significant reduction of total cholesterol than m18L, and this reduction by both peptides was minimized by pretreatment with heparinase [79]. However, when incorporated into the diet for 6 weeks there was a modest reduction of total cholesterol by mR18L, but not by m18L, and a reduction in aortic root lesion formation by mR18L.

6.2. Dual-Domain Peptides

The LDL receptor-binding domain and a major proteoglycan-binding domain of apoE had been identified as residues 141–150. This sequence was coupled with the class A amphipathic helix 18A (2F). The rationale for the construction of this dual-domain peptide (designated Ac-hE18A-NH$_2$) was that the 18A peptide will promote the association of the non-lipid associating apoE ligand-binding sequence with lipoproteins. The properties of this peptide have been compared to a control peptide (Ac-nhE18A-NH$_2$), in which the apoE sequence 151–160 is coupled to 18A. The dual domain peptide Ac-hE18A-NH$_2$ lowers plasma cholesterol in hyperlipidemic mouse models, while Ac-nhE18A-NH$_2$ and 18A on their own do not [80,81]. The cholesterol-lowering effect is thought to be attributable to the ligand-binding domain of apoE in the dual-domain peptide facilitating the clearance of lipoproteins. Indeed, in vitro Ac-hE18A-NH$_2$, but not Ac-nhE18A-NH$_2$, facilitated LDL uptake by HepG2 cells via a heparan sulfate–dependent pathway [82]. In in vitro studies, Ac-hE18A-NH$_2$ also decreased monocyte adhesion to bovine aortic endothelial cells in culture, attenuated LPS-induced inflammatory responses of HUVECs, and reduced lipid hydroperoxides in LDL [81,82], consistent with anti-inflammatory and antioxidative properties. Interestingly, Ac-nhE18A-NH$_2$ increased LDL hydroperoxides [81]. When these peptides were administered intravenously into apoE$^{-/-}$ mice for 6 weeks, three times per week, Ac-hE18A-NH$_2$ modestly reduced aortic sinus lesion, while the Ac-nhE18A-NH$_2$ peptide actually increased lesion area, perhaps related to its effect on LDL hydroperoxides. In this experiment, total cholesterol and triglyceride was decreased by Ac-hE18A-NH$_2$ due primarily to a reduction in VLDL levels, and there was an increase in paraoxonase antioxidant activity in the plasma. On the other hand, Ac-nhE18A-NH$_2$ had no effect on lipoproteins and actually slightly decreased paraoxonase activity [81]. A more robust effect of Ac-hE18A-NH$_2$ on atherosclerosis was observed in the *en face* analysis of mice treated with peptide for 4 weeks. The neutral effects of Ac-nhE18A-NH$_2$ are mostly understandable, but not its increase of atherosclerosis and decline of paraoxonase activity.

In two recent studies, the Ac-hE18A-NH$_2$ apoE dual-domain peptide has been compared with 4F on the one hand and with mR18L on the other for their effects on atherosclerosis and plasma lipids. In the first study [83], both Ac-hE18A-NH$_2$ and 4F peptides are shown to promote cholesterol efflux, to improve endothelial cell function, and to lower plasma lipid hydroperoxides. But they do have unique functions. 4F binds oxidized lipid with high affinity, while Ac-hE18A-NH$_2$ rapidly reduces plasma cholesterol levels, involving lowering of VLDL and LDL levels, though they gradually rise over time [80]. When Ac-hE18A-NH$_2$ was administered intravenously and 4F intraperitoneally 2–3 times a week once the apoE$^{-/-}$ mice were expected to have established lesions, both reduced atherosclerosis, as measured by *en face* analysis of the aorta, to the same extent. However, when both were administered intravenously at the same dose and frequency, the

Ac-hE18A-NH$_2$ was more effective than 4F, perhaps at least partly attributable to the reduction of the plasma lipids by the apoE dual-domain peptide. Both peptides were equally effective in reducing plasma hydroperoxides and HDL inflammatory indices. This is the first study that compares, in the same animal model and in the same laboratory, an apoE and an apoA-I mimetic peptide [84].

In a similar comparison by the same group, Ac-hE18A-NH$_2$ and mR18L were given to LDLR$^{-/-}$ mice [85]. There was a greater reduction of plasma total cholesterol by mR18L than Ac-hE18A-NH$_2$ peptide following the acute administration of a single dose. However with multiple doses administered intravenously twice a week for 8 weeks, both peptides limited the increase of plasma cholesterol induced by a high-fat, high-cholesterol diet in LDLR$^{-/-}$ mice. The peptides affected VLDL, LDL, and HDL levels. Reactive oxygen species in the plasma are more substantially reduced by the Ac-hE18A-NH$_2$ peptide than the mR18L peptide. But both significantly reduced atherosclerosis measured at either the aortic root or the whole aorta (*en face*), with Ac-hE18A-NH$_2$ being more effective.

6.3. Other Apoprotein-Related Mimetic Peptides

There are peptides other than those mimicking apoA-I and apoE that influence lipoprotein homeostasis and atherosclerosis. These include apoJ-derived peptides, short synthetic peptides, and SAA-derived peptides. These have been briefly discussed in our recent review [86].

7. CONCLUSION

Here we have reviewed the structure and pathophysiological function of apoA-I mimetic and apoE mimetic peptides, as well as variants and hybrids of these two peptide families. Many of the peptides attenuate atherogenesis, particularly early atherogenesis. The properties they generally exhibit are lipid association, promotion of cholesterol efflux, anti-inflammatory and antioxidative activities, and reduction in hepatic inflammation as manifested by reduced plasma levels of SAA. Some of the peptides also promote lipoprotein clearance. Any or all of these functional properties, either alone or in combination, could account for the attenuation of atherogenesis. The mechanism by which the peptides are anti-atherogenic may not be identical. It is clear, for example, from the studies of D'Souza and colleagues [49] that different peptide variants have different optima for different in vitro or in culture assays that could relate to atherogenesis. These variants have yet to be studied in vivo to ascertain which function most closely relates to the protection against atherogenesis.

In the in vivo experimental protocols that have been employed to assess the anti-atherogenic potential of the mimetic peptides, there are many differences, including the animal model; the variations in structure of the peptide; and the dose, route, frequency, and duration of treatment. Recent studies comparing comparable doses of two peptides administered by the same route at the same frequency for the same duration to the same animal model conducted in the same laboratory are encouraging [84]. The stability of the peptide and its pharmacodynamics in vivo require much further work. The recent protocol for oral administration of stable peptide offers potential as an effective means of delivering sufficient peptides to animal models for investigation of mechanism of action and for the delivery of peptides to humans as potential therapeutic agents [58].

Consideration needs to be given to each of the features of the experiments and properties of the peptides to arrive at optimal treatment protocols. The best design will depend upon the more complete understanding of the mechanisms involved in the attenuation of atherogenesis by the various mimetic peptides. This will probably require careful genetic modulation of the putative mechanisms. Such studies are ongoing in several laboratories, including our laboratory.

REFERENCES

[1] Gordon DJ, Probstfield JL, Garrison RJ, Neaton JD, Castelli WP, Knoke JD, et al. High-density lipoprotein cholesterol and cardiovascular disease. Four prospective American studies. Circulation 1989;79:8–15.

[2] Vaisar T, Pennathur S, Green PS, Gharib SA, Hoffnagle AN, Cheung MC, et al. HDL proteomics: shotgun proteomics implicates protease inhibition and complement activation in the antiinflammatory properties of HDL. J Clin Invest 2007;117:746–9.

[3] Davidson WS, Silva RA, Chantepie S, Lagor WR, Chapman MJ, Konush A. Proteomic analysis of defined HDL subpopulations reveals particle-specific protein clusters: relevance to antioxidative function. Arterioscler Thromb Vasc Biol 2009;29:870–6.

[4] Plump AS, Scott CJ, Breslow JL. Human apolipoprotein A-I gene expression increases high density lipoprotein and suppresses atherosclerosis in apolipoprotein E-deficient mice. Proc Natl Acad Sci USA 1994;91:9607–11.

[5] Duverger N, Kruth H, Emmanuel F, Caillaud JM, Viglietta C, Castro G, et al. Inhibition of atherosclerosis development in cholesterol-fed human apolipoproteins A-I transgenic rabbits. Circulation 1996;94(4):713–7.

[6] Voyiazikis E, Goldberg IJ, Plump AS, Rubin EM, Breslow JL, Huang LS. ApoA-I deficiency causes both hypertriglyceridemia and increased atherosclerosis in human apoB transgenic mice. J Lipid Res 1998;39:313–21.

[7] deGoma EM, Rader DJ. Novel HDL-directed pharmacotherapeutic strategies. Nat Rev Cardiol 2011;8:266–77.

[8] Heinecke JW. The not-so-simple HDL story: a new era for quantifying HDL and cardiovascular risk? Nat Med 2012;18:1346–7.

[9] Rothblat GH, de la Llera-Moya M, Favari E, Yancey PG, Kellner-Weibel G. Cellular cholesterol flux studies: methodological considerations. Atherosclerosis 2002;163:1–8.

[10] Sankaranarayanan S, Kellner-Weibel G, de la Llera-Moya M, Phillips MC, Asztalos BF, Bittman R, et al. A sensitive assay for ABCA1-mediated cholesterol efflux using BODIPY-cholesterol. J Lipid Res 2011;52:2332–40.

[11] Zhang Y, Zanotti I, Reilly MP, Glick JM, Rothblat GH, Rader DJ. Overexpression of apolipoproteins A-I promotes reverse cholesterol transport from macrophages to feces in vivo. Circulation 2003;108:661–3.

[12] Navab M, Imes SS, Hama SY, Hough GP, Ross LA, Bork RW, et al. Monocyte transmigration induced modification of low density lipoprotein in cocultures of human aortic wall cells is due to induction of monocyte chemotactic protein 1 synthesis and is abolished by high density lipoprotein. J Clin Invest 1991;88:2039–46.

[13] Navab M, Hama SY, Anantharamaiah GM, Hassan K, Hough GP, Watson AD, et al. Normal high density lipoprotein inhibits three steps in the formation of mildly oxidized low density lipoprotein: steps 2 and 3. J Lipid Res 2000;41:1495–508.

[14] Mackness B, Mackness M. Anti-inflammatory properties of paraoxonase-1 in atherosclerosis. Adv Exp Med Biol 2010;660:143–51.

[15] Ansell BJ, Navab M, Hama S, Kamranpour N, Fonarow G, Hough G, et al. Inflammatory/anti-inflammatory properties of high-density lipoprotein distinguish patients from control subjects better than high-density lipoprotein cholesterol levels and are favorably affected by simvastatin treatment. Circulation 2003;108:2751–6.

[16] Frank PG, Marcel YL, Apolipoprotein A-I. structure-function relationships. J Lipid Res 2000;41:853–72.

[17] Palgunachari MN, Mishra VK, Lund-Katz S, Phillips MC, Adeyeye SO, Alluri S, et al. Only the two end helixes of eight tandem amphipathic helical domains of human apoA-I have significant lipid affinity. Implications for HDL assembly. Arterioscler Thromb Vasc Biol 1996;16:328–38.

[18] Anantharamaiah GM, Jones JL, Brouillette CG, Schmidt CF, Chung BH, Hughes TA, et al. Studies of synthetic peptide analogs of the amphipathic helix. Structure of complexes with dimyristoyl phosphatidycholine. J Biol Chem 1985;260:10248–55.

[19] Yancey PG, Bielicki JK, Johnson WJ, Lund-Katz S, Palgunachari MN, Anantharamaiah GM, et al. Efflux of cellular cholesterol and phospholipid to lipid-free apolipoproteins and class A amphipathic peptides. Biochemistry 1995;34:7955–65.

[20] Venkatachalapathi YV, Phillips MC, Epand RM, Epand RF, Tytler EM, Segrest JP, et al. Effect of end group blockage on the properties of class A amphipathic helical peptides. Proteins 1993;15:349–59.

[21] Anantharamaiah GM, Brouillette CG, Engler JA, De Loof H, Venkatachalapathi YV, Boogaerts J, et al. Role of amphipathic helixes in HDL structure/function. Adv Exp Med Biol 1991;285:131–40.

[22] Mishra VK, Palgunachari MN, Segrest JP, Anantharamaiah GM. Interactions of synthetic peptide analogs of the class A amphipathic helix. J Biol Chem 1994;269:1785–91.

[23] Datta G, Chaddha M, Hama S, Navab M, Fogelman AM, Garber DW, et al. Effects of increasing hydrophobicity on the physical-chemical and biological properties of a class A amphipathic helical peptide. J Lipid Res 2001;42:1096–104.

[24] Datta G, Epand RF, Epand RM, Chaddha M, Kirksey MA, Garber DW, et al. Aromatic residue position on the nonpolar face of class A amphipathic helical peptides determines biological activity. J Biol Chem 2004;279:26509–17.

[25] Handattu SP, Garber DW, Horn DC, Hughes DW, Berno B, Bain AD, et al. ApoA-I mimetic peptides with differing ability to inhibit atherosclerosis also exhibit differences in their interactions with membrane bilayers. J Biol Chem 2007;282:1980–8.

[26] Anantharamaiah GM, Mishra VK, Garber DW, Datta G, Handattu SP, Palgunachari MN, et al. Structural requirements for antioxidative and anti-inflammatory properties of apolipoprotein A-I mimetic peptides. J Lipid Res 2007;48:1915–23.

[27] Van Lenten BJ, Wagner AC, Jung CL, Ruchalal P, Waring AJ, Lehrer RI, et al. Anti-inflammatory apoA-I-mimetic peptides bind oxidized lipids with much higher affinity than human apoA-I. J Lipid Res 2008;49:2302–11.

[28] Navab N, Anantharamaiah GM, Reddy ST, Hama S, Hough G, Grijalva VR, et al. Apolipoprotein A-I mimetic peptides. Arterioscler Thromb Vasc Biol 2005;25:1325–31.

[29] Navab M, Anantharamaiah GM, Hama S, Garber DW, Chaddha M, Hough G, et al. Oral administration of an apoA-I mimetic peptide synthesized from D-amino acids dramatically reduces atherosclerosis in mice independent of plasma cholesterol. Circulation 2002;105:290–2.

[30] Li X, Chyu KY, Faria Neto JR, Yano J, Nathwani N, Ferreira C, et al. Differential effects of apolipoprotein A-I-mimetic peptide on evolving and established atherosclerosis in apolipoprotein E-null mice. Circulation 2004;110:1701–5.

[31] Navab M, Anantharamaiah GM, Hama S, Hough G, Reddy ST, Frank JS, et al. D-4F and statins synergize to render HDL antiinflammatory in mice and monkeys and cause lesion regression in old apolipoprotein E-null mice. Arterioscler Thromb Vasc Biol 2005;25:1426–32.

[32] Van Lenten BJ, Wagner AC, Navab M, Anantharamaiah GM, Hama S, Reddy ST, et al. Lipoprotein inflammatory properties and serum amyloid A levels but not cholesterol levels predict lesion area in cholesterol-fed rabbits. J Lipid Res 2007;48:2344–53.

[33] Navab M, Ruchala P, Waring AJ, Lehrer RI, Hama S, Hough G, et al. A novel method for oral delivery of apolipoprotein mimetic peptides synthesized from all L-amino acids. J Lipid Res 2009;50:1538–47.

[34] Wool GD, Cabana VG, Lukens J, Shaw PX, Binder CJ, Witztum JL, et al. 4F peptide reduces atherosclerosis and induces natural antibody production in apolipoproteins E-null mice. FASEB J 2011;25:290–300.

[35] Getz GS, Wool GD, Reardon CA. HDL apolipoprotein-related peptides in the treatment of atherosclerosis and other inflammatory disorders. Curr Pharm Des 2010; 16:3173–84.

[36] Tang C, Vaughan AM, Anantharamaiah GM, Oram JF. Janus kinase 2 modulates the lipid-removing but not protein stabilizing interactions of amphipathic helices with ABCA1. J Lipid Res 2006;47:107–14.

[37] Wool GD, Reardon CA, Getz GS. Apolipoprotein A-I mimetic peptide helix number and helix linker influence potentially anti-atherogenic properties. J Lipid Res 2008;49:1268–83.

[38] Vedhachalam C, Duong PT, Nickel M, Nguyen D, Dhanasekaran P, Saito H, et al. Mechanism of ATP-binding cassette transporter A1-mediated cellular lipid efflux to apolipoprotein A-I and formation of high density lipoprotein particles. J Biol Chem 2007;282:25123–30.

[39] Tang C, Liu Y, Kessler PS, Vaughan AM, Oram JF. The macrophage cholesterol exporter ABCA1 functions as an anti-inflammatory receptor. J Biol Chem 2009;284:32336–43.

[40] Smythies LE, White CR, Maheshwari A, Palgunachari MN, Anantharamaiah GM, Chaddha M, et al. Apolipoprotein A-I mimetic 4F alters the function of human monocyte-derived macrophages. Am J Physiol Cell Physiol 2010;298:C1538–48.

[41] Gupta H, Dai L, Datta G, Garber DW, Grenett H, Li Y, et al. Inhibition of lipopolysaccharide-induced inflammatory responses by an apolipoprotein AI mimetic peptide. Circ Res 2005;97:236–43.

[42] Navab M, Berliner JA, Subbanagounder G, Hama S, Lusis AJ, Castellani LW, et al. HDL and the inflammatory response induced by LDL-derived oxidized phospholipids. Arterioscler Thromb Vasc Biol 2001;21:481–8.

[43] Mendez AJ, Anantharamaiah GM, Segrest JP, Oram JF. Synthetic amphipathic helical peptides that mimic apolipoprotein A-I in clearing cellular cholesterol. J Clin Invest 1994;94:1698–705.

[44] Madenspacher JH, Azzam KM, Gong W, Gowdy KM, Vitek MP, Laskowitz DT, et al. Apolipoproteins and apolipoprotein mimetic peptides modulate phagocyte trafficking through chemotactic activity. J Biol Chem 2012;287:43730–40.

[45] Wool GD, Vaisar T, Reardon CA, Getz GS. An apoA-I mimetic peptide containing a proline residue has greater in vivo HDL binding and anti-inflammatory ability than the 4F peptide. J Lipid Res 2009;50:1889–900.

[46] Sethi AA, Stonik JA, Thomas F, Demosky SJ, Amar M, Neufeld E, et al. Asymmetry in the lipid affinity of bihelical amphipathic peptides. A structural determinant for the specificity of ABCA1-dependent cholesterol efflux by peptides. J Biol Chem 2008; 283:32273–82.

[47] Tabet F, Remaley AT, Segaliny AI, Millet J, Yan L, Nakhal S, et al. The 5A apolipoprotein A-I mimetic peptide displays anti-inflammatory and antioxidant properties in vivo and in vitro. Arterioscler Thromb Vasc Biol 2010;30:246–52.

[48] Amar MJ, D'Souza W, Turner S, Demosky S, Sviridov D, Stonik J, et al. 5A apolipoproteins mimetic peptide promotes cholesterol efflux and reduces atherosclerosis in mice. J Pharmacol Exp Ther 2010;334:634–41.

[49] D'Souza W, Stonik JA, Murphy A, Demosky SJ, Sethi AA, Moore XL, et al. Structure/function relationships of apolipoprotein A-I mimetic peptides: implications for antiatherogenic activities of high-density lipoprotein. Circ Res 2010;107:217–27.

[50] Natarajan P, Forte RM, Chu B, Phillips MC, Oram JF, Bielicki JK. Identification of an apolipoprotein A-I structural element that mediates cellular cholesterol efflux and stabilizes ATP binding cassette transporter A1. J Biol Chem 2004;279:24044–52.

[51] Sontag TJ, Carnemolla R, Vaisar T, Reardon CA, Getz GS. Naturally occurring variant of mouse apolipoprotein A-I alters the lipid and HDL association properties of the protein. J Lipid Res 2012;53:951–63.

[52] Imaizumi S, Grijalva V, Navab M, Van Lenten BJ, Wagner AC, Anantharamaiah GM, et al. L-4F differentially alters plasma levels of oxidized fatty acids resulting in more anti-inflammatory HDL in mice. Drug Metab Lett 2010;4: 139–118.

[53] Navab M, Reddy ST, Anantharamaiah GM, Hough G, Buga GM, Danciger J, et al. D-4F-mediated reduction in metabolites of arachidonic and linoleic acids in the small intestine is associated with decreased inflammation in low-density lipoprotein receptor-null mice. J Lipid Res 2012;53:437–45.

[54] Navab M, Reddy ST, Van Lenten BJ, Buga GM, Hough G, Wagner AC, et al. High density lipoprotein and 4F peptide reduce systemic inflammation by modulating intestinal oxidized lipid metabolism: novel hypotheses and review of literature. Arterioscler Thromb Vasc Biol 2012;32:2553–60.

[55] Binder CJ, Chou MY, Fogelstrand L, Hartvigsen K, Shaw PX, Boullier A, et al. Natural antibodies in murine atherosclerosis. Curr Drug Targets 2008;9:190–5.

[56] Ou J, Wang J, Xu H, Ou Z, Sorci-Thomas MG, Jones DW, et al. Effects of D-4F on vasodilation and vessel wall thickness in hypercholesterolemic LDL receptor-null and LDL receptor/apolipoprotein A-I double-knockout mice on Western diet. Circ Res 2005;97:1190–7.

[57] Navab M, Reddy ST, Anantharamaiah GM, Imaizumi S, Hough G, Hama S, et al. Intestine may be a major site of action for the apoA-I mimetic peptide 4F whether administered subcutaneously or orally. J Lipid Res 2011;52:1200–10.

[58] Chattopadhyay A, Navab M, Hough G, Gao F, Meriwether D, Grijalva V, et al. A novel approach to oral apoA-I mimetic therapy. J Lipid Res 2013;54:995–1010.

[59] Getz GS, Reardon CA. Apo-I mimetics: tomatoes to the rescue. J Lipid Res 2013;54:878–80.

[60] Navab M, Anantharamaiah GM, Fogelman AM. The effect of apolipoprotein mimetic peptides in inflammatory disorders other than atherosclerosis. Trends Cardiovasc Med 2008;18:61–6.

[61] Van Lenten BJ, Wagner AC, Anantharamaiah GM, Navab M, Reddy ST, Buga GM, et al. Apolipoprotein A-I mimetic peptides. Curr Atheroscler Rep 2009;11:52–7.

[62] Van Lenten BJ, Wagner AC, Navab M, Anantharamaiah GM, Hui EKW, Nayak DP, et al. D-4F, an apolipoprotein A-I mimetic peptide, inhibits the inflammatory response induced by influenza A infection of human Type II pneumocytes. Circulation 2004;110:3252–8.

[63] DeLeve LD, Wang X, Kanel GC, Atkinson RD, McCuskey RS. Prevention of hepatic fibrosis in a murine model of metabolic syndrome with nonalcoholic steatohepatitis. Am J Pathol 2008;173:993–1001.

[64] Ou J, Ou Z, Jones DW, Holzhauer S, Hatoum OA, Ackerman AW, et al. L-4F, an apolipoprotein A-1 mimetic, dramatically improves vasodilation in hypercholesterolemia and sickle cell disease. Circulation 2003;107:2337–41.

[65] Handattu SP, Garber DW, Monroe CE, van Groen T, Kadish I, Nayyar G, et al. Oral apolipoprotein A-I mimetic peptide improves cognitive function and reduces amyloid burden in a mouse model of Alzheimer's disease. Neurobiol Dis 2009;34:525–34.

[66] Nandekar SD, Weihrauch D, Xu H, Shi Y, Feroah T, Hutchins W, et al. D-4F, an apoA-I mimetic, decreases airway hyperresponsiveness, inflammation, and oxidative stress in a murine model of asthma. J Lipid Res 2011;52:499–508.

[67] Yao X, Dai C, Fredriksson K, Dagur PK, McCoy JP, Qu X, et al. 5A, an apolipoproteins A-I mimetic peptide, attenuates the induction of house mite-induced asthma. J Immunol 2011;186:576–83.

[68] Su F, Kozak KR, Imaizumi S, Gao F, Amneus MW, Grijalva V, et al. Apolipoprotein A-I (apoA-I) and apoA-I mimetic peptides inhibit tumor development in a mouse model of ovarian cancer. Proc Natl Acad Sci USA 2010;107:19997–20002.

[69] Ganapathy E, Su F, Meriwether D, Devarajan A, Grijalva V, Gao F, et al. D-4F, an apoA-I mimetic peptide, inhibits proliferation and tumorigenicity of epithelial ovarian cancer cells by upregulating the antioxidant enzyme MnSOD. Int J Cancer 2012;130: 1071–81.

[70] Dai L, Datta G, Zhang Z, Gupta H, Patel R, Honavar J, et al. The apolipoprotein A-I mimetic peptide 4F prevents defects in vascular function in endotoxemic rats. J Lipid Res 2010;51:2695–705.

[71] Zhang Z, Datta G, Zhang Y, Miller AP, Mochon P, Chen Y-F, et al. Apolipoprotein A-I mimetic peptide treatment inhibits inflammatory responses and improves survival in septic rats. Am J Physiol Heart Circ Physiol 2009;297:H866–73.

[72] White CR, Smythies LE, Crossman DK, Palgunachari MN, Anantharamaiah GM, Datta G. Regulation of pattern recognition receptors by the apolipoprotein A-I mimetic peptide 4F. Arterioscler Thromb Vasc Biol 2012;32:2631–9.

[73] Getz GS, Reardon CA. Apoprotein E as a lipid transport and signaling protein in the blood, liver and artery wall. J Lipid Res 2009;50:S156–61.

[74] Vedhachalam C, Narayanaswami V, Neto N, Forte TM, Phillips MC, Lund-Katz S, et al. The C-terminal lipid-binding domain of apolipoprotein E is a highly efficient mediator of ABCA1-dependent cholesterol efflux that promotes the assembly of high-density lipoproteins. Biochemistry 2007;46:2583–93.

[75] Bielicki JK, Zhang H, Cortez Y, Zheng Y, Narayanaswami V, Patel A, et al. A new HDL mimetic peptide that stimulates cellular cholesterol efflux with high efficiency greatly reduces atherosclerosis in mice. J Lipid Res 2010;51:1496–503.

[76] Laskowitz DT, Lei B, Dawson HN, Wang H, Bellows ST, Christensen DJ, et al. The apoE-mimetic peptide, COG1410, improves functional recovery in a murine model of intracerebral hemorrhage. Neurocrit Care 2012;16:316–26.

[77] Wang H, Christensen DJ, Vitek MP, Sullivan PM, Laskowitz DT. APOE genotype affects outcome in a murine model of sepsis: implications for a new treatment strategy. Anaesth Intensive Care 2009;37:38–45.

[78] Tytler EM, Segrest JP, Epand RM, Nie SQ, Epand RF, Mishra VK, et al. Reciprocal effects of apolipoprotein and lytic peptide analogs on membranes. Cross-sectional molecular shapes of amphipathic alpha helixes control membrane stability. J Biol Chem 1993;268:22112–8.

[79] Handattu SP, Datta G, Epand RM, Epand RF, Palgunachari MN, Mishra VK, et al. Oral administration of L-mR18L, a single domain cationic amphipathic helical peptide, inhibits lesion formation in ApoE null mice. J Lipid Res 2010;51:3491–9.

[80] Garber DW, Handattu S, Aslan I, Datta G, Chaddha M, Anantharamaiah GM. Effect of an arginine-rich amphipathic helical peptide on plasma cholesterol in dyslipidemic mice. Atherosclerosis 2003;168:229–37.

[81] Nayyar G, Handattu SP, Monroe CE, Chaddha M, Datta G, Mishra VK, et al. Two adjacent domains (141-150 and 151-160) of apoE covalently linked to a class A amphipathic helical peptide exhibit opposite atherogenic effects. Atherosclerosis 2010;213:449–57.

[82] Datta G, Garber DW, Chung BH, Chaddha M, Dashti N, Bradley WA, et al. Cationic domain 141-150 of apoE covalently linked to a class A amphipathic helix enhances atherogenic lipoprotein metabolism in vitro and in vivo. J Lipid Res 2001;42:959–66.

[83] Nayyar G, Garber DW, Palgunachari MN, Monroe CE, Keenum TD, Handattu SP, et al. Apolipoprotein E mimetic is more effective than apolipoprotein A-I mimetic in reducing lesion formation in older female apoE null mice. Atherosclerosis 2012;224:326–31.

[84] Reardon CA. Apolipoprotein E mimetic is more effective than apolipoprotein A-I mimetic in reducing lesion formation in older female apoE null mice: a commentary. Atherosclerosis 2012;225:39–40.

[85] Handattu SP, Nayyar G, Garber DW, Palgunachari MN, Monroe CE, Keenum TD, et al. Two apolipoproteins E mimetic peptides with similar cholesterol reducing properties exhibit differential atheroprotective effects in LDL-R null mice. Atherosclerosis 2013;227:58–64.

[86] Getz GS, Wool GD, Reardon CA. Biological properties of apolipoprotein A-I mimetic peptides. Curr Atheroscler Rep 2010;12:96–104.

CHAPTER 10

Oxidized High-Density Lipoprotein: Friend or Foe

Toshiyuki Matsunaga*, Akira Hara*, Tsugikazu Komoda**

*Laboratory of Biochemistry, Gifu Pharmaceutical University, Gifu, Japan **Department of Urology, Toho University School of Medicine, Tokyo, Japan

Contents

Abstract

High-density lipoprotein (HDL) is considered to be a major anti-atherogenic particle that inhibits conversion of low-density lipoprotein (LDL) into its oxidized form (oxLDL), reduction in the vasoregulatory function, and formation of foam cells. In in vitro experiments, the HDL particle is found to be more vulnerable to oxidative modification than LDL. In addition, recent attentive investigations of clinical specimens have shown that HDL from coronary artery disease and dyslipidemia subjects has a higher sensitivity to lipid oxidation, and that levels of plasma-oxidized HDL (oxHDL) in patients with atherosclerosis, diabetes mellitus, and renal failure are significantly higher, compared with those from healthy donors. Numerous studies using potential oxidants moreover suggest that the oxidative modification of HDL reduces its intrinsic beneficial properties, such as reverse cholesterol transport and antioxidant property, and progressively participates in the development of atherosclerosis through appearance of the

247

pro-inflammatory and cytotoxic effects. By contrast, we recently found that 4-hydroxy-2-nonenal, which is generated during oxidation of HDL, induces cross-linking of apo-proteins and upregulates antioxidant enzymes such as aldo-keto reductase (AKR) 1C1 and AKR1C3. In this chapter, we introduce the current literature on clinical relevance of oxHDL to vascular-related diseases, and the structural and functional alterations in HDL due to treating with oxidants. In addition, on the basis of the oxidation-elicited functional changes in HDL and cellular responses to the oxHDL treatment, we discuss whether oxHDL is a causal factor, like oxLDL, or a protective factor for the development of atherosclerosis.

1. INTRODUCTION

Lipoproteins are heterogenous particles that contain hydrophilic compounds (apoproteins and phospholipids) in the outer surface and the inner hydrophobic lipids (triacylglycerols and cholesteryl esters). Based on their biochemical and compositional characteristics such as size, density, electrophoretic mobility, and function, they are divided into five major classes: chylomicron, very low–density lipoprotein (VLDL), intermediate-density lipoprotein (IDL), low-density lipoprotein (LDL), and high-density lipoprotein (HDL). Among them, the least dense lipoproteins, chylomicrons and VLDL, transport dietary and endogenous triglycerides, respectively, and then supply the triacylglycerols-derived fatty acids for energy supplementation against fasting or postprandial storage in cells. Most of the circulating VLDL is converted by lipolysis of hepatic lipase into IDL, which in turn forms a smaller particle LDL that directly confers cholesterol on peripheral tissues. In contrast, HDL, an apoprotein-rich and the smallest particle of the five classes, is generated in liver and small intestine, and is involved in reverse cholesterol transport from extra-hepatic tissues to the liver and in part delivery of cholesterol to steroidogenic tissues [1]. In addition to its ability to promote reverse cholesterol transport, HDL is capable of inhibiting LDL oxidation [2], vascular cell dysfunction [3,4], and macrophage-derived foam cell formation [5], all of which are key events for the formation and progression of atheromatous plaque. Therefore, HDL is considered to be an anti-atherogenic factor [6–11]. Based on multiple investigations, HDL has recently been reported to possess antiviral activity [12,13], antithrombotic effects [14], and prostacyclin stabilizing activity [15,16].

Abnormal lipid transport due to chronic alteration of the plasma lipoprotein profile (i.e., elevated and decreased levels of LDL and HDL, respectively) gives rise to an arterial cholesterol accumulation, which enhances the incidence of atherosclerosis [17]. The first step of the atherogenic processes is formation of oxidized LDL (oxLDL), which is known to result from low plasma HDL levels and/or dysfunction of the lipoprotein. One potential mechanism underlying the HDL abnormality is oxidative modification caused during progression of dyslipidemia and the resultant vascular disorders. Up to now, presence of the oxidized HDL (oxHDL) is verified by the following fundamental and clinical experiments: 1) HDL is oxidized more rapidly than LDL during in vitro oxidation [18,19], 2) HDL from patients with coronary artery spasm has a higher susceptibility to lipid peroxidative modification than LDL [20], and 3) an immunohistochemical analysis detected oxHDL in the intima of atheromatous plaques in the human abdominal aorta [21]. Much of the literature proposed myeloperoxidase [22–26], 12/15-lipoxygenases [27–30], 5-lipoxygenase [31–33], cyclooxygenase [34–36], and nicotinamide adenine dinucleotide phosphate (NADPH) oxidase [37,38] as predominant enzymes involved in oxidation of LDL. Among the proposed enzymes, recent research has revealed the contribution of myeloperoxidase to oxHDL formation and accentuated a possibility that oxidized lipoproteins including HDL and LDL are closely implicated in the onset and development of atherosclerosis [39,40]. In addition to the verification of in vivo presence of oxHDL, experiments using different oxidants found that oxidative modification of HDL raises multiple alterations of its components and conformation [41–43]. In this chapter, we introduce the current literature on clinical relevance of oxHDL to progression of vascular-related diseases and the structural alterations observed in HDL by treating with oxidants. In addition, on the basis of the oxidation-elicited functional changes in HDL and cellular responses to the oxHDL treatment, we discuss whether oxHDL is a causal factor, like oxLDL, or a protective factor for the development of atherosclerosis.

2. CLINICAL RELEVANCE OF oxHDL TO VASCULAR-RELATED DISEASES

2.1. Level of oxHDL in Healthy Subjects

High plasma concentrations of LDL, particularly its small, dense subclass, correlate with an increased risk of atherogenesis. This was assumed to be accounted for by the inverse relationship between the size of the LDL particle and the inflow rate into atheromatous plaque. Using an in vitro perfusion system of the experimental rabbit models, Björnheden and colleagues found that HDL has the highest inflow rate among the three lipoprotein classes (VLDL, LDL, and HDL), and its moderate amounts are distributed even in nonplaque lesions [44]. The data may imply that HDL can readily diffuse in subendothelial fraction because of its smaller particle size than that of LDL [44,45]. The long retention in the arterial intima might increase the frequency of HDL oxidation by the tissue- and/or interstitial fluid-located oxidases, even in healthy subjects. Indeed, high-performance liquid chromatography (HPLC) chemiluminescence analysis of HDL from healthy donors detected higher levels of lipid hydroperoxides (cholesteryl ester hydroperoxides and phospholipid hydroperoxides), compared to LDL [46]. In addition to the lipid peroxidation, oxidative modification of apoproteins in HDL isolated from control human specimens was demonstrated by a sensitive and specific enzyme-linked immunosorbent assay (ELISA) [47]. Our previous ELISA furthermore estimated that the plasma oxHDL concentration in 23 healthy volunteers was 127 µg/L, which is lower than the levels in patients with coronary artery disease and non-insulin-dependent diabetes mellitus described later [48].

2.2. Level of oxHDL in Patients with Vascular-Related Diseases

A growing body of evidence implicates oxHDL as a new diagnostic marker for vascular diseases, in particular coronary artery diseases and diabetes, and is listed in Table 10.1. Pennathur and colleagues focused on 3-nitrotyrosine formation as a potential indicator of HDL oxidation, and found that 3-nitrotyrosine contents in circulating HDL isolated from patients with established atherosclerosis are 2-fold higher than those in healthy humans [39]. Interestingly, they also showed that the 3-nitrotyrosine level in HDL isolated from human aortic atherosclerotic intima was 6-fold higher than that in circulating HDL, and demonstrated the outstanding colocalization of 3-nitrotyrosine with myeloperoxidase, which generates nitrotyrosine through nitrite oxidation of hydrogen peroxide [49]. Based on their other report, levels of 3-chlorotyrosine, another index of oxHDL, appeared to be 13-fold higher in HDL from plasma of subjects with coronary artery disease, compared with those of healthy subjects [40]. In diabetes mellitus, plasma levels of oxHDL in the patients were reported to be higher than that in healthy individuals [50,51]. Consistent with these data, our analyses using an oxHDL-specific antibody showed that oxHDL is present in the intima of atheromatous plaques in the human abdominal aorta [21], and its level is increased in patients with coronary

Table 10.1 Relationship Between Plasma oxHDL Level and Progression of Diseases

Change	Disease	Reference
Increase	Coronary artery diseases	
	Coronary artery diseases	[39,40,48]
	Diabetes mellitus	
	Gestational diabetes	[50]
	Non-insulin-dependent diabetes	[48]
	Diabetes mellitus	[51]
	Renal failure	
	Chronic renal failure	[53]
	Protein-energy wasting	[54]
	Coronary artery disease complication	[55]
	Inflammatory disease	
	Inflammatory disease	[51]
	Juvenile dermatomyositis	[56]
Decrease	Alzheimer's disease	[57]

artery diseases (200 μg/L) or Type 2 diabetes mellitus (191 μg/L) [48]. Thus, oxHDL is considered to be a potential marker that sensitively reflects progression of coronary artery disease and diabetes mellitus. Oxidation of lipidic components is detected in each particle of HDL and LDL separated from human atherosclerotic lesions, and the amounts of the oxidized lipids are positively correlated with the severity of the disease [52]. In fact, Ohmura and colleagues [20] showed that levels of thiobarbituric acid reactive substances in plasma and HDL are high in patients with coronary artery spasm, suggesting that the HDL sensitization to lipid peroxidation contributes to the pathogenic mechanism of coronary artery spasm. Thereupon, our experiments using plasma from normal and dyslipidemia subjects infer the correlation of HDL sensitivity to Cu^{2+}-mediated oxidation with triacylglycerol level in blood (unpublished data). Thus, a measurement of the HDL sensitization to oxidation might be useful for evaluating the development of vascular diseases, including coronary artery disease and dyslipidemia.

Besides the atherosclerosis-related diseases, recent studies have shown increased plasma oxHDL levels in patients with renal failure [53–55]. Tsumura and colleagues reported that HDL susceptibility to oxidation is high in patients on hemodialysis, diabetic nephropathy before initiation of dialysis, and chronic glomerulonephritis in the conservative stage [53]. On the other hand, Honda and colleagues suggest that a high oxHDL state is associated with protein-energy wasting in maintenance hemodialysis patients [54]. More recently, they have also proposed oxHDL as a risk factor for cardiovascular events in prevalent hemodialysis patients [55]. Thus, the oxidative modification of HDL could occur in patients with renal diseases, although it remains to be determined whether oxHDL promotes or inhibits the development of kidney diseases. To date, it is reported that levels of oxHDL increase and decrease in patients with inflammatory diseases [51,56] and Alzheimer's disease [57], respectively. Since there is less evidence to support the clinical significance and alteration of oxHDL, further large population-based investigations are needed to clarify the pathophysiological role of oxHDL in diseases.

3. STRUCTURAL ALTERATIONS IN HDL COMPONENTS BY OXIDATION

Molecular mechanisms of oxidation of HDL particles appear to be different among oxidants. In the system including metal ions and lipoxygenase, oxidants primarily attack lipidic components, especially phospholipids and cholesterols, in HDL particles, and subsequent oxidation of apoproteins is mediated by the oxidized lipids [43,58,59]. In contrast, low doses of tyrosyl radical, which is generated by the myeloperoxidase system, oxidize lipid-free apoproteins in HDL [60], and the apoproteins in HDL are oxidized more highly than the lipidic components [60,61]. The structural alterations of apoproteins and lipid moieties in HDL particles by various oxidants are summarized as follows.

3.1. Alteration in Apoproteins by Oxidation

Cu^{2+}-mediated oxidation of HDL particles increases a net negative charge, which is speculated to be due to modification of positively charged residues such as lysine in apoproteins [41]. The modification of lysine and tryptophan in HDL during oxidation appears to be involved in the reduction of the binding affinity of HDL for receptor and in its ability to remove cholesterol from the cells [41,62]. Indeed, Duell and colleagues demonstrated the importance of lysine residues in the HDL receptor-dependent efflux of intracellular cholesterol [63]. In addition, mild oxidation of HDL by 2,2′-azo-bis(2-amidinopropane) dihydrochloride (AAPH), a generator of aqueous peroxyl radical, is capable of converting methionine (Met) residues of HDL apoproteins into Met sulfoxide (MetO) [58,64,65]. Lipid hydroperoxides formed during the AAPH-induced oxidation of HDL convert Met86 and Met112 of three Met residues in apoprotein A-I (apoA-I) to MetO in a step-wise manner, while Met148 is not oxidized. In the case of homodimeric apoA-II, two Met26 residues of the subunits are oxidized to MetO. These studies also suggest the clinical importance of apoA-I-containing MetO as a new marker for coronary artery diseases, because there is a relationship between its circulating concentration and the risk for vascular diseases. It should be noted that the high molar ratio of hypochlorite to HDL induces oxidative modification of amino acids other than Met [61], although at its low ratios of the oxidant/HDL ($< 6:1$) the modification is limited to MetO formation [66].

3.2. Alteration in Lipids by Oxidation

Lipidic components in HDL are oxidized rapidly and preferably, compared with those in LDL, during in vitro oxidation [18,19,46]. In the HDL subclass, HDL_2 is more susceptible to Cu^{2+}-mediated oxidation than HDL_3 [67]. Accordingly, determination of oxidized lipids in HDL_2 particles might be a useful marker for an earlier stage of atherogenic events, while in vivo oxidation of HDL_2 lipids has barely been investigated.

Characterization of the lipid moiety in native HDL revealed that phosphatidylcholine (PC) is a major phospholipid occupying up to 90% of total phospholipids, whereas the remaining species are composed of phosphatidylinositol, phosphatidic acid, phosphatidylserine, and phosphatidylethanolamine. In general, PC plays an important role in functional properties of the cell membrane of all living organisms, and is metabolized into lysophosphatidylcholine (lyso-PC) and polyunsaturated fatty acid by phospholipase A_2. HDL oxidation decreases PC content and in parallel increases the amount of lyso-PC, in addition to the disappearance of phosphatidylethanolamine [19,43,68,69]. The enhanced formation of lyso-PC is similarly observed in HDL treated with peroxynitrite donor and AAPH [58,68]. In contrast, it was reported that the oxidation of HDL by increasing concentrations of hypochlorous acid (HOCl) solely results in a slight decrease of PC [70]. These pieces of evidence suggest that increased formation of lyso-PC is one of the most reasonable indicators observed in HDL particles during oxidation. Cu^{2+}-mediated oxidation of HDL markedly reduces the amount of polyunsaturated fatty acids such as linoleic acid, arachidonic acid, docosahexaenoic acid, and linolenic acid in esterified fatty acids, whereas it does not alter the contents of saturated and monounsaturated fatty acids [71]. These data may imply that two-electron oxidation by the incubation with Cu^{2+} specifically impairs polyunsaturated fatty acids at the site of *sn*-2 in PC. It should be noted that PC hydroperoxides are formed as reactive metabolites of PC during oxidation of HDL [72–74].

Compositional studies showed that HDL oxidation alters the moiety of cholesteryl ester and unesterified cholesterol form through a reduction of the ability to esterify cholesterols, and these alterations might affect its role in reverse cholesterol transport [60,75]. The cholesterols in HDL particles are readily oxidized by incubation with oxidant molecules, and then converted into cholesterol hydroperoxides and its hydroxides, like phospholipids [58,64]. Nuclear magnetic resonance spectroscopy revealed that several kinds of oxidized derivatives of cholesterol are identified in an inner fraction of the Cu^{2+}-oxidized HDL particle [69].

Increased levels of proteins modified by the reactive lipid metabolites are detected in the plasma of patients with atherosclerosis [76,77]. 4-Hydroxy-2-nonenal (HNE), one of the reactive lipid metabolites, is generated by undergoing alkoxyl radical formation and its β-scission [78], and is a reactive electrophile that elicits multiple cellular functions as illustrated in Figure 10.1. It also forms Michael adducts with side chains of lysine, histidine and cysteine, and a pyrrole adduct with lysine in proteins [79,80]. The modification of LDL with HNE appears to be responsible for an enhanced recognition by scavenger receptors on macrophages, because the binding affinities for the LDL and scavenger receptors are critically dependent on free and modified forms, respectively, of lysine residues in LDL [81–83]. In order to evaluate formation of HNE during HDL oxidation, we have prepared Cu^{2+}-oxidized HDL from plasma of healthy volunteers as previously described [19], and then quantitated the HNE contents in the oxHDL by sequential methods using TLC and HPLC (Figure 10.2A). The formation of HNE in the oxidized particles is increased in a dose-depend manner on the concentrations of $CuSO_4$ used. In addition, when the protein-HNE adducts in the Cu^{2+}-oxidized HDL were detected by Western blotting using the anti-HNE antibody, the amounts of adducts formed were elevated similarly to the pattern of free HNE, as evident from the results shown in Figure 10.2B. Therefore, we suggested that HNE plays an important role in oxidant-induced molecular alterations in the HDL particle. Apoprotein-lipid association in Cu^{2+}-oxidized HDL is more stable than that in native HDL and may be due to an oxidant-induced cross-linking of apoproteins [84]. Cross-linking between apoA-I and apoA-II (apoA-I dimer, apoA-I trimer, and apoA-I-[apoA-II]$_2$ heterotrimer) was previously detected in HDL particles modified by other oxidants

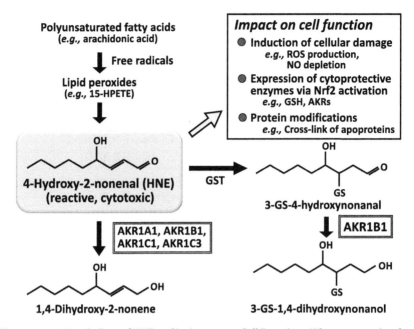

Figure 10.1 *Metabolism of HNE and its Impact on Cell Function.* When exposed to free radicals such as reactive oxygen and nitrogen species, polyunsaturated fatty acids in phospholipids are oxidized into various lipid peroxides, some of which generate a reactive and cytotoxic aldehyde, HNE, as a decomposition product. In the cytosol, most of the aldehyde is metabolized by glutathione transferase (GST) into the glutathione conjugate 3-GS-4-hydroxynonanal, which is followed by the NADPH-linked reduction of AKR1B1 into the stable metabolite 3-GS-1,4-dihydroxynonanol. Although HNE formed is in part reduced by some AKRs (1A1, 1B1, 1C1, 1C3) into its less toxic metabolite 1,4-dihydroxy-2-nonene without binding to glutathione, the residual HNE influences multiple cellular functions (reactive oxygen species [ROS] formation, NO depletion, and induction of antioxidant enzymes via Nrf2 activation) through its binding to protein thiols. In oxidation of circulating HDL, HNE generated is thought to mainly participate in cross-linking of the proteins such as apoA-I and apoA-II. 15-HPETE, 15-hydroperoxyeicosatetraenoic acid.

such as hypochlorite [70], peroxynitrite [68,85], and aldehydes (e.g., malondialdehyde, acrolein, and HNE) [42,86]. These findings indicate that the active metabolites of lipid peroxides also act as cross-linkers of apoproteins.

4. FUNCTIONAL ALTERATIONS IN HDL BY OXIDATION

In the literature, several functional alterations of HDL by its oxidative modification have been detected (Table 10.2).

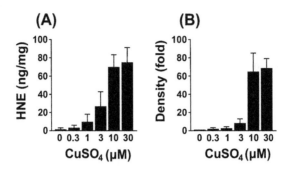

(A) **(B)**

Figure 10.2 *Amounts of Free (A) and Protein-Bound HNE (B) in Cu²⁺-Oxidized HDL.* Human HDL fraction (1.063–1.210 g/mL) was prepared by sequential ultracentrifugation as previously reported [19], and then incubated at 37°C for 18 hours with the indicated concentrations of $CuSO_4$. To detect the free HNE, 2 N HCl containing 10 M 2, 4-dinitrophenylhydrazine was added to the oxHDL solution immediately after termination of the oxidation, and the mixture was incubated at room temperature for 48 hours under anaerobic conditions to change HNE into the dinitrophenyl derivative. After the TLC separation, the derivative was applied to HPLC analysis using a μBondasphere 5 μ C_{18} column, which was eluted isocratically with a mobile phase of 80% methanol, and then monitored using a UV detector at 378 nm. Value is expressed as a ratio (mean ± SD, $n = 4$) of free HNE (ng)/HDL (mg). For determination of the HNE-apoprotein adduct, the Cu²⁺-oxidized HDL (20 μg) was subjected to Western blotting using the anti-HNE antibody, and density of the immunopositive band was analyzed. Value is expressed as a fold increase (mean ± SD, $n = 3$) of the band density of the adduct in oxHDL to that in unoxidized HDL.

Table 10.2 Functional Alterations in HDL by Oxidation

Alteration	Reference
Reduction of cholesterol transport	[41–43,63,70,75,88,90–92,156,157]
Reduction of antioxidant activity	[68,95,109–111,114,116]
Decrease in nitric oxide production and its function	[85,125,129]
Induction of pro-inflammatory effect	[87,130–135,152]
Promotion of breast cancer metastasis	[136,137]
Modulation of aldosterone release	[138]
Induction of cytotoxic effect	[18,85,129,130,139]

4.1. Reverse Cholesterol Transport

As described earlier, the anti-atherogenic effect of HDL is mainly attributed to transport of the excess cholesterol from peripheral tissues to the liver. Many results concerning effects of HDL oxidation on its capacity to mediate cellular cholesterol efflux have been documented. Oxidative modification of HDL reduces cholesterol efflux from cells [41,42,87] and induces the subsequent accumulation of unesterified cholesterol in macrophages and fibroblasts [75,88]. Morel showed that short incubation time (< 6 hours) of human HDL with Cu^{2+} decreases its phospholipid content and the ability to mediate cholesterol efflux from cells and instead increases free cholesterol, whereas long exposure (> 12 hours) diminishes efflux of cellular cholesterol without alteration of apoproteins [43]. In contrast, tyrosylation is unable to interfere with the ability of intrinsic HDL to take up cholesterol efflux, because there is no alteration in lipidic components [89]. Inactivation of lecithin:cholesterol acyltransferase (LCAT) observed during HDL oxidation is thought to be responsible for the reduction in the cholesterol efflux from the cells. McCall and colleagues [86] and Maziere and colleagues [90] independently found that short time exposure of aldehydes with HDL particle induces both inactivation and reduction of LCAT prior to initiation of cross-linking of apoproteins. These results imply that inactivation of LCAT during HDL oxidation results from the lipid peroxidation rather than apoprotein modification. Experiments using foam cells found that Cu^{2+}-mediated oxidation of HDL reduces cholesterol efflux from the cells, leading to a loss of macrophage membrane fluidity [91], and the decrease in cholesterol efflux results from both apoprotein cross-linking and the resulting less binding affinity of HDL for cell surface [41,70]. In addition, oxHDL disrupts cholesterol synthesis through inhibition of hydroxymethylglutaryl-CoA reductase by lipid peroxides [88]. Moreover, oxidation of HDL influences cholesterol transports mediated by both scavenger receptor-BI and ATP-binding cassette-A1 [70,92]. Thus, oxHDL is suggested to act as a less effective transporter for cholesterol than native HDL and to induce disruption of cholesterol homeostasis.

4.2. Antioxidant Property

HDL plays an important role in inhibiting the generation of oxLDL [2,93–95]. The inhibitory mechanism is thought to be coordinately mediated by apoA-I and several enzymes, such as paraoxonase (PON), LCAT, platelet-activating factor acetylhydrolase, and glutathione peroxidase, in addition to antioxidant molecules in HDL particles. ApoA-I is solely capable of reducing peroxides of both phospholipids and cholesteryl esters, and removing eicosanoid hydroperoxides that cause the nonenzymatic oxidation of LDL phospholipids [58,96–99]. On the other hand, it is suggested that a calcium-dependent HDL-associated ester hydrolase, PON, catalyzes the hydrolysis of organic phosphates, aromatic carboxylic acid esters, and carbamates, and mainly participates in antioxidant function of HDL toward LDL oxidation [100]. Antioxidant ability of PON seems to be mediated by hydrolyzing long-chain oxidized phospholipids formed during the oxidation process of lipoproteins [101]. In animal models, downregulation of PON raises the susceptibility for the formation of atherosclerotic lesions [102], and its genetic overexpression diminishes the lesions [103]. Thus, it is well accepted that both apoA-I and PON play key roles in the prevention of lipoprotein oxidation by degrading the reactive lipid products formed [46,58,64,101,102,104–108].

It has been shown that HDL isolated from Cu^{2+}-treated plasma significantly loses its ability to inhibit LDL oxidation, suggesting that the antioxidant ability of HDL is quenched when HDL itself is oxidized [95]. The loss of antioxidant function of HDL by oxidation is inferred to be due to a decrease in HDL-associated PON activity, but not in free PON activity [109,110]. Although the inactivation rate of PON depends on both the oxidation extent of HDL and the oxidation system used, a significant correlation between the PON inactivation and HDL oxidation extent is observed [108,110–113]. Recent studies have found that arylesterase activity of PON is markedly inhibited during HDL oxidation, whereas its phospholipase A_2 activity remains unaffected [68,114]. The findings suggest that the unaltered phospholipase-A_2 activity of PON under oxidative conditions participates in the cleavage of peroxidized PC into lyso-PC and peroxidized fatty acids detected in oxidized lipoproteins, because incubation of phospholipid with PON in the presence of apoA-I generates lyso-PC [115]. Thus, the formation of lyso-PC appears to be mainly responsible for the interaction of apoA-I and PON in HDL particles. PON translocates from cell membrane to lipoproteins during association of HDL with hepatocyte. In the particle of oxHDL, the abilities of HDL to promote release of PON and to stabilize its activity are markedly decreased, indicating that the depletion of active PON in oxHDL particles leads to impairment of antioxidant property of intrinsic HDL [116].

4.3. Nitric Oxide (NO) Generation

NO, a potent vasodilator produced in macrophages and endothelial cells is considered to act as a mediator for both atherogenic and anti-atherogenic events, depending on its source of production. The literature shows that HDL enhances the production of endothelial constitutive NO synthase (ecNOS)-derived NO through the pathway including scavenger receptor-BI and a protein kinase, Akt, leading to vasorelaxation [117–124]. These studies suggest that the increased level of NO plays an important role in the anti-atherogenic ability of HDL to preserve normal endothelial function. OxLDL and the increased content of lyso-PC in the oxLDL particle are suggested to play a crucial role in alteration of endothelium-derived arterial relaxation via reduced expression of ecNOS and the subsequent decrease in NO level [125–128]. We have shown that, like oxLDL, highly oxidized HDL by Cu^{2+} reduces the expression of ecNOS and the release of NO in human endothelial cells [129], whereas such alteration is not observed in the cells treated with peroxynitrite-modified HDL [85]. Treatment with low concentrations of $oxHDL_3$ or its lipidic fraction almost blocks NO-stimulated cGMP accumulation, suggesting that the lipidic components of oxHDL quench NO-mediated function after its release from endothelial cells [125]. It is speculated that the reduced generation of NO by HDL oxidation loses a lot of its beneficial properties to prevent the development of atherosclerosis and its complications, because NO inhibits platelet aggregation, leukocyte adhesion, and vascular smooth muscle cell proliferation, in addition to maintaining vasodilation.

4.4. Anti-Inflammatory Effect

Increased binding of circulating leukocytes on the endothelial surface is often encountered in the early stage of atherosclerosis. OxLDL enhances adhesion of macrophages to the endothelial monolayer, whereas HDL can completely inhibit it. Cell adhesion assays found that oxHDL is unable to improve the oxLDL-induced adhesion of monocytes to aortic endothelium, and itself rather increases the attached cell number compared to the control, indicating that oxidation makes HDL more pro-inflammatory [130]. Recent investigations have shown that Cu^{2+}-oxidized HDL promotes inflammatory response in human platelets [131], aortic endothelial cells [87] and monocytes [132], and rat mesangial cells [133]. Our recent study has shown that Cu^{2+}-oxidized HDL activates signaling, including nuclear factor-kappa B (NF-κB), which is triggered by enhanced generation of reactive oxygen species and subsequent inhibitor-κB ubiquitination, as observed in oxLDL [134]. This fact permits us to speculate that oxHDL particle includes certain initiating factor(s) for transcriptional activation of adhesion molecules via the NF-κB signaling pathway, because genes for adhesion molecules including vascular cell adhesion molecule-1 are conserved downstream of the transcriptional region activated by NF-κB. Other literature shows that oxHDL inhibits the secretion of tumor necrosis factor-α from macrophages, and the inhibition is mainly due to reactive low molecular weight aldehydes

(2,4-heptadienal, hexanal, 2-nonenal, 2-octenal, 2,4-decadienal), but not hydroperoxides of fatty acids [135]. Thus, oxHDL, particularly some reactive products derived from its lipid peroxidation, is likely to play a crucial role in the modulation of the inflammatory response by vascular cells early in the process of atherogenesis.

4.5. Other Effects

Recent studies have shown contribution of oxHDL to breast cancers metastasis, which is confirmed by using both chemically oxidized HDL and HDL from Type 2 diabetes mellitus patients [136,137]. Based on these reports, the metastasis appears to be due to induction of oxidative stress and activation of protein kinase C-dependent signaling pathway. On the other hand, Saha and colleagues recently found that modified HDL modulates aldosterone release via extracellular signal-regulated kinase and Janus kinase-dependent pathways [138]. Thus, these exciting data underscore significance of oxHDL as a potential factor involved in development of the oxidative stress-related diseases such as cancer metastasis and hypertension. It should be mentioned that oxidation not only loses antioxidant activity of HDL, but also converts HDL into a cytotoxic particle [18,139]. In an experiment using cultured macrophages, the lethal effect of oxHDL is equally or more cytotoxic than oxLDL when both lipoproteins were compared at the same level of lipids [18]. We have also shown that Cu^{2+}-mediated oxidation of HDL induced endothelial apoptosis, and suggested that oxHDL participates in the onset and progression of atherosclerosis [129]. As a possible lethal pathway of oxHDL, it is proposed that, similar to oxLDL, oxHDL increases intracellular levels of reactive oxygen species and the related compounds, including HNE, and that the oxidative stress is associated with activation of NADPH oxidase and downregulation of antioxidant enzymes (superoxide dismutase, catalase, and glutathione peroxidase) [85,129,130].

5. IS oxHDL FRIEND OR FOE?

Oxidative modification loses the anti-atherogenic functions distinctive of HDL, and, like oxLDL, resultant oxHDL acts as an atherogenic molecule, as mentioned earlier. This allows us to suggest that the HDL oxidation is a disadvantageous event against the health-related quality of life. Conversely, some of the previously accumulated data and our experimental results propose the following two beneficial roles of HDL oxidation: first, in the artery wall, HDL is oxidized instead of LDL in order to inhibit generation of oxLDL that is a more potent toxicant to vascular cells than oxHDL. This hypothesis is supported by the fact that oxHDL shows longer retention into the atheromatous plaque [44], faster clearance rate than native HDL [140], and less toxicity to aortic endothelial cells than oxLDL [134]. In addition to these properties of oxHDL, Gao and colleagues recently found that mild oxidation of HDL elevates the remodeling rate of this lipoprotein, probably due to removal of oxidatively damaged HDL species and supply of the fresh particle into the circulation [141]. Second, OxHDL enhances antioxidant activity of aortic endothelial cells through induction of aldo-keto reductase (AKR) 1C1 and AKR1C3. AKRs are NADPH-dependent reductases involved in metabolism of carbohydrates, steroids, prostaglandins, and other endogenous aldehydes and ketones as well as xenobiotic compounds (http://www.med.upenn.edu/akr). Among the four AKRs (1A1, 1B1, 1C1, and 1C3) expressed in human endothelial cells [142], AKR1B1, AKR1C1, and AKR1C3, which are known as aldose reductase, 20α-hydroxysteroid dehydrogenase, and Type 5 17β-hydroxysteroid dehydrogenase, respectively [143], efficiently reduce HNE into less toxic 1,4-dihydroxy-2-nonene [144,145] (see Figure 10.1).

We currently found that treatment of human aortic endothelial cells with low concentrations of oxHDL or its lipid extract highly increases the expression of AKR1C1 and AKR1C3 (Figure 10.3), without influence on AKR1B1 expression (data not shown). In addition, overexpression of AKR1C1 and AKR1C3 into bovine artery endothelial cells significantly abolished the damage elicited by HNE as well as the lipid extract (Figures 10.4A and 4B). These results imply that the reactive aldehyde generated during the HDL oxidation is detoxified by upregulated AKR1C1 and AKR1C3. Expressions of the two AKRs and other antioxidant enzymes, such as NAD(P)H:quinone oxidoreductase-1, hemeoxygenase-1, and glutathione peroxidase, are known to be regulated by a transcription factor nuclear factor erythroid 2-related factor 2, which is activated by HNE and reactive oxygen species [146,147]. Therefore, formation of oxHDL might inhibit several atherogenic processes, including oxidation of LDL and its induced oxidative stress, through upregulation of a variety of antioxidant genes.

To date, there is growing evidence in the literature suggesting the presence of oxHDL-specific receptors on the cell surface. Cu^{2+}-mediated oxidation of HDL reduces its affinity for the specific receptor [148], and oxHDL is recognized and endocytosed by scavenger receptor for oxLDL on the macrophages [149,150]. On the other hand, it is reported that oxHDL

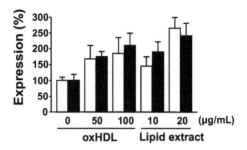

Figure 10.3 *Induction of AKR1C1 and AKR1C3 in Aortic Endothelial Cells by Treatment with oxHDL and its Lipid Extract.* Lipidic fraction was prepared from the Cu^{2+}-oxidized HDL (oxHDL) according to the Folchi method, and its dimethylsulfoxide-soluble fraction (containing aldehydes such as HNE and acrolein, but excluding phospholipids, triglycerides, and cholesterol esters) was used as the lipid extract. Human aortic endothelial cells were suspended in EBM2 medium supplemented with 1 mg/mL lipoprotein-deficient human serum and antibiotics, and seeded into a type I collagen-coated 60-mm dish. After 24-hour culture, the cells were treated for 24 hours with the indicated concentrations of oxHDL and the lipid extract, and then the expression of transcripts for AKR1C1 (□) and AKR1C3 (■) in the cells were determined by semiquantitative PCR using their specific primers [142]. The expression levels of the two genes in the treated cells are expressed as percentages (means ± SD, $n = 3$) of those in the control cells without the treatments.

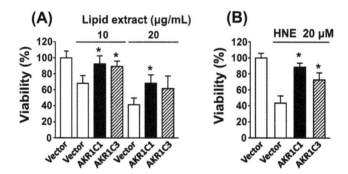

Figure 10.4 *Overexpression of AKR1C1 and AKR1C3 Protects Endothelial Cells from Damage Provoked by the Lipid Extract from oxHDL (A) and HNE (B).* Bovine artery endothelial cells were transfected with pGW1 expression vector harboring for the cDNA for AKR1C1 (■) or AKR1C3 (▨) or with the empty vector (□) as previously reported [142], and then treated for 48 hours with 10 or 20 μg/mL lipid extract of oxHDL (prepared in Figure 10.3) or 20 μM HNE. The cell viability was estimated by a dye-based cytotoxicity assay using WST1. The values (means ± SD, $n = 3$) are normalized to those in the vector-transfected cells treated with vehicle dimethylsulfoxide alone. *Significant difference from the vector cells treated with the same concentration of the compound, $p < 0.05$.

does not apparently behave as a ligand for the scavenger receptor [93]. CD36 has recently been reported to be a receptor for oxHDL, but not for native HDL and LDL [151,152]. More recently, Nofer and van Eck have shown that scavenger receptor-BI is a receptor of native and moderately oxidized HDL for inhibiting platelet activation [153]. We previously showed that oxHDL reduces a binding affinity for HDL receptor and instead enhances the affinity for a 130-kDa protein, distinct from the native HDL receptor on the surface of human endothelial cells, suggesting that a novel receptor for oxHDL is present on human endothelial cells [21]. In addition, the lethal effects of oxHDL on cultured human endothelial cells were markedly suppressed by concomitant incubation with an antibody against lectin-like oxLDL receptor 1, which is involved in initiation and progression of atherosclerotic plaque [134,154]. These apparently conflicting observations concerning oxHDL receptor(s) may depend on the type of oxidants experimentally used and the degree of oxidative modification of HDL.

Immunohistochemical study by Vollmer and colleagues revealed that apoA-I is present in lumen–adjacent layers of the intima in the earlier stage and in deeper layers of the vascular wall of patients in the advanced stages of the diseases [155]. Within the intimal layer, apoproteins in HDL are detected either in an intracellular (mainly in foam cells) or extracellular location, depending on the stage of atherosclerosis. Our study revealed that oxHDL is localized in the surface of the endothelial layer and arterial intima of atheromatous plaques of the human abdominal aorta, in contrast to the localization of oxLDL in the whole area of the plaques [21]. This different localization between oxHDL and oxLDL may be of crucial importance in order to clarify their roles in the atherogenic events. In order to prove how and why HDL is oxidized in vivo, it would be required to further investigate the localization of oxHDL in clinical samples and to identify the specific receptor(s) on the vascular cells.

6. CONCLUSION

Collectively, localization of oxHDL as a nonfunctional HDL has been a gradually obvious incident in the plasma from patients with atherosclerosis and its related vascular diseases. In addition, oxidation of HDL particles is likely to have important consequences not only for cholesterol homeostasis in peripheral tissue, but also for the development of oxidative stress–associated vascular diseases, including inflammatory diseases. In previous experiments using oxidant molecules, functional alterations in oxidized components in

HDL particles have also become clear by continuous investigations by numerous researchers. However, little is reported about the direct patho-physiological role of oxHDL in vascular diseases, including atherosclerosis and diabetes mellitus. Therefore, further fundamental and clinical investigations will be needed in order to accurately judge whether oxHDL is a friend or foe for our healthy lives and to exploit a new method of medication targeted at oxHDL.

REFERENCES

[1] Frohlich JJ, Pritchard PH. The clinical significance of serum high density lipoproteins. Clin Biochem 1989;22:417–23.
[2] Ohta T, Takata K, Horiuchi S, Morino Y, Matsuda I. Protective effect of lipoproteins containing apoprotein A-I on Cu^{2+}-catalyzed oxidation of human low density lipo-protein. FEBS Lett 1989;257:435–8.
[3] O'Connell BJ, Genest J. High-density lipoproteins and endothelial function. Circula-tion 2001;104:1978–83.
[4] Rohrer L, Hersberger M, von Eckardstein A. High density lipoproteins in the inter-section of diabetes mellitus, inflammation and cardiovascular disease. Curr Opin Lipidol 2004;15:269–78.
[5] Brown MS, Ho YK, Goldstein JL. The cholesteryl ester cycle in macrophage foam cells. Continual hydrolysis and re-esterification of cytoplasmic cholesteryl esters. J Biol Chem 1980;255:9344–52.
[6] Rader DJ. Regulation of reverse cholesterol transport and clinical implications. Am J Cardiol 2003;92:42–9.
[7] Assmann G, Gotto AM. HDL cholesterol and protective factors in atherosclerosis. Circulation 2004;109:8–14.
[8] Steinberg D. The rediscovery of high density lipoprotein: a negative risk factor in atherosclerosis. Eur J Clin Invest 1978;8:107–9.
[9] Tall AR. Plasma high density lipoproteins. Metabolism and relationship to atherogen-esis. J Clin Invest 1990;86:379–84.
[10] Johnson WJ, Mahlberg FH, Rothblat GH, Phillips MC. Cholesterol transport between cells and high-density lipoproteins. Biochim Biophys Acta 1991;1085:273–98.
[11] Collins T, Cybulsky MI. NF-kappaB: pivotal mediator or innocent bystander in ath-erogenesis? J Clin Invest 2001;107:255–64.
[12] Singh IP, Baron S. Innate defences against viremia. Rev Med Virol 2000;10:395–403.
[13] Van Lenten BJ, Wagner AC, Nayak DP, Hama S, Navab M, Fogelman AM. High-density lipoprotein loses its anti-inflammatory properties during acute influenza A infection. Circulation 2001;103:2283–8.
[14] Saku K, Ahmad M, Glas-Greenwalt P, Kashyap ML. Activation of fibrinolysis by apo-lipoproteins of high density lipoproteins in man. Thromb Res 1985;39:1–8.
[15] Fleisher LN, Tall AR, Witte LD, Miller RW, Cannon PJ. Stimulation of arterial endo-thelial cell prostacyclin synthesis by high density lipoproteins. J Biol Chem 1982;257:6653–5.
[16] Yui Y, Aoyama T, Morishita H, Takahashi M, Takatsu Y, Kawai C. Serum prostacyclin stabilizing factor is identical to apolipoprotein A-I (ApoAI). A novel function of ApoAI. J Clin Invest 1988;82:803–7.
[17] Reichl D, Miller NE. Pathophysiology of reverse cholesterol transport. Insights from inherited disorders of lipoprotein metabolism. Arteriosclerosis 1989;9:785–97.

[18] Hurtado I, Fiol C, Gracia V, Caldu P. In vitro oxidised HDL exerts a cytotoxic effect on macrophages. Atherosclerosis 1996;125:39–46.

[19] Nakajima T, Sakagishi Y, Katahira T, Nagata A, Kuwae T, Nakamura H, et al. Characterization of a specific monoclonal antibody 9F5-3a and the development of assay system for oxidized HDL. Biochem Biophys Res Commun 1995;217:407–11.

[20] Ohmura H, Watanabe Y, Hatsumi C, Sato H, Daida H, Mokuno H, et al. Possible role of high susceptibility of high-density lipoprotein to lipid peroxidative modification and oxidized high-density lipoprotein in genesis of coronary artery spasm. Atherosclerosis 1999;142:179–84.

[21] Nakajima T, Origuchi N, Matsunaga T, Kawai S, Hokari S, Nakamura H, et al. Localization of oxidized HDL in atheromatous plaques and oxidized HDL binding sites on human aortic endothelial cells. Ann Clin Biochem 2000;37:179–86.

[22] Carr AC, McCall MR, Frei B. Oxidation of LDL by myeloperoxidase and reactive nitrogen species oxidation of LDL by myeloperoxidase and reactive nitrogen species. Arterioscler Thromb Vasc Biol 2000;20:1716–23.

[23] Zhang R, Brennan ML, Shen Z, MacPherson JC, Schmitt D, Molenda CE, et al. Myeloperoxidase functions as a major enzymatic catalyst for initiation of lipid peroxidation at sites of inflammation. J Biol Chem 2002;277:46116–22.

[24] Brennan ML, Penn MS, Van Lente F, Nambi V, Shishehbor MH, Aviles RJ, et al. Prognostic value of myeloperoxidase in patients with chest pain. N Engl J Med 2003;349:1595–604.

[25] Brennan ML, Hazen SL. Emerging role of myeloperoxidase and oxidant stress markers in cardiovascular risk assessment. Curr Opin Lipidol 2003;14:353–9.

[26] Gaut JP, Byun J, Tran HD, Lauber WM, Carroll JA, Hotchkiss RS, et al. Myeloperoxidase produces nitrating oxidants in vivo. J Clin Invest 2002;109:1311–9.

[27] Cyrus T, Pratico D, Zhao L, Witztum JL, Rader DJ, Rokach J, et al. Absence of 12/15-lipoxygenase expression decreases lipid peroxidation and atherogenesis in apolipoprotein E-deficient mice. Circulation 2001;103:2277–82.

[28] George J, Afek A, Shaish A, Levkovitz H, Bloom N, Cyrus T, et al. 12/15-lipoxygenase gene disruption attenuates atherogenesis in LDL receptor-deficient mice. Circulation 2001;104:1646–50.

[29] Zhao L, Cuff CA, Moss E, Wille U, Cyrus T, Klein EA, et al. Selective interleukin-12 synthesis defect in 12/15-lipoxygenase deficient macrophages associated with reduced atherosclerosis in a mouse model of familial hypercholesterolemia. J Biol Chem 2002;277:35350–6.

[30] Harats D, Shaish A, George J, Mulkins M, Kurihara H, Levkovitz H, et al. Overexpression of 15-lipoxygenase in vascular endothelium accelerates early atherosclerosis in LDL receptor-deficient mice. Arterioscler Thromb Vasc Biol 2000;20:2100–5.

[31] Mehrabian M, Allayee H, Wong J, Shi W, Wang XP, Shaposhnik Z, et al. Identification of 5-lipoxygenase as a major gene contributing to atherosclerosis susceptibility in mice. Circ Res 2002;91:120–6.

[32] Mehrabian M, Allayee H. 5-lipoxygenase and atherosclerosis. Curr Opin Lipidol 2003;14:447–57.

[33] Dwyer JH, Allayee H, Dwyer KM, Fan J, Wu H, Mar R, et al. Arachidonate 5-lipoxygenase promoter genotype, dietary arachidonic acid, and atherosclerosis. N Engl J Med 2004;350:29–37.

[34] Linton MF, Fazio S. Cyclooxygenase-2 and inflammation in atherosclerosis. Curr Opin Pharmacol 2004;4:116–23.

[35] Natarajan R, Nadler JL. Lipid inflammatory mediators in diabetic vascular disease. Arterioscler Thromb Vasc Biol 2004;24:1542–8.

[36] Vila L. Cyclooxygenase and 5-lipoxygenase pathways in the vessel wall: role in atherosclerosis. Med Res Rev 2004;24:399–424.

[37] Sorescu D, Szocs K, Griendling KK. NAD(P)H oxidases and their relevance to atherosclerosis. Trends Cardiovasc Med 2001;11:124–31.

[38] Cathcart MK. Regulation of superoxide anion production by NADPH oxidase in monocytes/macrophages. Contributions to atherosclerosis. Arterioscler Thromb Vasc Biol 2004;24:23–8.

[39] Pennathur S, Bergt C, Shao B, Byun J, Kassim SY, Singh P, et al. Human atherosclerotic intima and blood of patients with established coronary artery disease contain high density lipoprotein damaged by reactive nitrogen species. J Biol Chem 2004;279:42977–83.

[40] Bergt C, Pennathur S, Fu X, Byun J, O'Brien K, McDonald TO, et al. The myeloperoxidase product hypochlorous acid oxidizes HDL in the human artery wall and impairs ABCA1-dependent cholesterol transport. Proc Natl Acad Sci USA 2004;101; 13032–7.

[41] Nagano Y, Arai H, Kita T. High density lipoprotein loses its effect to stimulate efflux of cholesterol from foam cells after oxidative modification. Proc Natl Acad Sci USA 1991;88:6457–61.

[42] Salmon S, Mazière C, Auclair M, Theron L, Santus R, Mazière JC. Malondialdehyde modification and copper-induced autooxidation of high-density lipoprotein decrease cholesterol efflux from human cultured fibroblasts. Biochim Biophys Acta 1992;1125:230–5.

[43] Morel DW. Reduced cholesterol efflux to mildly oxidized high density lipoprotein. Biochem Biophys Res Commun 1994;200:408–16.

[44] Björnheden T, Babyi A, Bondjers G, Wiklund O. Accumulation of lipoprotein fractions and subfractions in the arterial wall, determined in an in vitro perfusion system. Atherosclerosis 1996;123:43–56.

[45] Sloop CH, Dory L, Roheim PS. Interstitial fluid lipoproteins. J Lipid Res 1987;28: 225–37.

[46] Bowry VW, Stanley KK, Stocker R. High density lipoprotein is the major carrier of lipid hydroperoxides in human blood plasma from fasting donors. Proc Natl Acad Sci USA 1992;89:10316–20.

[47] Wang XS, Shao B, Oda MN, Heinecke JW, Mahler S, Stocker R. A sensitive and specific ELISA detects methionine sulfoxide-containing apolipoprotein A-I in HDL. J Lipid Res 2009;50:586–94.

[48] Nakajima T, Matsunaga T, Kawai S, Hokari S, Inoue I, Katayama S, et al. Characterization of the epitopes specific for the monoclonal antibody 9F5-3a and quantification of oxidized HDL in human plasma. Ann Clin Biochem 2004;41:309–15.

[49] van der Vliet A, Eiserich JP, Halliwell B, Cross CE. Formation of reactive nitrogen species during peroxidase-catalyzed oxidation of nitrite. A potential additional mechanism of nitric oxide-dependent toxicity. J Biol Chem 1997;272:7617–25.

[50] Santra D, Sawhney H, Aggarwal N, Majumdar S, Vasishta K. Lipid peroxidation and vitamin E status in gestational diabetes mellitus. J Obstet Gynaecol Res 2003;29:300–4.

[51] Ueda M, Hayase Y, Mashiba S. Establishment and evaluation of 2 monoclonal antibodies against oxidized apolipoprotein A-I (apoA-I) and its application to determine blood oxidized apoA-I levels. Clin Chim Acta 2007;378:105–11.

[52] Frei B, Yamamoto Y, Niclas D, Ames BN. Evaluation of an isoluminol chemiluminescence assay for the detection of hydroperoxides in human blood plasma. Anal Biochem 1988;175:120–30.

[53] Tsumura M, Kinouchi T, Ono S, Nakajima T, Komoda T. Serum lipid metabolism abnormalities and change in lipoprotein contents in patients with advanced-stage renal disease. Clin Chim Acta 2001;314:27–37.

[54] Honda H, Ueda M, Kojima S, Mashiba S, Suzuki H, Hosaka N, et al. Oxidized high-density lipoprotein is associated with protein-energy wasting in maintenance hemodialysis patients. Clin J Am Soc Nephrol 2010;5:1021–8.

[55] Honda H, Ueda M, Kojima S, Mashiba S, Michihata T, Takahashi K, et al. Oxidized high-density lipoprotein as a risk factor for cardiovascular events in prevalent hemodialysis patients. Atherosclerosis 2012;220:493–501.

[56] Eimer MJ, Brickman WJ, Seshadri R, Ramsey-Goldman R, McPherson DD, Smulevitz B, et al. Clinical status and cardiovascular risk profile of adults with a history of juvenile dermatomyositis. J Pediatr 2011;159:795–801.

[57] Bergt C, Nakano T, Ditterich J, DeCarli C, Eiserich JP. Oxidized plasma high-density lipoprotein is decreased in Alzheimer's disease. Free Radic Biol Med 2006;41:1542–7.

[58] Garner B, Waldeck AR, Witting PK, Rye KA, Stocker R. Oxidation of high density lipoproteins. II. Evidence for direct reduction of lipid hydroperoxides by methionine residues of apolipoproteins AI and AII. J Biol Chem 1998;273:6088–95.

[59] Marcel YL, Jewer D, Leblond L, Weech PK, Milne RW. Lipid peroxidation changes the expression of specific epitopes of apolipoprotein A-I. J Biol Chem 1989;264: 19942–50.

[60] Wang WQ, Merriam DL, Moses AS, Francis GA. Enhanced cholesterol efflux by tyrosyl radical-oxidized high density lipoprotein is mediated by apolipoprotein AI-AII heterodimers. J Biol Chem 1998;273:17391–8.

[61] Panzenboeck U, Raitmayer S, Reicher H, Lindner H, Glatter O, Malle E, et al. Effects of reagent and enzymatically generated hypochlorite on physicochemical and metabolic properties of high density lipoproteins. J Biol Chem 1997;272:29711–20.

[62] Reyftmann JP, Santus R, Mazière JC, Morlière P, Salmon S, Candide C, et al. Sensitivity of tryptophan and related compounds to oxidation induced by lipid autoperoxidation. Application to human serum low- and high-density lipoproteins. Biochim Biophys Acta 1990;1042:159–67.

[63] Duell PB, Oram JF, Bierman EL. Nonenzymatic glycosylation of HDL and impaired HDL-receptor-mediated cholesterol efflux. Diabetes 1991;40:377–84.

[64] Garner B, Witting PK, Waldeck AR, Christison JK, Raftery M, Stocker R. Oxidation of high density lipoproteins. I. Formation of methionine sulfoxide in apolipoproteins AI and AII is an early event that accompanies lipid peroxidation and can be enhanced by alpha-tocopherol. J Biol Chem 1998;273:6080–7.

[65] Pankhurst G, Wang XL, Wilcken DE, Baernthaler G, Panzenbock U, Raftery M, et al. Characterization of specifically oxidized apolipoproteins in mildly oxidized high density lipoprotein. J Lipid Res 2003;44:349–55.

[66] Bergt C, Oettl K, Keller W, Andreae F, Leis HJ, Malle E, et al. Reagent or myeloperoxidase-generated hypochlorite affects discrete regions in lipid-free and lipid-associated human apolipoprotein A-I. Biochem J 2000;346:345–54.

[67] Stojanović N, Krilov D, Herak JN. Slow oxidation of high density lipoproteins as studied by EPR spectroscopy. Free Radic Res 2006;40:135–40.

[68] Ahmed Z, Ravandi A, Maguire GF, Emili A, Draganov D, La Du BN, et al. Apolipoprotein A-I promotes the formation of phosphatidylcholine core aldehydes that are hydrolyzed by paraoxonase (PON-1) during high density lipoprotein oxidation with a peroxynitrite donor. J Biol Chem 2001;276:24473–81.

[69] Bradamante S, Barenghi L, Giudici GA, Vergani C. Free radicals promote modifications in plasma high-density lipoprotein: nuclear magnetic resonance analysis. Free Radic Biol Med 1992;12:193–203.

[70] Marsche G, Hammer A, Oskolkova O, Kozarsky KF, Sattler W, Malle E. Hypochlorite-modified high density lipoprotein, a high affinity ligand to scavenger receptor class B, type I, impairs high density lipoprotein-dependent selective lipid uptake and reverse cholesterol transport. J Biol Chem 2002;277:32172–9.

[71] Nakajima T. Evaluation of the oxidized HDL-specific monoclonal antibody 9F5-3a epitope and identification of possible oxidized HDL receptor in atherosclerotic lesions. J Saitama Med School 1998;25:255–66.

[72] Mashima R, Yamamoto Y, Yoshimura S. Reduction of phosphatidylcholine hydroperoxide by apolipoprotein A-I: purification of the hydroperoxide-reducing proteins from human blood plasma. J Lipid Res 1998;39:1133–40.

[73] Wu CY, Peng YN, Chiu JH, Ho YL, Chong CP, Yang YL, et al. Characterization of in vitro modified human high-density lipoprotein particles and phospholipids by capillary zone electrophoresis and LC ESI-MS. J Chromatogr B Analyt Technol Biomed Life Sci 2009;877:3495–505.

[74] Hui SP, Taguchi Y, Takeda S, Ohkawa F, Sakurai T, Yamaki S, et al. Quantitative determination of phosphatidylcholine hydroperoxides during copper oxidation of LDL and HDL by liquid chromatography/mass spectrometry. Anal Bioanal Chem 2012;403:1831–40.

[75] Musanti R, Ghiselli G. Interaction of oxidized HDLs with J774-A1 macrophages causes intracellular accumulation of unesterified cholesterol. Arterioscler Thromb 1993;13:1334–45.

[76] Palinski W, Yla-Herttuala S, Rosenfeld ME, Butler SW, Socher SA, Parthasarathy S, et al. Antisera and monoclonal antibodies specific for epitopes generated during oxidative modification of low density lipoprotein. Arteriosclerosis 1990;10:325–35.

[77] Holvoet P, Perez G, Zhao Z, Brouwers E, Bernar H, Collen D. Malondialdehyde-modified low density lipoproteins in patients with atherosclerotic disease. J Clin Invest 1995;95:2611–9.

[78] Esterbauer H, Schaur RJ, Zollner H. Chemistry and biochemistry of 4-hydroxynonenal, malonaldehyde and related aldehydes. Free Radic Biol Med 1991;11:81–128.

[79] Uchida K, Toyokuni S, Nishikawa K, Kawakishi S, Oda H, Hiai H, et al. Michael addition-type 4-hydroxy-2-nonenal adducts in modified low-density lipoproteins: markers for atherosclerosis. Biochemistry 1994;33:12487–94.

[80] O'Neil J, Hoppe G, Sayre LM, Hoff HF. Inactivation of cathepsin B by oxidized LDL involves complex formation induced by binding of putative reactive sites exposed at low pH to thiols on the enzyme. Free Radic Biol Med 1997;23:215–25.

[81] Hoff HF, O'Neil J, Chisolm GM, Cole TB, Quehenberger O, Esterbauer H, et al. Modification of low density lipoprotein with 4-hydroxynonenal induces uptake by macrophages. Arteriosclerosis 1989;9:538–49.

[82] Esterbauer H, Gebicki J, Puhl H, Jurgens G. The role of lipid peroxidation and antioxidants in oxidative modification of LDL. Free Radic Biol Med 1992;13:341–90.

[83] Steinbrecher UP, Lougheed M, Kwan WC, Dirks M. Recognition of oxidized low density lipoprotein by the scavenger receptor of macrophages results from derivatization of apolipoprotein B by products of fatty acid peroxidation. J Biol Chem 1989;264:15216–23.

[84] Shoukry MI, Gong EL, Nichols AV. Apolipoprotein-lipid association in oxidatively modified HDL and LDL. Biochim Biophys Acta 1994;1210:355–60.

[85] Matsunaga T, Nakajima T, Sonoda M, Koyama I, Kawai S, Inoue I, et al. Modulation of reactive oxygen species in endothelial cells by peroxynitrite-treated lipoproteins. J Biochem 2001;130:285–93.

[86] McCall MR, Tang JY, Bielicki JK, Forte TM. Inhibition of lecithin-cholesterol acyltransferase and modification of HDL apolipoproteins by aldehydes. Arterioscler Thromb Vasc Biol 1995;15:1599–606.

[87] Undurti A, Huang Y, Lupica JA, Smith JD, DiDonato JA, Hazen SL. Modification of high density lipoprotein by myeloperoxidase generates a pro-inflammatory particle. J Biol Chem 2009;284:30825–35.

[88] Ghiselli G, Giorgini L, Gelati M, Musanti R. Oxidatively modified HDLs are potent inhibitors of cholesterol biosynthesis in human skin fibroblasts. Arterioscler Thromb 1992;12:929–35.

[89] Francis GA. High density lipoprotein oxidation: in vitro susceptibility and potential in vivo consequences. Biochim Biophys Acta 2000;1483:217–35.

[90] Mazière JC, Myara I, Salmon S, Auclair M, Haigle J, Santus R, et al. Copper- and malondialdehyde-induced modification of high density lipoprotein and parallel loss of lecithin cholesterol acyltransferase activation. Atherosclerosis 1993;104:213–9.

[91] Girona J, LaVille AE, Solà R, Motta C, Masana L. HDL derived from the different phases of conjugated diene formation reduces membrane fluidity and contributes to a decrease in free cholesterol efflux from human THP-1 macrophages. Biochim Biophys Acta 2003;1633:143–8.

[92] Shao B, Bergt C, Fu X, Green P, Voss JC, Oda MN, et al. Tyrosine 192 in apolipoprotein A-I is the major site of nitration and chlorination by myeloperoxidase, but only chlorination markedly impairs ABCA1-dependent cholesterol transport. J Biol Chem 2005;280:5983–93.

[93] Parthasarathy S, Barnett J, Fong LG. High-density lipoprotein inhibits the oxidative modification of low-density lipoprotein. Biochim Biophys Acta 1990;1044:275–83.

[94] Mackness MI, Abbott C, Arrol S, Durrington PN. The role of high-density lipoprotein and lipid-soluble antioxidant vitamins in inhibiting low-density lipoprotein oxidation. Biochem J 1993;294:829–34.

[95] Hahn M, Subbiah MT. Significant association of lipid peroxidation products with high density lipoproteins. Biochem Mol Biol Int 1994;33:699–704.

[96] Ohta T, Takata K, Horiuchi S, Morino Y, Matsuda I. Protective effect of lipoproteins containing apolipoprotein A-I on Cu^{2+}-catalyzed oxidation of human low density lipoprotein. FEBS Lett 1989;257:435–8.

[97] Navab M, Hama SY, Anantharamaiah GM, Hassan K, Hough GP, Watson AD, et al. Normal high density lipoprotein inhibits three steps in the formation of mildly oxidized low density lipoprotein: steps 2 and 3. J Lipid Res 2000;41:1495–508.

[98] Navab M, Berliner JA, Subbanagounder G, Hama S, Lusis AJ, Castellani LW, et al. HDL and the inflammatory response induced by LDL-derived oxidized phospholipids. Arterioscler Thromb Vasc Biol 2001;21:481–8.

[99] Assmann G, Nofer JR. Atheroprotective effects of high-density lipoproteins. Annu Rev Med 2003;54:321–41.

[100] Mackness MI, Durrington PN. HDL, its enzymes and its potential to influence lipid peroxidation. Atherosclerosis 1995;115:243–53.

[101] Watson AD, Berliner JA, Hama SY, La Du BN, Faull KF, Fogelman AM, et al. Protective effect of high density lipoprotein associated paraoxonase. Inhibition of the biological activity of minimally oxidized low density lipoprotein. J Clin Invest 1995;96:2882–91.

[102] Shih DM, Gu L, Xia YR, Navab M, Li WF, Hama S, et al. Mice lacking serum paraoxonase are susceptible to organophosphate toxicity and atherosclerosis. Nature 1998;394:284–7.

[103] Tward A, Xia YR, Wang XP, Shi YS, Park C, Castellani LW, et al. Decreased atherosclerotic lesion formation in human serum paraoxonase transgenic mice. Circulation 2002;106:484–90.

[104] Mackness MI, Durrington PN, Mackness B. How high-density lipoprotein protects against the effects of lipid peroxidation. Curr Opin Lipidol 2000;11:383–8.

[105] Laplaud PM, Dantoine T, Chapman MJ. Paraoxonase as a risk marker for cardiovascular disease: facts and hypotheses. Clin Chem Lab Med 1998;36:431–41.

[106] Mackness MI, Arrol S, Abbott C, Durrington PN. Protection of low-density lipoprotein against oxidative modification by high-density lipoprotein associated paraoxonase. Atherosclerosis 1993;104:129–35.

[107] Mackness MI, Arrol S, Durrington PN. Paraoxonase prevents accumulation of lipoperoxides in low-density lipoprotein. FEBS Lett 1991;286:152–4.

[108] Aviram M, Rosenblat M, Bisgaier CL, Newton RS, Primo-Parmo SL, La Du BN. Paraoxonase inhibits high-density lipoprotein oxidation and preserves its functions. A possible peroxidative role for paraoxonase. J Clin Invest 1998;101:1581–90.

[109] Nguyen SD, Sok DE. Oxidative inactivation of paraoxonase1, an antioxidant protein and its effect on antioxidant action. Free Radic Res 2003;37:1319–30.

[110] Nguyen SD, Kim JR, Kim MR, Jung TS, Soka DE. Copper ions and hypochlorite are mainly responsible for oxidative inactivation of paraoxon-hydrolyzing activity in human high density lipoprotein. Toxicol Lett 2004;147:201–8.

[111] Jaouad L, Milochevitch C, Khalil A. PON1 paraoxonase activity is reduced during HDL oxidation and is an indicator of HDL antioxidant capacity. Free Radic Res 2003;37:77–83.

[112] Ferretti G, Bacchetti T, Moroni C, Savino S, Liuzzi A, Balzola F, et al. Paraoxonase activity in high-density lipoproteins: a comparison between healthy and obese females. J Clin Endocrinol Metab 2005;90:1728–33.

[113] Aviram M, Rosenblat M, Billecke S, Erogul J, Sorenson R, Bisgaier CL, et al. Human serum paraoxonase (PON 1) is inactivated by oxidized low density lipoprotein and preserved by antioxidants. Free Radic Biol Med 1999;26:892–904.

[114] Karabina SA, Lehner AN, Frank E, Parthasarathy S, Santanam N. Oxidative inactivation of paraoxonase—implications in diabetes mellitus and atherosclerosis. Biochim Biophys Acta 2005;1725:213–21.

[115] Ahmed Z, Ravandi A, Maguire GF, Emili A, Draganov D, La Du BN, et al. Multiple substrates for paraoxonase-1 during oxidation of phosphatidylcholine by peroxynitrite. Biochem Biophys Res Commun 2002;290:391–6.

[116] Deakin S, Moren X, James RW. HDL oxidation compromises its influence on paraoxonase-1 secretion and its capacity to modulate enzyme activity. Arterioscler Thromb Vasc Biol 2007;27:1146–52.

[117] Kuhn FE, Mohler ER, Satler LF, Reagan K, Lu DY, Rackley CE. Effects of high-density lipoprotein on acetylcholine-induced coronary vasoreactivity. Am J Cardiol 1991;68:1425–30.

[118] Tauber JP, Cheng J, Gospodarowicz D. Effect of high and low density lipoproteins on proliferation of cultured bovine vascular endothelial cells. J Clin Invest 1980;66:696–708.

[119] Murugesan G, Sa G, Fox PL. High-density lipoprotein stimulates endothelial cell movement by a mechanism distinct from basic fibroblast growth factor. Circ Res 1994;74:1149–56.

[120] Zeiher AM, Schächlinger V, Hohnloser SH, Saurbier B, Just H. Coronary atherosclerotic wall thickening and vascular reactivity in humans. Elevated high-density lipoprotein levels ameliorate abnormal vasoconstriction in early atherosclerosis. Circulation 1994;89:2525–32.

[121] Uittenbogaard A, Shaul PW, Yuhanna IS, Blair A, Smart EJ. High density lipoprotein prevents oxidized low density lipoprotein-induced inhibition of endothelial nitric-oxide synthase localization and activation in caveolae. J Biol Chem 2000;275:11278–83.

[122] Yuhanna IS, Zhu Y, Cox BE, Hahner LD, Osborne-Lawrence S, Lu P, et al. High-density lipoprotein binding to scavenger receptor-BI activates endothelial nitric oxide synthase. Nat Med 2001;7:853–7.

[123] Mineo C, Yuhanna IS, Quon MJ, Shaul PW. High density lipoprotein-induced endothelial nitric-oxide synthase activation is mediated by Akt and MAP kinases. J Biol Chem 2003;278:9142–9.

[124] Li XA, Titlow WB, Jackson BA, Giltiay N, Nikolova-Karakashian M, Uittenbogaard A, et al. High density lipoprotein binding to scavenger receptor, Class B, type I activates endothelial nitric-oxide synthase in a ceramide-dependent manner. J Biol Chem 2002;277:11058–63.

[125] Chin JH, Azhar S, Hoffman BB. Inactivation of endothelial derived relaxing factor by oxidized lipoproteins. J Clin Invest 1992;89:10–8.

[126] Simon BC, Cunningham LD, Cohen RA. Oxidized low density lipoproteins cause contraction and inhibit endothelium-dependent relaxation in the pig coronary artery. J Clin Invest 1990;86:75–9.

[127] Mangin EL, Kugiyama K, Nguy JH, Kerns SA, Henry PD. Effects of lysolipids and oxidatively modified low density lipoprotein on endothelium-dependent relaxation of rabbit aorta. Circ Res 1993;72:161–6.

[128] Nuszkowski A, Gräbner R, Marsche G, Unbehaun A, Malle E, Heller R. Hypochlorite-modified low density lipoprotein inhibits nitric oxide synthesis in endothelial cells via an intracellular dislocalization of endothelial nitric-oxide synthase. J Biol Chem 2001;276:14212–21.

[129] Matsunaga T, Nakajima T, Miyazaki T, Koyama I, Hokari S, Inoue I, et al. Glycated high-density lipoprotein regulates reactive oxygen species and reactive nitrogen species in endothelial cells. Metabolism 2003;52:42–9.

[130] Sharma N, Desigan B, Ghosh S, Sanyal SN, Ganguly NK, Majumdar S. The role of oxidized HDL in monocyte/macrophage functions in the pathogenesis of atherosclerosis in Rhesus monkeys. Scand J Clin Lab Invest 1999;59:215–25.

[131] Assinger A, Schmid W, Eder S, Schmid D, Koller E, Volf I. Oxidation by hypochlorite converts protective HDL into a potent platelet agonist. FEBS Lett 2008;582(5): 778–84.

[132] Soumyarani VS, Jayakumari N. Oxidatively modified high density lipoprotein promotes inflammatory response in human monocytes-macrophages by enhanced production of ROS, TNF-α, MMP-9, and MMP-2. Mol Cell Biochem 2012;366:277–85.

[133] Zhang M, Gao X, Wu J, Liu D, Cai H, Fu L, et al. Oxidized high-density lipoprotein enhances inflammatory activity in rat mesangial cells. Diabetes Metab Res Rev 2010;26:455–63.

[134] Matsunaga T, Hokari S, Koyama I, Harada T, Komoda T. NF-kappa B activation in endothelial cells treated with oxidized high-density lipoprotein. Biochem Biophys Res Commun 2003;303:313–9.

[135] Girona J, La Ville AE, Heras M, Olivé S, Masana L. Oxidized lipoproteins including HDL and their lipid peroxidation products inhibit TNF-alpha secretion by THP-1 human macrophages. Free Radic Biol Med 1997;23:658–67.

[136] Pan B, Ren H, Lv X, Zhao Y, Yu B, He Y, et al. Hypochlorite-induced oxidative stress elevates the capability of HDL in promoting breast cancer metastasis. J Transl Med 2012;10:65.

[137] Pan B, Ren H, Ma Y, Liu D, Yu B, Ji L, et al. High-density lipoprotein of patients with type 2 diabetes mellitus elevates the capability of promoting migration and invasion of breast cancer cells. Int J Cancer 2012;131:70–82.

[138] Saha S, Graessler J, Schwarz PE, Goettsch C, Bornstein SR, Kopprasch S. Modified high-density lipoprotein modulates aldosterone release through scavenger receptors via extra cellular signal-regulated kinase and Janus kinase-dependent pathways. Mol Cell Biochem 2012;366:1–10.

[139] Alomar Y, Nègre-Salvayre A, Levade T, Valdiguié P, Salvayre R. Oxidized HDL are much less cytotoxic to lymphoblastoid cells than oxidized LDL. Biochim Biophys Acta 1992;1128:163–6.

[140] Racek J, Holecek V, Trefil L. Indicators of oxidative stress in cardiovascular diseases. I. Lipoperoxidation. Vnitr Lek 1999;45:367–72.

[141] Gao X, Jayaraman S, Gursky O. Mild oxidation promotes and advanced oxidation impairs remodeling of human high-density lipoprotein in vitro. J Mol Biol 2008;376:997–1007.

[142] Matsunaga T, Arakaki M, Kamiya T, Endo S, El-Kabbani O, Hara A. Involvement of an aldo-keto reductase (AKR1C3) in redox cycling of 9,10-phenanthrenequinone leading to apoptosis in human endothelial cells. Chem Biol Interact 2009;181:52–60.

[143] Matsunaga T, Shintani S, Hara A. Multiplicity of mammalian reductases for xenobiotic carbonyl compounds. Drug Metab Pharmacokinet 2006;21:1–18.

[144] Burczynski ME, Sridhar GR, Palackal NT, Penning TM. The reactive oxygen species—and Michael acceptor-inducible human aldo-keto reductase AKR1C1 reduces the α,β-unsaturated aldehyde 4-hydroxy-2-nonenal to 1,4-dihydroxy-2-nonene. J Biol Chem 2001;276:2890–7.

[145] Srivastava S, Liu SQ, Conklin DJ, Zacarias A, Srivastava SK, Bhatnagar A. Involvement of aldose reductase in the metabolism of atherogenic aldehydes. Chem Biol Interact 2001;130-132:563–71.

[146] MacLeod AK, McMahon M, Plummer SM, Higgins LG, Penning TM, Igarashi K, et al. Characterization of the cancer chemopreventive NRF2-dependent gene battery in human keratinocytes: demonstration that the KEAP1-NRF2 pathway, and not the BACH1-NRF2 pathway, controls cytoprotection against electrophiles as well as redox-cycling compounds. Carcinogenesis 2009;30:1571–80.

[147] Reuter S, Gupta SC, Chaturvedi MM, Aggarwal BB. Oxidative stress, inflammation, and cancer: how are they linked? Free Radic Biol Med 2010;49:1603–16.

[148] Sakai M, Miyazaki A, Sakamoto Y, Shichiri M, Horiuchi S. Cross-linking of apolipoproteins is involved in a loss of the ligand activity of high density lipoprotein upon Cu^{2+}-mediated oxidation. FEBS Lett 1992;314:199–202.

[149] Steinberg D, Parthasarathy S, Carew TE, Khoo JC, Witztum JL. Beyond cholesterol. Modifications of low-density lipoprotein that increase its atherogenicity. N Engl J Med 1989;320:915–24.

[150] La Ville AE, Sola R, Balanya J, Turner PR, Masana L. In vitro oxidised HDL is recognized by the scavenger receptor of macrophages: implications for its protective role in vivo. Atherosclerosis 1994;105:179–89.

[151] Thorne RF, Mhaidat NM, Ralston KJ, Burns GF. CD36 is a receptor for oxidized high density lipoprotein: implications for the development of atherosclerosis. FEBS Lett 2007;581:1227–32.

[152] Assinger A, Koller F, Schmid W, Zellner M, Babeluk R, Koller E, et al. Specific binding of hypochlorite-oxidized HDL to platelet CD36 triggers proinflammatory and procoagulant effects. Atherosclerosis 2010;212:153–60.

[153] Nofer JR, van Eck M. HDL scavenger receptor class B type I and platelet function. Curr Opin Lipidol 2011;22:277–82.

[154] Vohra RS, Murphy JE, Walker JH, Ponnambalam S, Homer-Vanniasinkam S. Atherosclerosis and the lectin-like oxidized low-density lipoprotein scavenger receptor. Trends Cardiovasc Med 2006;16:60–4.

[155] Vollmer E, Brust J, Roessner A, Bosse A, Burwikel F, Kaesberg B, et al. Distribution patterns of apolipoproteins A1, A2, and B in the wall of atherosclerotic vessels. Virchows Arch A Pathol Anat Histopathol 1991;419:79–88.

[156] Francis GA, Mendez AJ, Bierman EL, Heinecke JW. Oxidative tyrosylation of high density lipoprotein by peroxidase enhances cholesterol removal from cultured fibroblasts and macrophage foam cells. Proc Natl Acad Sci USA 1993;90:6631–5.

[157] Francis GA, Oram JF, Heinecke JW, Bierman EL. Oxidative tyrosylation of HDL enhances the depletion of cellular cholesteryl esters by a mechanism independent of passive sterol desorption. Biochemistry 1996;35:15188–97.

CHAPTER 11

Current Aspects of Paraoxonase-1 Research

Mike Mackness, Bharti Mackness

AVDA Princip D'Espanya No 152, Miami Playa, Tarragona, Spain

Contents

Abstract

Paraoxonase-1 (PON1) is an HDL-associated lipolactonase capable of preventing LDL and cell membrane oxidation and is therefore considered to be atheroprotective. PON1 contributes to the antioxidative function of HDL and reductions in HDL-PON1 activity, prevalent in a wide variety of diseases with an inflammatory component, and is believed to lead to dysfunctional HDL, which can promote inflammation and atherosclerosis. However, PON1 is multifunctional and may contribute to other HDL functions such as in innate immunity, preventing infection by quorum-sensing Gram-negative bacteria by destroying acyl lactone mediators of quorum sensing. In this chapter we explore the role of PON1 in atherosclerosis and other inflammatory diseases, its role in determining the toxicity of certain pesticides, prevention of bacterial infection, putative new roles for PON1 in cancer development, the noninflammatory removal of toxic membrane products of apoptosis, and the promotion of healthy aging.

Key Concepts

- In section 1 of this chapter we provide a historical background and basic characteristics of paraoxonase-1 (PON1). We then review the evidence that PON1 protects against atherosclerosis development by antioxidative/anti-inflammatory mechanisms and how PON1 contributes to the atheroprotective function of HDL (section 2). The concept of dysfunctional HDL and its consequences are discussed in section 3, while the contribution of low serum PON1 activity to the dysfunctional HDL in a wide variety of diseases with an inflammatory component is discussed in section 4.

- We then turn to non-antioxidative/non-anti-inflammatory functions of PON1, beginning with a review of the evidence that PON1 protects against the toxic effects of certain pesticides in section 5, before describing three relatively new and exciting functions of PON1 in cancer development (section 6), in protecting against bacterial infection (section 7), and as a possible determinant of healthy aging (section 8) before summing up and concluding in section 9.

1. INTRODUCTION

Human serum paraoxonase-1 (PON1) is a Ca^{2+}-dependent, high-density lipoprotein (HDL)-associated lactonase capable of hydrolyzing a wide variety of lactones, thiolactones, aryl esters, cyclic carbonates, and organophosphate pesticides [1]. PON1 is currently classified as an aryldialkylphosphatase (EC 3.1.8.1) by the Enzyme Commission of the International Union of Biochemistry and Molecular Biology [2].

PON1 was first described in the 1940s when Mazur reported an enzyme activity found in mammalian tissues which was capable of hydrolyzing organophosphate pesticides [3]. The enzymes were further classified by Aldridge [4] as "A"-esterases (esterases capable of hydrolyzing organophosphates, as opposed to "B"-esterases which are inhibited by organophosphates). However, the widespread use of paraoxon as substrate for the enzyme led to the almost universal adoption of the name paraoxonase.

PON1 is a glycoprotein of 354 amino acids and approximate molecular mass of 43 KDa. It retains its hydrophobic signal sequence in the N-terminal region (with the exception of the initial methionine), which enables its association with HDL [2]. PON1 associates with a specific HDL subspecies which also contains apolipoprotein A-I (apoA-I) and clusterin. On ultracentrifugation, the majority of PON1 (and HDL antioxidant activity) resides on the small dense HDL_3 subfraction [5].

The gene for PON1 is located between q21.3 and q22.1 on the long arm of chromosome 7 in humans (chromosome 6 in mice). An X-ray crystallography study has indicated the structure of PON1 to be a 6-bladed propeller, with a lid covering the active site passage and containing two Ca^{2+}, one essential for activity and one essential for stability [6].

PON1 is the first discovered member of the paraoxonase (PON) multigene family, which is comprised of three members, PON1, PON2 and

PON3, the genes for which are located adjacent to each other [7]. The genes for all three members of the family are widely expressed in mammalian tissues [8], however, PON1 and PON3 are predominantly located in the plasma associated with HDL, while PON2 is not found in the plasma but has a wide cellular distribution [9]. PON1, PON2, and PON3 all retard the pro-atherogenic oxidative modification of low-density lipoprotein (LDL) and cell membranes and are therefore considered to be anti-atherogenic [10]. PON1 is now considered to be a major factor in the antioxidative activity of HDL [11].

Although many nutritional, lifestyle, and pharmaceutical modulators of PON1 are known [12,13], by far the biggest effect on PON1 activity levels, which can vary by more than 40-fold between individuals, is through PON1 genetic polymorphisms [2]. The coding region PON1-Q192R polymorphism determines a substrate-dependent effect on activity. Some substrates, such as paraoxon, are hydrolyzed faster by the R isoform, while others such as diazoxon are hydrolyzed more rapidly by the Q isoform [2]. Both the coding region PON1-L55M and the promoter region PON1-T-108C polymorphisms are associated with different serum concentrations and therefore different activities. The 55L allele results in significantly higher PON1 mRNA and serum protein levels and therefore more activity compared to the 55M allele [14]. The -108C allele has greater promoter activity than the -108T allele, which results in different serum activities [15]. Several other polymorphisms affect serum PON1 activity to a lesser extent [16].

The PON1-Q192R polymorphism also determines the efficacy with which PON1 inhibits LDL oxidation, with the Q isoform being the most efficient and the R isoform least efficient [17,18]. These observations resulted in a plethora of genetic epidemiological studies to link the PON1 polymorphisms with coronary heart disease (CHD) presence to little or no effect, with meta-analyses showing a marginal relationship at best [19].

2. PON1 AND ATHEROSCLEROSIS

The concentration of LDL is directly related to the risk of developing atherosclerosis [20]. The current theory to explain the development of fatty streaks in the artery wall, which are believed to initiate atherosclerosis, states that the oxidation of LDL is critical [21,22]. Several cell types present in the artery wall, including endothelial cells, smooth muscle cells, and macrophages, can oxidize LDL [21,22]. Monocyte-derived macrophages present in the subintimal space of arteries avidly take up oxidized LDL to become foam cells. This uptake is not mediated by the LDL receptor, instead macrophage uptake of oxidized LDL involves other receptors, the so-called "scavenger" or "oxidized LDL" receptors [21–24]. The foam cells formed in this way secrete monocyte-chemotactic protein 1 (MCP-1) and macrophage colony-stimulating factor, causing the further recruitment and retention of lipid-laden macrophages which aggregate to form the fatty streak [22]. Thus, oxidized LDL is believed to play a central role in events that initiate atherosclerosis.

Epidemiological studies have shown a strong inverse relationship between serum HDL cholesterol concentration and the development of atherosclerosis [25], indicating that HDL is atheroprotective. The mechanism of this protection has been the subject of intense research, with the majority of this research directed at the central role of HDL in reverse cholesterol transport (RCT), the process of transporting excess cholesterol from peripheral tissues, particularly arterial wall macrophages, to the liver for disposal. Recently, however, attention has turned to other pleiotropic mechanisms whereby HDL exerts its atheroprotective effects. These include antioxidative, anti-inflammatory, anti-apoptopic, vasorelaxative, and anti-thrombotic effects, as well as promoting the normalization of endothelial function and stimulating endothelial progenitor cell function [26].

Recently, however, a number of human pharmaceutical intervention studies with HDL-raising agents failed to show clinical benefit. Unfortunately, one using the cholesteryl ester transfer protein (CETP) inhibitor torcetrapib was terminated due to off-target toxicity of the drug, while another with niacin had a flawed design [27]. A genetic Mendelian randomization study also failed to find a relationship between myocardial infarction and common genetic variants only associated with HDL cholesterol (HDL-C) levels [27]. Although these studies have cast some doubt on the "HDL Hypothesis", it is not known whether HDL function was affected in these studies (see the next section of this chapter, "Dysfunctional HDL"), and many more studies will be required before the hypothesis is cast aside.

In vitro studies attempting to discover the mechanism by which HDL prevented LDL oxidation found that the mechanism was saturable and therefore potentially enzymatic [28]. At that time, several potential candidate enzymes were known to be associated with HDL such as lecithin:cholesterol acyltransferase (LCAT), CETP, PON1, and lipoprotein-associated phospholipase A_2 (Lp-PLA$_2$), also known as platelet-activating factor acetyltransferase, that could be responsible [29]. It has turned out that all four can in some way prevent/retard LDL oxidation along with several other HDL-associated proteins [30].

PON1 was first shown to retard the oxidation of LDL in vitro by Mackness and colleagues [31,32], results which were subsequently confirmed by many other laboratories [17,33–35] and extended to include HDL and cell membranes [36,37]. In studies of macrophages in cell culture, PON1 exhibits a variety of potentially atheroprotective properties such as reducing macrophage oxidative stress and the ability of macrophages to oxidize LDL, inhibit cholesterol synthesis, and promote cholesterol efflux [9,38].

The mechanism by which PON1 retards LDL oxidation is unproven but appears to involve the hydrolysis of the truncated oxidized fatty acids from phospholipid, cholesteryl ester, and triglyceride hydroperoxides, resulting in the production of lysolipids, cholesterol, diglyceride, and oxidized fatty acids. A recent excellent study by Tavori and colleagues identified a specific triglyceride found in the atherosclerotic plaque and containing palmitic, oleic, and linoleic acids with 0.3% of the linoleate oxidized as a natural substrate for PON1 [39]. The lysolipids and oxidized fatty acids produced by PON1 are themselves potentially atherogenic but do not appear to be so when produced on HDL [29]. Unmetabolized lipid hydroperoxides are highly inflammatory, inducing the production of MCP-1 by arterial cells, which attracts monocytes into the arterial intima at the very start of the atherosclerotic process [21,22]. In the presence of PON1, lipid hydroperoxide concentrations are reduced, MCP-1 production is inhibited [40,41], and the atherosclerotic process is attenuated.

It is worth noting that a recent study has indicated that HDL can prevent the glycation of LDL (another pro-atherosclerotic modification of LDL) and that this is very probably due to PON1 [42].

The transgenic or adenovirus-mediated expression of human PON1 in various mouse models of atherosclerosis has been shown to retard or reverse atherosclerosis by mechanisms which include a reduction in circulating and aortic oxidized-LDL (oxLDL), a reduction in macrophage oxidative stress and foam cell formation, an increase in reverse cholesterol transport, and a normalization of endothelial function [43–47]. Interestingly, in human aortas, immunostaining for PON1 progressively increases as atherosclerosis develops [48], and the presence of both PON1 and PON3 in aortic macrophages indicates a cellular protective effect of these enzymes [49]. Recently, it has been shown that the expression of human PON1 can prevent diabetes development in mice through its antioxidant properties and the stimulation of beta-cell insulin release, suggesting a possible role for PON1 in insulin biosynthesis [50,51].

Low levels of serum PON1 have been consistently associated with susceptibility to CHD development in case-control studies. Several studies have previously shown prospectively that PON1 activity is a risk factor for CHD development independently of HDL concentration [52–54], including a study in Type 2 diabetes [55] although the finding is not universal [56,57]. However, a recent meta-analysis of the relationship between PON1 activity and CHD susceptibility which studied 9853 cases and 11,408 controls showed that on pooled analysis CHD patients had 19% lower PON1 activity than did controls ($P < 10^{-5}$). The same results were found in all subgroup analyses, including coronary stenosis, myocardial infarction, ethnicity, age, and sample size, amongst others [58]. Low PON1 concentration predicts cardiovascular mortality in hemodialysis patients [59].

Many studies have also investigated several PON1 polymorphisms as risk factors for CHD with positive associations being seen in some but not all studies. Meta-analyses have shown at best a marginal significance of the PON1-Q192R polymorphism as a risk factor for CHD but no relationship of other PON1 polymorphisms and CHD [19,60].

In a study of 3668 subjects undergoing elective coronary angiography without acute coronary syndrome, each had serum paraoxon and phenylacetate hydrolysis measured and were followed for 3 years for major adverse cardiovascular events (MACE; i.e., death, MI, stroke). Low PON1 paraoxonase and arylesterase activities were both associated with increased MACE risk [61]. Low arylesterase activity predicted future development of MACE in both primary and secondary prevention cohorts and even reclassified some subjects into higher risk categories. A genome-wide association study (GWAS) of SNPs in the PON1 gene associated with PON1 activity were not associated with MACE risk in an angiographic cohort of 2136, or with history of CHD or MI in the Coronary Artery Disease Genome-Wide Replication and Meta-Analysis Consortium of some 80,000 subjects [61]. These authors were therefore able to replicate the findings of several previous studies (PON1 activity is an important determinant of CHD development, genotype is not [60,62]), with far greater numbers of subjects.

It would appear, therefore, on the balance of current evidence, that PON1 activity is atheroprotective. The putative mechanisms explaining this atheroprotection are shown in Table 11.1. The prevention of LDL and cell membrane oxidation, reduction in macrophage oxidative stress, and promotion of RCT by PON1 are well documented [31–35,37,43,53]. The prevention of LDL glycation and diabetes development are new and require confirmation [42,50,51]. Nonetheless, they are exciting developments.

Table 11.1 Potential Anti-Atherosclerotic Mechanisms of PON1

Mechanism	Reference
Prevention of LDL and cell membrane oxidation	[31–35,37,42,52]
Prevention of LDL Glycation	[41]
Prevention of diabetes development	[49,50]
Reduction of macrophage oxidative stress	[37]
Promotion of macrophage RCT	[37,106]
Normalization of endothelial function	[44,64]
Metabolism of homocysteine thiolactones	[60–62]
Prevention of LCAT oxidative inactivation	[63]
Disposal of toxic apoptosis products	[65]

Diabetes is associated with a greatly increased burden of atherosclerosis, therefore preventing diabetes development would greatly decrease this burden.

The overexpression of PON1 in aged apoE-deficient mice normalized endothelial function by restoring cellular Ca^{2+} flux [45]. This finding requires confirmation in other models of atherosclerosis and in humans.

Homocysteine thiolactone (HcyTL) is a consequence of Hcy editing by methionyl-tRNA synthase and is associated with the atherosclerosis development related to hyperhomocysteinemia [63]. HcyTL is very reactive and modifies protein lysine residues by N–homocysteinylation. This in turn affects protein structure and function. Damage to proteins on lipoproteins or in cells of the vessel wall by N–homocysteinylation leads to cell death, autoimmune responses, inflammation, and, eventually, atherosclerosis [64]. However, HcyTL is a PON1 substrate (one of the few known to occur naturally). By metabolising HcyTL, PON1 reduces its concentration thereby limiting N–homocysteinylation and retarding atherosclerosis development [65]. This may be a major atheroprotective mechanism of PON1.

The HDL-associated enzyme LCAT plays a major role in RCT by esterifying cholesterol on HDL, thus creating a gradient for the movement of free cholesterol from cell membranes to HDL [20]. However, LCAT also has the ability to prevent LDL oxidation [30]. Unfortunately, LCAT is also inhibited by oxidative stress. A recent in vitro study has reported that PON1 prevented the oxidative inactivation of LCAT when incubated together in conditions of oxidative stress, and that this in turn prolonged the time HDL was able to prevent LDL oxidation [66]. If this finding can be confirmed, this atheroprotective function of PON1 could prolong protection against LDL oxidation while at the same time preserving normal RCT in conditions of inflammation and oxidative stress.

Recently, a number of studies have further confirmed a role for PON1 in atheroprotection. HDL isolated from healthy individuals increases NO bioavailability (improving endothelial function) via interaction with the SR-B1 receptor when incubated with cultured endothelial cells. In contrast, HDL isolated from individuals with CHD contained more malondialdehyde, was less able to interact with SR-B1, and caused no increase in (and in some cases inhibited) NO bioavailability. Adding malondialdehyde to healthy HDL or using HDL from PON1-deficient mice blunted NO production. The reason for this difference between healthy and CHD HDL was due to much lower PON1 activity associated with CHD HDL and hence a lower ability to remove the malondialdehyde [67].

Enrichment of human monocyte/macrophages with unesterified cholesterol causes the release of highly procoagulant microvesicles (UCMV) via the induction of apoptosis [68]. MVs contain damage-associated molecular patterns (DAMPs) which are endogenous danger signals that stimulate an immune response. UCMVs induce lymphocyte rolling and adhesion to post-capillary venules in rats in vivo, and augment adhesion of human monocytes to mouse aortic explants and cultured human endothelial cells via induction of intercellular adhesion protein-1. UCMVs induce mitochondrial production of superoxide and peroxides and contain malondialdehyde-like peroxidized epitopes. The incubation of UCMVs with HDL or purified PON1 detoxifies the malondialdehyde DAMPs and prevents the immune response. Thus, this is a potentially novel atheroprotective role of PON1, but also has wider implications of PON1 as a noninflammatory remover of apoptosis induced cellular debris [68].

3. DYSFUNCTIONAL HDL

The concept of dysfunctional HDL arose from observations that some individuals with high or normal HDL-C but low PON1 activity were susceptible to CHD development, while others with low HDL-C but high PON1 activity were not [69]. Since that time, many studies have been performed to show how and why HDL becomes dysfunctional [27]. Dysfunctionality of HDL can take the form of reduced cholesterol efflux capacity, but is most commonly measured by the loss of anti-inflammatory/antioxidative function. LDL added to endothelial cells in co-culture models is oxidized, inducing the production of monocyte chemotactic factors that increase monocyte binding and migration. The addition of HDL to the co-culture prevents the oxidation of the LDL and impairs the inflammatory response [40].

However, HDL from patients with a wide variety of diseases with an inflammatory component does not inhibit monocyte chemotaxis but may actually increase it, thus this dysfunctional HDL is often pro-inflammatory, promoting inflammation and CHD development (see also Chapter 4).

4. PON1 AND OTHER INFLAMMATORY DISEASES

Low serum PON1 is associated with many diseases with a large inflammatory component, including diabetes mellitus, rheumatoid arthritis, systemic lupus erythematosus, various hepatic and renal diseases, psoriasis, and macular degeneration [70]. These diseases are also characterized by having dysfunctional HDL believed to be caused by the low PON1 activity [71].

At present it is not known whether the low PON1 in these diseases is a cause or a consequence of disease presence. Large prospective epidemiological studies will be required to determine this and to determine the future direction of PON1 research in these diseases, either as a causal agent and potential therapeutic target or as a potential diagnostic tool.

5. PON1 AND ORGANOPHOSPHATE TOXICITY

This is the subject of excellent recent reviews and will be dealt with only briefly here [72,73].

Organophosphorus compounds (OPs) are widely used in both rural and urban settings, leading to widespread exposure. OPs are activated in the body by the process known as oxidative desulphuration to produce the toxic oxon forms. Some, but not all, parent or activated OPs are PON1 substrates, and of those that are (which include some of the most widely used including diazinon and chlorpyriphos [CP] oxons), most are hydrolyzed at different rates by the PON1-Q and R isoforms. Therefore, the majority of studies in this area have concentrated on PON1 as a genetic determinant of OP toxicity [72].

Animal studies have consistently shown that PON1 protects against OP toxicity. The administration of exogenous PON1 to rats and mice protects against OP toxicity, and administration of the PON1 isoform that hydrolyses the OP at the greatest rate affords most protection [74]. PON1 knockout mice are dramatically more susceptible to diazoxon and CP oxon toxicity, and the administration of exogenous PON1 restores resistance to these OPs [47].

In contrast to animal studies, there have been relatively few studies on the role of PON1 in OP toxicity in humans. Military personnel deployed in the Persian GulfWar of 1990–1991 were exposed to low levels of the OP nerve gas sarin and various OP insecticides as well as other chemical and biological agents [72]. In U.S. veterans, low activity of the PON1-192Q isoform correlated better with neurological symptoms of GulfWar Illness (GWS) than did the PON1-192R isoform or PON1 genotype [75]. In U.K. deployed veterans, serum PON1 activity was 25–35% lower than in nondeployed veterans, which were not due to differences in PON1 genotype [76]. However, neither PON1 activity nor genotype was associated with specific symptoms of illness. Further studies are required to show PON1 as a risk factor for GWS.

Several studies have investigated the role of PON1 in modulating the chronic central and peripheral nervous system abnormalities, the so-called "dippers flu", associated with exposure of sheep dippers to diazinon [77–79]. On the basis of the findings reported, it appears that diazinon has contributed to ill health in sheep dippers who had a lower capacity to detoxify diazoxon. Several other studies of OP-exposed agricultural workers have indicated individuals with PON1 genotypes associated with low activity (Q/M) had a greater frequency of various indices of OP toxicity (chronic toxicity, genotoxicity, impaired thyroid function) [80–82]. Unfortunately, in these studies exposure was to a large number of pesticides including OPs, not all of which were PON1 substrates, limiting the significance of the results.

Residential exposure to OPs (mainly diazinon and CP) was associated with an increase in Parkinson's disease risk in PON1-55M carriers [83] and to an increased risk of brain tumors in children with low PON1 activity [84]. In utero exposure to OPs of fetuses of mothers with low PON1 resulted in babies born with smaller head circumference and shorter gestational age [85,86]. In children of OP-exposed Mexican American women living in California and children in NewYork exposed in utero, low PON1 was associated with poorer scores in various tests of mental development [87,88].

Amyotrophic lateral sclerosis (ALS) is an adult-onset progressive and fatal neurodegenerative disease with unknown etiology. Recent evidence has, however, suggested a link between exposure to toxic environmental compounds, including OPs and sporadic ALS [89]. Population studies have indicated an association of PON1 polymorphisms and ALS, however, the findings are inconsistent between populations [90–92]. In a large multipopulation study, Wills and colleagues were unable to find any reductions in PON1 OP hydrolysis and ALS [93], clearly demonstrating the problems in defining a role for PON1 in human toxicology.

Although available human studies provide some evidence that low PON1 activity may increase susceptibility to various adverse effects of some OPs, doubts remain, and further studies in which OP exposure is more carefully characterized and monitored are required.

6. PON1 AND CANCER

Although PON1 activity changes or PON1 SNPs have been linked to a number of cancers, research is patchy and often poorly conducted [70]. However, two recent studies on PON1 and breast cancer merit mention. Saadat [94] conducted a meta-analysis of PON1 genetic polymorphisms and breast cancer susceptibility covering six eligible studies. The PON1-192R allele was associated with a decreased risk of breast cancer (OR = 0.57, 95% CI 0.49–0.67, P < 0.001). However, both PON1-55LM and PON1-55MM genotypes were associated with increased risk (OR = 1.32, 95% CI 1.10–1.58, $P = 0.002$ and OR = 2.16, 95% CI 1.75–2.68, P < 0.001, respectively). There was also a significant linear trend in risk-associated 0,1 and 2 PON1-55M alleles ($X^2 = 54.2, P < 0.001$).

In a second study of patients with recurrent breast cancer, PON1 arylesterase activity was negatively associated with early death ($P = 0.0109$). In a multiple logistic regression model, arylesterase activity was independently associated with early death (OR = 0.12, $P = 0.047$), the time interval between first diagnosis and recurrence (OR = 0.54, $P = 0.078$), and undernutrition (OR = 3.95, $P = 0.088$), indicating arylesterase activity to be a potential survival marker in these patients [95].

Although small, these studies offer some hope that rigorously conducted research may indicate a role for PON1 in the development or diagnosis of some cancers.

7. PON1 AND QUORUM QUENCHING

Quorum sensing (QS) is a bacterial cell-to-cell signaling system that controls the production of virulence factors in many pathological bacteria [96,97]. QS is mediated by the production of autoinducers, small signal molecules that activate or repress gene expression when a minimal threshold concentration is reached [96,97]. N-acyl homoserine lactones (AHLs) are used as autoinducers by many Gram-negative pathogens, such as *Pseudomonas aeruginosa* to control the expression of virulence factors [96,97]. AHLs are degraded by lactonases, which is termed *quorum quenching*, and the production of these enzymes is an effective way to interfere with QS and prevent bacterial virulence [98].

Both murine and human PON1 are capable of hydrolyzing AHLs and therefore of modulating *P. aeruginosa* QS [99]. Using a *Drosophila melanogaster* (which have no intrinsic PON1) model of infection, the transgenic expression of human PON1 protected *Drosophila* from *P. aeruginosa* lethality [100]. Further studies in this model have indicated that in addition to quorum quenching, PON1 expression decreased superoxide anion levels and altered the expression of multiple genes related to oxidative stress. In addition, the profile of bacterial species colonizing the *Drosophila* gut was dramatically altered [101]. These and other studies indicate that PON1 (and PON2 and PON3, which share these quorum-quenching properties) could be useful tools in combating infections by QS bacteria, warranting further studies.

8. PON1 AND AGING

Several studies have shown a reduction in serum PON1 activity and consequent loss of protective functions of HDL, with advancing age [102,103]. On the other hand, polymorphisms within the PON1 gene have been associated with successful aging, leading to the suggestion that PON1 is a longevity-associated protein [104].

In an exciting recent advance, Lee and colleagues [105] investigated the effects of silencing the PON1 gene in cultured human dermal microvascular endothelial cells (HDMECs). The expression of endogenous PON1 was knocked down using small interfering RNA. The authors found that 1) cell viability was decreased, 2) the expression of cellular senescence biomarkers such as moesin and Rho-GDI were significantly decreased, and 3) aging of the HDMECs was triggered. The findings suggest that cellular PON1 has a functional role in cellular senescence and also functions as an aging-related protein. With an aging world population, healthy aging is an important goal, therefore further study into this new and potentially exciting function of PON1 is warranted.

9. CONCLUSION

It appears increasingly obvious that serum PON1 contributes to the atheroprotective function of HDL by decreasing lipid peroxidation in a variety of diseases with an inflammatory component. Much more research has been and is being conducted into PON1 and atherosclerosis than into other diseases, although there is still a need to determine exactly how PON1 contributes to HDL function. It is not known whether PON1 acts directly on lipid peroxides on molecules other than HDL, such as LDL, or whether lipid peroxides are transferred to HDL for disposal (although there is some evidence to suggest this [106,107] or whether PON1 acts indirectly by protecting HDL from oxidative damage and therefore maintaining other normal functions. This uncertainty needs to be removed before PON1 becomes a therapeutic target in atherosclerosis.

It is also true to say that a definitive role for PON1 in protecting humans from the toxic effects of OP pesticides has not been proven despite decades of research. Research into other putative roles for PON1 in cancer development, bacterial infection, and aging are really in their infancy but could nevertheless prove fruitful and exciting avenues of research.

In conclusion, PON1 contributes to the antioxidative function of HDL and is atheroprotective. The putative role of PON1 in preventing bacterial infection may contribute to HDL's role in innate immunity. More basic and clinical investigations of all the possible functions of PON1 are required. In particular, although much is known about genetic, nutritional, lifestyle, and pharmacological factors that affect human serum PON1 activity levels in vivo and some information is available on receptors and signaling pathways affecting PON1 synthesis in vitro [13,108,109], almost nothing is known about the regulatory pathways which operate in humans in vivo. There is, therefore, a large hole in our knowledge base that hampers the discovery of small molecule effectors of PON1 for therapeutic interventions.

REFERENCES

[1] Rajkovic MG, Rumora L, Barisic K. The paraoxonase 1, 2 and 3 in humans. Biochemia Medica 2011;21:122–30.
[2] Mackness B, Durrington PN, Mackness MI. Human serum paraoxonase. Gen Pharmac 1998;31:329–36.
[3] Mazur A. An enzyme in the animal organism capable of hydrolysing the phosphorus-fluorine bond of alkyl fluorophosphates. J Biol Chem 1946;164:271–89.
[4] Aldridge WN. Serum esterases 2. An enzyme hydrolysing diethyl p-nitrophenylphosphate (E600) and its identity with the A-esterase of mammalian sera. Biochem J 1953;53:117–24.
[5] Davidson WS, Silva RAGD, Chantepie S, Lagor WR, Chapman MJ, Kontush A. Proteomic analysis of defined HDL subpopulations reveals particle-specific protein clusters—Relevance to antioxidative function. Arterioscler Thromb Vasc Biol 2009;29:870–6.

[6] Harel M, Aharoni A, Gaidukov L, Brumshtein B, Khersonsky O, Maged R, et al. Structure and evolution of the serum paraoxonase family of detoxifying and anti-atherosclerotic enzymes. Nat Struct Mol Biol 2004;11:412–9.

[7] Primo-Parma SL, Sorenson RC, Teiber J, La Du BN. The human serum paraoxonase/arylesterase gene (PON1) is one member of a multigene family. Genomics 1996;33:498–509.

[8] Rodriguez-Sanabria F, Rull A, Beltran-Debon R, Aragones G, Camps J, Mackness B, et al. Tissue distribution and expression of paraoxonases and chemokines in the mouse: the ubiquitous and joint localisation suggest a systemic and coordinated role. J Mol Histol 2010;41:379–86.

[9] Aviram M, Rosenblat M. Paraoxonases 1, 2, and 3, oxidative stress, and macrophage foam cell formation during atherosclerosis development. Free Radic Biol Med 2004;37:1304–16.

[10] Reddy ST, Devarajan A, Bourquard N, Shih D, Fogelman AM. Is it just paraoxonase 1 or are other members of the paraoxonase gene family implicated in atherosclerosis? Curr Opin Lipidol 2008;19:405–8.

[11] Deakin SP, James RW. Genetic and environmental factors modulating serum concentrations and activities of the antioxidant enzyme paraoxonase-1. Clin Sci 2004;107:435–47.

[12] Costa LG, Vitalone A, Cole TB, Furlong CE. Modulation of paraoxonase activity. Biochem Pharmacol 2005;69:541–50.

[13] Schrader C, Rimbach G. Determinants of paraoxonase 1 status: genes, drugs and nutrition. Curr Med Chem 2011;18; 5624–3.

[14] Leviev I, Negro F, James RW. Two alleles of the human paraoxonase gene produce different amounts of mRNA. An explanation for differences in serum concentrations of paraoxonase associated with the (Leu-Met54) polymorphism. Arterioscler Thromb Vasc Biol 1997;17:2935–9.

[15] Leviev I, James RW. Promoter polymorphisms of human paraoxonase PON1 gene and serum paraoxonase activities and concentrations. Arterioscler Thromb Vasc Biol 2000;20:516–21.

[16] Kim DS, Burt AA, Ranchalis JE, Richter RJ, Marshall JK, Eintracht JF, et al. Additional common polymorphisms in the PON gene cluster predict PON1 activity but not vascular disease. J Lipids 2012;2012:476316.

[17] Aviram M, Billecke S, Sorenson R, Bisgaier C, Newton R, Rosenblat M, et al. Paraoxonase active site required for protection against LDL oxidation involves its free sulphydryl group and is different from that required for its arylesterase/paraoxonase activities: selective action of human paraoxonase alloenzymes Q and R. Arterioscl Thromb Vasc Biol 1998;10:1617–24.

[18] Mackness B, Mackness MI, Arrol S, Turkie W, Durrington PN. Effect of the human serum paraoxonase 55 and 192 genetic polymorphisms on the protection by high density lipoprotein against low density lipoprotein oxidative modification. FEBS Letts 1998;423:57–60.

[19] Wheeler JG, Keavney BD, Watkins H, Collins R, Danesh J. Four paraoxonase gene polymorphisms in 11212 cases of coronary heart disease and 12786 controls: meta-analysis of 43 studies. Lancet 2004;363:689–95.

[20] Durrington PN. Hyperlipidaemia. Diagnosis and management. 2nd ed. London: Butterworth Heinemann; 1995.

[21] Steinberg D, Parthasarathy S, Carew TE, Khoo JC, Witztum JL. Beyond cholesterol modifications of low-density lipoprotein that increase its atherogenicity. New Engl J Med 1989;320:915–24.

[22] Lusis AJ. Atherosclerosis. Nature 2000;407:233–41.

[23] Chisholm GM, Penn MS. Oxidised lipoproteins and atherosclerosis. In: Fuster V, Ross R, Topol EJ, editors. Atherosclerosis and coronary artery disease. Philadelphia: Lippincott-Raven; 1996. p. 129–49.

[24] Endemann G, Stanton LW, Madden KS, Bryant CM, White RT, Protter AA. CD36 is a receptor for oxidised low density lipoprotein. J Biol Chem 1993;268:11811–6.

[25] Gordon DJ, Probstfield JL, Garrison RJ, Neaton JD, Castelli WP, Knoke JD, et al. High-density lipoprotein cholesterol and cardiovascular disease. Four prospective American studies. Circulation 1989;79:8–15.

[26] Shah PK. Evolving concepts on benefits and risks associated with therapeutic strategies to raise HDL. Curr Opin Cardiol 2010;25:603–8.

[27] Fisher EA, Feig JE, Hewing B, Hazen SL, Smith JD. High-density lipoprotein function, dysfunction and reverse cholesterol transport. Arterioscler Thromb Vasc Biol 2012;32:2813–20.

[28] Mackness MI, Abbott CA, Arrol S, Durrington PN. The role of high density lipoprotein and lipid-soluble antioxidant vitamins in inhibiting low-density lipoprotein oxidation. Biochem J 1993;294:829–35.

[29] Mackness MI, Durrington PN. High density lipoprotein, its enzymes and its potential to influence lipid peroxidation. Atherosclerosis 1995;115:243–53.

[30] Mackness B, Mackness M. The antioxidant properties of high-density lipoproteins in atherosclerosis. Panminerva Med 2012;54:83–90.

[31] Mackness MI, Arrol S, Durrington PN. Paraoxonase prevents accumulation of lipoperoxides in low-density lipoprotein. FEBS Letts 1991;286:152–4.

[32] Mackness MI, Arrol S, Abbott CA, Durrington PN. Protection of low-density lipoprotein against oxidative modification by high-density lipoprotein associated paraoxonase. Atherosclerosis 1993;104:129–35.

[33] Watson AD, Berliner JA, Hama SY, La Du BN, Fault KF, Fogelman AM, et al. Protective effect of high density lipoprotein associated paraoxonase—inhibition of the biological activity of minimally oxidised low-density lipoprotein. J Clin Invest 1995;96:2882–91.

[34] Ahmed Z, Ravandi A, Maguire GF, Emili A, Draganov D, La Du BN, et al. Apolipoprotein AI promotes the formation of phosphatidylcholine core aldehydes that are hydrolysed by paraoxonase (PON1) during high density lipoprotein oxidation with a peroxynitrite donor. J Biol Chem 2001;276:24473–81.

[35] Draganov DI, La Du BN. Pharmacogenetics of paraoxonase: a brief review. Naunyn-Schmiedeberg's Arch Pharmacol 2004;369:78–88.

[36] Aviram M, Rosenblat M, Bisgaier CL, Newton RS, Primo-Parmo SL, La Du BN. Paraoxonase inhibits high-density lipoprotein oxidation and preserves its functions. J Clin Invest 1998;101:1581–90.

[37] Deakin SP, Bioletto S, Bochaton-Piallat ML, James RW. HDL-associated paraoxonase-1 can redistribute to cell membranes and influence sensitivity to oxidative stress. Free Radic Biol Med 2011;50:102–9.

[38] Berrougui H, Loued S, Khalil A. Purified human paraoxonase-1 interacts with plasma membrane lipid rafts and mediates cholesterol efflux from macrophages. Free Rad Biol Med 2012;52:1372–81.

[39] Tavori H, Aviram M, Khatib S, Musa R, Mannheim D, Karmeli R, et al. Paraoxonase 1 protects macrophages from atherogenicity of a specific triglyceride isolated from human carotid lesion. Free Rad Biol Med 2011;51:234–42.

[40] Navab M, Imes SS, Hama SY, Hough GP, Ross LA, Bork RW, et al. Monocyte transmigration induced by modification of low density lipoprotein in cocultures of human aortic wall cells is due to induction of monocyte chemotactic protein 1 synthesis and is abolished by high density lipoprotein. J Clin Invest 1991;88:2039–46.

[41] Mackness B, Hine D, Liu Y, Mastorikou M, Mackness M. Paraoxonase 1 inhibits oxidised LDL-induced MCP-1 production by endothelial cells. BBRC 2004;318:680–3.

[42] Younis NN, Soran H, Charlton-Menys V, Sharma R, Hama S, Pemberton P, et al. High-density lipoprotein impedes glycation of low-density lipoprotein. Diabet Vasc Dis Res 2013;10:152–60.

[43] Rozenberg O, Shih DM, Aviram M. Paraoxonase 1 (PON1) attenuates macrophage oxidative status: studies in PON1 transfected cells and in PON1 transgenic mice. Atherosclerosis 2005;181:9–18.

[44] Mackness B, Quarck R, Verreth W, Mackness M, Holvoet P. Human paraoxonase-1 overexpression inhibits atherosclerosis in a mouse model of metabolic syndrome. Arterioscler Thromb Vasc Biol 2006;26:1545–50.

[45] Guns P-J, Van Assche T, Verreth W, Fransen P, Mackness B. Mackness, M, Holvoet P, Bull H. Paraoxonase 1 gene transfer lowers vascular oxidative stress and improves vasomotor function in apolipoprotein E-deficient mice with pre-existing atherosclerosis. Br J Pharmacol 2008;153:508–16.

[46] Tward A, Xia YR, Wang XP, Shi YS, Park C, Castellani LW, et al. Decreased atherosclerotic lesion formation in human serum paraoxonase transgenic mice. Circulation 2002;106:484–90.

[47] Shih DM, Gu L, Xia Y-R, Navab M, Li W-F, Hama S, et al. Mice lacking serum paraoxonase are susceptible to organophosphate toxicity and atherosclerosis. Nature 1998;394:284–7.

[48] Mackness B, Hunt R, Durrington PN, Mackness MI. Increased immunolocalisation of paraoxonase, clusterin and apolipoprotein AI in the human artery wall with progression of atherosclerosis. Arterioscler Thromb Vasc Biol 1997;17:1233–8.

[49] Marsillach J, Camps J, Beltran-Debon R, Rull A, Aragones G, Maestre-Martinez C, et al. Immunohistochemical analysis of paraoxonases 1 and 3 in human atheromatous plaques. Eur J Clin Invest 2011;41:308–14.

[50] Rozenberg O, Shiner M, Aviram M, Hayek T. Paraoxonase 1 (PON1) attenuates diabetes development in mice through its antioxidative properties. Free Rad Biol Med 2008;44:1951–9.

[51] Koren-Gluzer M, Aviram M, Meilin E, Hayek T. The antioxidant HDL-associated paraoxonase-1 (PON1) attenuates diabetes development and stimulates β-cell insulin release. Atherosclerosis 2011;219:532–7.

[52] Mackness B, Durrington P, McElduff P, Yarnell J, Azam N, Watt M, et al. Low paraoxonase activity predicts coronary events in the Caerphilly Prospective Study. Circulation 2003;107:2775–9.

[53] Bhattacharyya T, Nicholls SJ, Topol EJ, Zhang R, Yang X, Schmitt D, et al. Relationship of paraoxonase 1 (PON1) gene polymorphisms and functional activity with systemic oxidative stress and cardiovascular risk. JAMA 2008;299:1265–76.

[54] van Himbergen TM, van der Schouw YT, Voorbij HAM, van Tits LJH, Stalenhoef AFH, Peeters PHM, et al. Paraoxonase (PON1) and the risk for coronary heart disease and myocardial infarction in a general population of Dutch women. Atherosclerosis 2008;198:408–14.

[55] Ikeda Y, Inoue M, Suehiro T, Arii K, Kumon Y, Hashimoto K. Low human paraoxonase predicts cardiovascular events in Japanese patients with type 2 diabetes. Acta Diabetol 2009;46:239–42.

[56] Troughton JA, Woodside JV, Yarnell JWG, Arveiler D, Amouyel P, Ferrieres J, et al. Paraoxonase activity and coronary heart disease in healthy middle-aged males: the PRIME study. Atherosclerosis 2008;197:556–63.

[57] Birjmohun RS, Vergeer M, Stroes ES, Sandhu MS, Ricketts SL, Tanck MW, et al. Both paraoxonase-1 genotype and activity do not predict the risk of future coronary artery disease; the EPIC- Norfolk prospective population study. PLoS One 2009;4:e6809.

[58] Wang M, Lang X, Cui S, Zou L, Wang S, Wu X. Quantitative assessment of the influence of paraoxonase 1 activity and coronary heart disease. DNA Cell Biol 2012;31:975–82.

[59] Ikeda Y, Suehiro T, Itahara T, Inui Y, Chikazawa H, Inoue M, et al. Human serum paraoxonase concentration predicts cardiovascular mortality in hemodialysis patients. Clin Nephrol 2007;67:358–65.

[60] Mackness B, Davies GK, Turkie W, Lee E, Roberts DH, Hill E, et al. Paraoxonase status in coronary heart disease. Are activity and concentration more important than genotype? Arterioscler Thromb Vasc Biol 2001;21:1451–7.

[61] Tang WHW, Hartiala J, Fan Y, Wu Y, Stewart AFR, Erdmann J, et al. Clinical and genetic association of serum paraoxonase and arylesterase activities with cardiovascular risk. Arterioscler Thromb Vasc Biol 2012;32:2803–12.

[62] Jarvik GP, Rozek LS, Brophy VH, Hatsukami TS, Richter RJ, Schellenberg GD. Furlong, C.E. Paraoxonase (PON1) phenotype is a better predictor of vascular disease than is $PON1_{192}$ or $PON1_{55}$ genotype. Arterioscl Thromb Vasc Biol 2000;20:2441–7.

[63] Jakubowski H. Protein homocysteinylation: possible mechanisms underlying pathological consequences of elevated homocysteine levels. FASEB J 1999;13:2277–83.

[64] Jakubowski H. Homocysteine thiolactone: metabolic origin and protein homocysteinylation in humans. J Nutr 2000;130:377S–81S.

[65] Jakubowski H. Calcium-dependent human serum homocysteine thiolactone hydrolase—a protective mechanism against protein s-homocysteinylation. J Biol Chem 2000;275:3957–62.

[66] Hine D, Mackness B, Mackness M. Co-incubation of PON1, APO A1 and LCAT increases the time HDL is able to prevent LDL oxidation. IUBMB Life 2012;64:157–61.

[67] Besler C, Heinrich K, Rohrer L, Doerries C, Riwanto M, Shih DM, et al. Mechanisms underlying adverse effects of HDL on eNOS-activating pathways in patients with coronary artery disease. J Clin Invest 2011;121:2693–708.

[68] Liu M-L, Scalia R, Mehta JL, Williams KJ. Cholesterol-induced membrane microvesicles as novel carriers of damage-associated molecular patterns. Arterioscler Thromb Vasc Biol 2012;32:2113–21.

[69] Navab M, Hama-Levy S, Van Lenten BJ, Fonarow GC, Carelinez CJ, Castellani LW, et al. Mildly oxidised LDL induces an increased apolipoprotein J/paraoxonase ratio. J Clin Invest 1997;99:2005–19.

[70] Goswami B, Tayal D, Gupta N, Mallika V. Paraoxonase: a multifaceted biomolecule. Clin Chim Acta 2009;410:1–12.

[71] Navab M, Anantharamaiah GM, Reddy ST, Van Lenten BJ, Ansell BJ, Fogelman AM. Mechanisms of disease: proatherogenic HDL—an evolving field. Nat Clin Pract Endocrinol Metab 2006;2:504–11.

[72] Costa LG, Giordano G, Cole TB, Marsillach J, Furlong CE. Paraoxonase 1 (PON1) as a genetic determinant of susceptibility to organophosphate toxicity. Toxicology 2013;307:115–22.

[73] Costa LG, Cole TB, Jarvik GP, Furlong CE. Functional genomics of the paraoxonase (PON1) polymorphisms: effect on pesticide sensitivity, cardiovascular disease and drug metabolism. Ann Rev Med 2003;54:371–92.

[74] Li WF, Costa LG, Richter RJ, Hagen T, Shih DM, Tward A, et al. Catalytic efficiency determines the in vivo efficacy of PON1 for detoxifying organophosphates. Pharmacogenetics 2000;10:767–79.

[75] Haley RW, Billecke S, La Du BN. Association of low PON1 type Q (type A) arylesterase activity with neurologic symptom complexes in Gulf War veterans. Toxicol Appl Pharmacol 2000;157:227–33.

[76] Hotopf M, Mackness MI, Nikolau V, Collier DA, Curtis C, David A, et al. Paraoxonase in Persian Gulf War veterans. J Occup Environ Med 2003;45:668–75.

[77] Mackenzie Ross SJ, Brewin CR, Curran HV, Furlong CE, Abraham-Smith KM, Harrison V. Neuropsychological and psychiatric functioning in sheep farmers exposed to low levels of organophosphate pesticides. Neurotoxicol Teratol 2010;32:452–9.

[78] Cherry N, Mackness M, Mackness B, Dippnall M, Povey A. "Dippers' flu" and its relationship to PON1 polymorphisms. Occ Environm Med 2011;68:211–7.

[79] Mackness B, Durrington P, Povey A, Thomson S, Dippnall M, Mackness M, et al. Paraoxonase and susceptibility to organophosphorus poisoning in farmers dipping sheep. Pharmacogenetics 2003;13:81–8.

[80] da Silva J, Moraes CR, Heuser VD, Andrade VM, Silva FR, Kvitko K, et al. Evaluation of genetic damage in a Brazilian population occupationally exposed to pesticides and its correlation with polymorphisms in metabolising genes. Mutagenesis 2008;23: 415–22.

[81] Lee BW, London L, Poulauskis J, Myers J, Christiani DC. Association between human paraoxonase gene polymorphism and chronic symptoms in pesticide-exposed workers. J Occup Environ Med 2003;45:118–22.

[82] Lacasana M, Lopez-Flores I, Rodriguez-Barranco M, Aguilar-Garduno C, Blanco-Munoz J, Perez-Mendez O, et al. Interaction between organophosphate pesticide exposure and PON1 activity on thyroid function. Toxicol Appl Pharmacol 2010;249:16–24.

[83] Manthripragada AD, Costello S, Cokburn MG, Bronstein JM, Ritz B. Paraoxonase 1, agricultural organophosphate exposure and Parkinson disease. Epidemiology 2010;21:87–94.

[84] Searles Nielsen S, Mueller BA, DeRoos AJ, Viernes HM, Farin FM, Checkoway H. Risk of brain tumors in children and susceptibility to organophosphorus insecticide; the potential role of paraoxonase (PON1). Environ Health Perspect 2005;113: 909–13.

[85] Berkowitz GS, Wetmur JG, Birman-Deych E, Obel J, Lapinski RH, Godbold JH, et al. In utero pesticide exposure, maternal paraoxonase activity and head circumference. Environ Health Perspect 2004;112:388–91.

[86] Harley KG, Huen K, Schall RA, Holland NT, Bradman A, Barr DB, et al. Association of organophosphate pesticide exposure and paraoxonase with birth outcome in Mexican American Women. PLoS One 2011;6:e23923.

[87] Engel SM, Wetmur J, Chen J, Zhu C, Barr DB, Canfield RL, et al. Prenatal exposure to organophosphates, paraoxonase 1 and cognitive development in childhood. Environ Health Perspect 2011;119:1182–8.

[88] Eskenazi B, Huen K, Marks A, Harley KG, Bradman A, Barr DB, et al. PON1 and neurodevelopment in children from the CHAMACOS study exposed to organophosphate pesticides in utero. Environ Health Perspect 2010;118:1775–81.

[89] Garliardi S, Abel K, Bianchi M, Milani P, Bernuzzi S, Corato M, et al. Regulation of FMO and PON detoxication systems in ALS human tissues. Neurotox Res 2013;23:370–7.

[90] Ricci C, Battistini S, Cozzi L. Lack of association of PON polymorphisms with sporadic ALS in an Italian population. Neurobiol Aging 2011;32:552; e7-552.e13.

[91] Slowik A, Tomik B, Wolkow PP, Partyka D, Turaj W, Malecki MT, et al. Paraoxonase gene polymorphisms and sporadic ALS. Neurology 2006;67:766–70.

[92] Ticozzi N, LeClerc AL, Keagle PJ, Glass JD, Wills AM, van Blitterswijk M, et al. Paraoxonase gene mutations in amyotrophic lateral sclerosis. Ann Neurol 2010;68:102–7.

[93] Wills AM, Landers JE, Zhang H, Richter RJ, Caraganis AJ, Cudkowicz ME, et al. Paraoxonase 1 (PON1) organophosphate hydrolysis is not reduced in ALS. Neurology 2008;70:929–34.

[94] Saadat M. Paraoxonase 1 genetic polymorphisms and susceptibility to breast cancer: a meta-analysis. Cancer Epidemiol 2012;36:e101–3.

[95] Bobin-Dubigeon C, Jaffre I, Joalland M-P, Classe J-M, Campone M, Herve M, et al. Paraoxonase 1 (PON1) as a marker of short term death in breast cancer recurrence. Clin Biochem 2012;45:1503–5.

[96] Camps J, Pujol I, Ballester F, Joven J, Simo JM. Paraoxonases as potential antibiofilm agents: their relationship with quorum-sensing signals in gram-negative bacteria. Antimocrob Agents Chemother 2011;55:1325–31.

[97] Simanski M, Babucke S, Eberl L, Harder J. Paraoxonase 2 acts as a quorum-quenching factor in human keratinocytes. J Invest Dermatol 2012;132:2296–9.

[98] Teiber JF, Horke S, Haines DC, Chowdhary PK, Xiao J, Kramer JK, et al. Dominant role of paraoxonases in inactivation of the pseudomonas aeruginosa quorum-sensing signal N-(3-oxododecanoyl)-L-homoserine lactone. Infect Immun 2008;76:2512–9.

[99] Ozer EA, Pezzulo A, Shih DM, Chun C, Furlong C, Lusis AJ, et al. Human and murine paraoxonase 1 are host modulators of *Pseudomonas aeruginosa* quorum-sensing. FEMS Microbiol Lett 2005;253:29–37.

[100] Stoltz DA, Ozer EA, Taft PJ, Barry M, Liu L, Kiss PJ, et al. *Drosophila* are protected from *Pseudomonas aeruginosa* lethality by transgenic expression of paraoxonase-1. J Clin Invest 2008;118:3123–31.

[101] Pezzulo AA, Hornick EE, Rector MV, Estin M, Reisetter AC, Taft PJ, et al. Expression of human paraoxonase 1 decreases superoxide levels and alters bacterial colonisation in the gut of *Drosophila melanogaster*. PLoS One 2012;7:e43777.

[102] Seres I, Paragh G, Deschene E, Fulop Jr T, Khalil A. Study of factors influencing the decreased HDL associated PON1 activity with aging. Exp Gerontol 2004;39:59–66.

[103] Jaouad L, de Guise C, Berrougui H, Cloutier M, Isabelle M, Fulop T, et al. Age-related decrease in high-density lipoproteins antioxidant activity is due to an alteration in the PON1's free sulfhydryl groups. Atherosclerosis 2006;185:191–200.

[104] Lescai F, Marchegiani F, Franceschi C. PON1 is a longevity gene: results of a meta-analysis. Ageing Res Rev 2009;8:277–84.

[105] Lee YS, Park CO, Noh JY, Jin S, Lee NR, Noh S, et al. Knockdown of paraoxonase 1 expression influences ageing of human dermal microvascular endothelial cells. Exp Dermatol 2012;21:682–7.

[106] Christison JK, Rye KA, Stocker R. Exchange of oxidised cholesteryl-linoleate between LDL and HDL mediated by cholesteryl ester transfer protein. J Lipid Res 1995;36:2017–26.

[107] Klimov AN, Nikiforova AA, Kuzmin AA, Kuznetsov AS, Mackness MI. Is high density lipoprotein a scavenger for oxidised phospholipids of low density lipoprotein? Advances in Lipoprotein and Atherosclerosis Research. Diagnostics and Treatment. Jena: Gustav Fischer Verlag; 1998, 78–82.

[108] Camps J, Marsillach J, Joven J. Pharmacological and lifestyle factors modulating serum paraoxonase-1 activity. Mini Rev Med Chem 2009;9:911–20.

[109] Costa LG, Giordano G, Furlong CE. Pharmacological and dietary modulators of paraoxonase 1 (PON1) activity and expression: the hunt goes on. Biochem Pharmacol 2011;81:337–44.

CHAPTER *12*

Proteomic Diversity in HDL: A Driving Force for Particle Function and Target for Therapeutic Intervention

Scott M. Gordon
National Heart, Lung and Blood Institute, National Institutes of Health, Bethesda, MD, USA

Contents

Abstract

High-density lipoproteins (HDL) are circulating particles composed of phospholipids, cholesterol, and proteins. There is a well established inverse correlation between plasma levels of HDL-associated cholesterol (HDL-C) and the incidence of cardiovascular disease. Attempts at therapeutic interventions to raise plasma HDL-C levels have been successful, raising it by 30% to more than 100%. However, so far, pharmacological raising of HDL-C has provided no benefit in terms of clinical outcomes. In an attempt to understand why, researchers have been looking deeper into the composition and functional properties of HDL. Recent analyses of the protein composition of HDL have discovered that upwards of 70 different plasma proteins are consistently identified on isolated particles. The known functions associated with these proteins are diverse and span physiological roles far beyond the traditional roles for HDL in lipid metabolism, suggesting that novel functional roles for HDL may exist. This information has sparked a new area of research aimed at identifying proteomic subspecies of HDL and understanding their functionality in health and disease. Here we discuss the proteomic diversity of HDL and the concept of subspeciation as well as the influence of disease on HDL-associated proteins and the potential for therapeutic manipulation of the HDL proteome to treat cardiovascular and other inflammatory diseases.

Key concepts

- HDL exists as a heterogeneous and proteomically diverse population of particles in the blood. More than 70 different proteins have been repeatedly found to associate with HDL.
- HDL particles are functionally diverse and the functionality of individual particle species is driven by their protein composition.
- The protein composition of HDL is influenced by the environment of the blood, which can change in response to various factors such as diet, disease, or therapeutic intervention.
- Further advancement in the understanding of HDL proteomics will require new methodologies for HDL isolation, quantitative analysis of individual proteins, and detection of posttranslational modifications.

1. INTRODUCTION TO HDL-ASSOCIATED PROTEINS

High-density lipoproteins (HDL), like other lipoproteins, are particles composed of a variety of lipid and protein components. Their high density relative to the other lipoprotein classes (i.e., chylomicrons, very low-density lipoprotein [VLDL], and low-density lipoprotein [LDL]) is a result of HDL's higher protein-to-lipid ratio. Figure 12.1 displays the lipid and protein compositions of the major lipoprotein classes. HDL are the smallest and densest of the lipoproteins and differ significantly in both lipid and protein composition. As much as 60% of HDL mass can come from associated protein components, much more than any other class of lipoprotein. HDL not only differs from other lipoprotein classes in total protein content but in the identities and variety of associated proteins. All lipoproteins rely on some protein component for stability and solubility in the blood; the proteins responsible for this are known as apolipoproteins. Apolipoproteins are amphipathic in nature, meaning that they have distinct regions that are

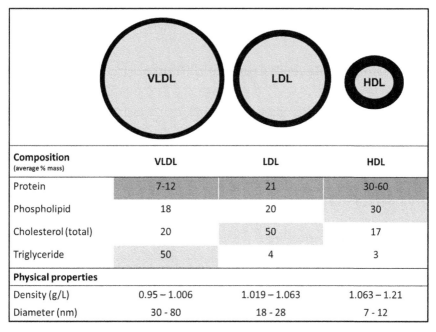

Composition (average % mass)	VLDL	LDL	HDL
Protein	7-12	21	30-60
Phospholipid	18	20	30
Cholesterol (total)	20	50	17
Triglyceride	50	4	3
Physical properties			
Density (g/L)	0.95 – 1.006	1.019 – 1.063	1.063 – 1.21
Diameter (nm)	30 - 80	18 - 28	7 - 12

Figure 12.1 *Composition and Physical Properties of Major Lipoprotein Classes.* Circular representations of lipoprotein particles depict an outer protein component (black) and major lipid compositions for each lipoprotein (gray). Average values representing the percentage of total particle mass of individual components are listed in the table as well as the distinguishing physical properties of density and particle diameter.

hydrophilic and prefer interactions with water and also regions that are hydrophobic, preferring interaction with lipid. This amphipathic nature of apolipoproteins allows them to shield hydrophobic components of lipid particles contributing to their solubility in the aqueous environment of the blood. Apolipoprotein B (apoB) is synthesized by cells of the liver and small intestine and is responsible for the formation of lower density lipoproteins. HDL, also formed by the liver and small intestine, is secreted into the blood as a nascent phospholipid disc stabilized by apolipoprotein A-I (apoA-I). As nascent HDL discs circulate throughout the body, they acquire a core of cholesteryl ester through mechanisms of cholesterol efflux (discussed in Chapter 5). As these nascent HDL mature, they take on a larger and more spherical shape and the resulting increase in surface area allows for the association of additional proteins with the particle. In mature HDL, apoA-I is the most common protein component comprising 50–70% of total protein mass, while the remainder comes from a number of peripherally associated proteins. The focus of this chapter is on how these proteins were identified and how they interact with HDL to influence its function. To begin, two brief examples of complex biological processes performed by HDL and driven by its protein components will be described.

1.1. Proteins of Reverse Cholesterol Transport

Since HDL was first isolated in the 1950s, many studies have examined its protein composition. Traditionally, proteins that associate with HDL have been grouped into three categories largely related to HDL's best described function which is as a transporter of lipids. These categories include structural proteins, lipid-modifying enzymes, and lipid transfer proteins. These traditional proteins are the best understood HDL-associated proteins in terms of their roles on HDL, and they work together to orchestrate a symphony of events driving what is commonly thought to be the primary function of HDL in humans—reverse cholesterol transport (RCT, discussed in Chapter 4). This is the process by which excess cholesterol is removed from peripheral tissues and transported to the liver where it can be converted to bile acids and excreted in the feces or recycled. In this process, structural proteins on the HDL particle (e.g., apoA-I) interact with cellular receptors such as ATP-binding cassette A1 (ABCA1) or ABCG1 to facilitate the efflux of free cholesterol from a cell to the HDL particle. In the next act, the free cholesterol, which is now embedded in the outer phospholipid shell of the HDL particle, is acted upon by a lipid-modifying enzyme on HDL called lecithin:cholesteryl acyltransferase (LCAT, further discussed in Chapter 7), which esterifies an acyl chain to the free cholesterol. The increased hydrophobicity of this new cholesteryl ester (CE) forces it to the hydrophobic core of the HDL particle. Accumulation of CE in the core of HDL results in the increased particle size and maturation to a more spherical morphology. The fate of this CE is also determined by HDL-associated proteins. One pathway involves a lipid transfer protein on HDL called cholesteryl ester transfer protein (CETP), which transfers CE to VLDL or LDL in exchange for triglyceride. The other involves interactions between apoA-I on HDL and surface proteins on hepatocytes, resulting in the net transfer of cholesterol from HDL to the liver where enzymes convert the cholesterol to bile acids, which are ultimately excreted in the feces or reabsorbed.

1.2. Proteins of the Trypanosome Lytic Factor

In 1902 a factor in human plasma (whole blood with cellular components removed) was identified as having a specific lytic activity against a class of protozoan parasites known as trypanosomes. It was not until nearly 80 years later that this "factor" was identified as an HDL particle with a very specific protein composition that conferred this activity [1,2]. This HDL particle has come to be known as the trypanosome lytic factor (TLF) and consists of the proteins apoA-I, haptoglobin-related protein (Hpr), and apoL1. Today, the mechanism of this specific lytic event has been largely worked out and demonstrates an instrumental role for each of these primary protein components. ApoA-I acts as a structural scaffold for the association of Hpr and apoL1 on the same HDL particle that is required for this lytic mechanism. First, Hpr on TLF binds to hemoglobin (Hb) in the circulation, forming a TLF-Hpr-Hb complex. This complex then binds a receptor in the flagellar pocket of the trypanosome resulting in the endocytic uptake of the TLF into the trypanosome. Upon acidification of the endosome, apoL1 undergoes a conformational change allowing it to dissociate from the TLF and insert in the lysosome membrane where it forms an ionic pore, allowing the influx of Cl^- ions and water into the lysosome. This then leads to lysosomal rupture and ultimately results in death of the trypanosome [3].

These two examples highlight the importance of HDL-associated proteins in driving physiological processes in the body related to both lipid metabolism and immune defense. It is easy to see how a seemingly random group of proteins can come together on an HDL particle and work in concert to perform rather complicated tasks. Today, the study of HDL-associated proteins and the physiological processes they mediate is still young and rapidly expanding with the development of new methods for the isolation of HDL and more sensitive mass spectrometry (MS) techniques. Over the next few sections of this chapter we will discuss the proteomic diversity of HDL and how this is influenced by disease and could potentially be modified therapeutically to prevent disease.

2. THE PROTEOMIC DIVERSITY OF HDL

Early studies of HDL isolated from plasma by ultracentrifugation techniques quickly identified several of the most commonly known HDL-associated proteins, the apolipoproteins. These proteins comprise the majority of total protein on HDL, and, in addition to acting as vital structural components as mentioned earlier, they also perform many other physiological roles.

The more recent application of modern high-resolution MS technologies to the identification of proteins on HDL has proven to be a fruitful field of research, identifying many new, low-abundance protein components that were undetectable in previous biochemical studies. Several MS-based studies have been done, each utilizing slightly different techniques for HDL isolation or types of MS analysis. The results of these studies are compiled in Table 12.1 [4–16]. Overall, 188 proteins have been reported to associate with HDL across 13 MS-based proteomics studies, 73 of these are common to at least three studies and are likely to be agreed upon by the majority of those in the field. This large number of HDL-associated proteins suggests a much more vast diversity in HDL composition than was previously supposed and along with this came new insights into HDL function. Most of these new HDL-associated proteins were previously known to be present in plasma but had known functions that are very different from the traditional role of HDL in lipid metabolism (e.g., protease inhibition and complement regulation). The presence of these proteins on HDL suggests the possibility of new, as yet undiscovered, physiological roles for HDL. The following section highlights some key features of many of the most commonly detected HDL-associated proteins. They are sorted into categories based on broad functional classifications but many have roles overlapping two or more categories.

2.1. Apolipoproteins

2.1.1. Apolipoprotein A-I (apoA-I)

ApoA-I is the most abundant protein component of HDL, making up from 60–70% of total protein mass on HDL. The mature protein consists of 243 amino acids (aa's) and its secondary structure is characterized by repeating amphipathic α-helical domains, a common feature of many apolipoproteins

Table 12.1 HDL Associated Proteins Identified by MS Based Studies

Protein name	Accession	Hits	Protein name	Accession	Hits
apolipoprotein A-I	P02647	13	prothrombin	P00734	6
apolipoprotein A-IV	P06727	13	retinol binding protein 4	P02753	6
apolipoprotein C-III	P02656	13	vitronectin	P04004	6
apolipoprotien L1	O14791	13	alpha-1-acid glycoprotein 1	P02763	5
apolipoprotein A-II	P02652	12	antithrombin-III	P01008	5
apolipoprotein E	P02649	12	apolipoprotein(a)	P08519	5
apolipoprotein C-I	P02654	11	haptoglobin	P00738	5
apolipoprotein C-II	P02655	11	hemoglobin subunit alpha	P69905	5
apolipoprotein M	O95445	11	kininogen-1	P01042	5
paraoxonase 1	P27169	11	phosholipid transfer protein	P55058	5
transthyretin	P02766	11	plasma protease C1 inhibitor	P05155	5
alpha-1-antitrypsin	P01009	10	alpha-1-antichymotrypsin	P01011	4
apolipoprotein D	P05090	10	alpha-2-antiplasmin	P08697	4
apolipoprotein J	P10909	10	alpha-2-macroglobulin	P01023	4
serum amyloid A4	P35542	10	angiotensinogen	P01019	4
albumin	P02768	9	complement factor B	P00751	4
apolipoprotein B	P04114	9	complement factor H	P08603	4
fibrinogen alpha chain	P02671	9	hemoglobin subunit beta	P68871	4
haptoglobin related protein	P00739	9	heparin cofactor 2	P05546	4
serotransferrin	P02787	9	lecithin:cholesterol acyltransferase	P04180	4
serum amyloid A1 &2	P02735	9	paraoxonase 3	Q15166	4
vitamin D binding protein	P02774	9	platelet basic protein	P02775	4
alpha-1B-glycoprotein	P04217	8	afamin	P43652	3
alpha-2-HS-glycoprotein	P02765	8	ceruloplasmin	P00450	3
complement C3	P01024	8	cholesteryl ester transfer protein	P11597	3
apolipoprotein F	Q13790	7	complement C1s subcomponent	P09871	3
apolipoprotein H	P02749	7	complement C9	P02748	3
hemopexin	P02790	7	fibrinogen beta chain	P02675	3
alpha-1-acid glycoprotein 2	P19652	6	fibronectin	P02751	3
apolipoprotein C-IV	P55056	6	gelsolin	P06396	3
complement C4-B	P0C0L5	6	platelet factor 4	P02776	3
inter alpha trypsin inhibitor H2	P19823	6	prenylcystein oxidase	Q9UHG3	3
inter alpha trypsin inhibitor H4	Q14624	6	serum amyloid P	P02743	3
pigment epithelium-derived factor	P36955	6	zinc-alpha-2-glycoprotein	P25311	3
protein AMBP	P02760	6			

which favors their association with lipids in the aqueous environment of the blood. It has been found to play a variety of functional roles primarily involved in the process of RCT. One primary function of this protein is interaction with transporters on the cellular surface to mediate cholesterol efflux as described above. ApoA-I also acts as a cofactor for LCAT, increasing the rate of CE formation.

2.1.2. Apolipoprotein A-II (apoA-II)

The second most abundant protein on HDL, apoA-II, is much smaller than apoA-I, only 77 aa, but shares the structural amphipathic α-helix motif. Synthesis occurs primarily in the liver and, to a lesser extent, in the small intestine. In the plasma of normal individuals, apoA-II is present at about 0.3 mg/mL and can be found as a monomer or homo/heterodimer by forming disulfide bonds with itself, apoE, or apoD [17,18]. It is mainly associated with HDL but can also bind chylomicrons and VLDL. ApoA-II plays a role in HDL remodeling by influencing HDL-associated enzymes, including CETP [19], PLTP [20], LCAT [21], and peripheral enzymes that act on HDL such as hepatic lipase [22]. HDL function is also affected by apoA-II. HDL isolated from apoA-II transgenic mice has impaired ability to promote cholesterol efflux and is unable to protect LDL from oxidation, limiting two major protective roles for HDL in cardiovascular disease [23,24]. Because of these and other observations, apoA-II has gained a reputation as a pro-atherogenic apolipoprotein.

2.1.3. Apolipoprotein A-IV (apoA-IV)
ApoA-IV is a 376-aa (46-kDa) protein synthesized by the intestine which can associate with chylomicrons as well as HDL or remain lipid free in the plasma as a monomer or homodimer. ApoA-IV plays roles in RCT, satiety in response to feeding [25], and can act as a potent antioxidant [26].

2.1.4. Apolipoprotein A-V (apoA-V)
ApoA-V is expressed by the liver and can associate with VLDL. This protein has been shown to influence plasma triglyceride levels by activating the enzyme lipoprotein lipase [27].

2.1.5. Apolipoprotein B (apoB)
ApoB is the key protein component of non-HDL lipoproteins, however proteomics studies often identify apoB-derived peptides in HDL preparations. It is synthesized as two isoforms: apoB100 (4563 aa), which is made by the liver and found on LDL and VLDL, and apoB48 (2152 aa), representing the amino terminal portion of full-length apoB100, which is produced by the small intestine and is the primary component of chylomicrons. The primary role of this protein is to act as a ligand for the LDL receptor.

2.1.6. Apolipoprotein Cs (apoC-I–apoC-III)
The physical and expression properties of the apoC proteins on HDL are displayed in Table 12.2 [28]. Functionally, these proteins play very important roles in lipoprotein metabolism and can exchange easily between the lipoprotein classes. All three of these apoC proteins act to inhibit interaction of lower density lipoproteins with one or more of their respective receptors, including the LDL receptor, LDL receptor–related protein, or VLDL receptor. This inhibition most often occurs through shielding of the lipoprotein-associated ligands (i.e., apoB or apoE) [28]. On HDL, they can modulate the functions

Table 12.2 Properties of apoC Proteins

Properties	apoC-I	apoC-II	apoC-III
Amino acids	57	79	79
MW(kDa)	6.6	8.8	8.8
Tissue expression	L/S/Sk/Ln/T	L/I	L/I
Plasma concentration (mg/mL)	0.06	0.04	0.12

L = liver; S = spleen; Sk = skin; Ln = lung; T = testis; I = intestine
Table derived from Jong et al [28]

of the lipid transfer enzymes LCAT and CETP. LCAT activity is strongly activated by apoC-I and moderately reduced by apoC-II and apoC-III. CETP is affected differently by these proteins with its activity being decreased by apoC-I and increased by apoC-II [28].

2.1.7. Apolipoprotein D (apoD)

Mature apoD is a 169-aa protein that shows little structural similarity to other apolipoproteins, containing few α-helical domains and mostly beta sheets [29,30]. It is, however, highly similar to the lipocalin family of proteins [31]. The apparent molecular weight (MW) of apoD can range from 19–32 kDa depending on the degree of glycosylation at various sites [30]. Interestingly, apoD contains a cysteine that has been demonstrated to form bonds with multiple other plasma proteins [32]. ApoD also varies from many other apolipoproteins in its tissue-expression profile, with little expression by the liver and intestine. ApoD is produced mostly by the adrenal glands and kidney, but also by numerous other tissues [29]. Normal plasma levels of apoD are between 0.05 and 0.23 mg/mL [33]. On HDL, apoD has been proposed to play a role in stabilizing LCAT and increasing cholesterol esterification [34].

2.1.8. Apolipoprotein E (apoE)

A classic apolipoprotein synthesized primarily by the liver (~75%), mature apoE is 299 aa (34 kDa), forming a four-helix bundle structure. Normal plasma levels of apoE are about 0.05 mg/mL. The C-terminal domain of apoE has high affinity for lipid and is responsible for its binding to chylomicrons, VLDL, and HDL [35]. This domain can also interact with ABCA1 to promote the assembly of HDL particles [36]. A cluster of positively charged amino acids in helix 4 of the N-terminal domain are responsible for binding of apoE to the LDL receptor. This interaction is responsible for uptake of apoE containing lipoproteins by all of the body's cells (except red blood cells) but the majority of uptake occurs in the liver. ApoE is generally considered to be anti-atherogenic. Knockout of apoE in mice models results in markedly increased plasma cholesterol and triglyceride levels (due to lack of clearance) and atherosclerosis [37]. On HDL, apoE can act to inhibit platelet aggregation [38]. Interestingly, apoE is present as three major isoforms (E2–E4) which strongly correlate with risk for and severity of Alzheimer's disease [39].

2.1.9. Apolipoprotein F (apoF or Lipid Transport Inhibitor Protein)

The liver is the primary site of synthesis for this o-linked glycoprotein consisting of 162 aa's (29 kDa) after cleavage of a large proprotein region. Plasma levels are typically around 0.08 mg/mL and apoF can associate with both HDL and LDL. On HDL, which is where the majority of plasma apoF resides (~75%) [40], its function is not known but Lagor and colleagues estimate that about 1 in 9 HDL particles could contain apoF [41]. On LDL, apoF acts as an inhibitor of CETP-mediated lipid exchange [40]. ApoF does not have well defined α-helical regions like many other apolipoproteins. Knockout of apoF has little effect on plasma lipid levels in mice or HDL protein composition, however, a decrease in ABCA1-mediated cholesterol efflux to apoB-depleted serum of male apoF KO mice was observed [41]. Overexpression of apoF resulted in a 25% reduction of HDL-C due to accelerated clearance from plasma [42].

2.1.10. Apolipoprotein H (apoH or Beta-2-glycoprotein)

Another glycoprotein, apoH is a 326-aa (50-kDa) protein synthesized by the liver. Although structurally different from the classical apolipoproteins, apoH can still bind lipids. The presence of apoH on lipoprotein particles has been questioned [43], however abundant evidence supports its binding to low- and high-density lipoproteins, including its identification by seven different proteomic studies of HDL (Table 12.1). Several functional properties have been described for apoH, including activation of lipoprotein lipase, role as a cofactor in binding of some antiphospholipid antibodies, inhibition of platelet prothrombinase and ADP-induced platelet aggregation, and involvement in macrophage-mediated clearance of immune complexes [44–46]. Interestingly, plasma levels of apoH are increased in patients with diabetes and correlate positively with cholesterol levels, but the causes and consequences of this are not understood [47].

2.1.11. Apolipoprotein J (apoJ or Clusterin)

The secretory form of apoJ is a glycosylated 427-aa (70-kDa) protein that is found in most body fluids and has considerable functional homology to the small heat-shock protein family [48,49]. In the plasma it is primarily associated with HDL. Little is known about its roles on HDL but the protein has been widely studied for potential functions in cellular protection [50], cancer [51], and Alzheimer's disease [52].

2.1.12. Apolipoprotein L (apoL)

The apoL family consists of six members (apoL-I to -VI) with significant structural and functional similarity to the Bcl-2 family of proteins and are thought to play roles as intracellular membrane channels and function in programmed cell death. Only apoL-I has been detected in association with HDL and has been identified as a critical component of the TLF described above. Additionally, plasma levels of apoL have been demonstrated to correlate with plasma triglyceride and cholesterol levels [53]. The mature apoL-I protein is 371 aa's long (44 kDa) and contains N-linked carbohydrate.

2.1.13. Apolipoprotein M (apoM)

Most similar to apoD, apoM is also a member of the lipocalin family. Interestingly, the 188-aa (25-kDa) protein is expressed in the liver and kidneys [54] and is secreted with its hydrophobic 20-aa signal peptide still intact. It has been suggested that this may be responsible for apoM's ability to associate with lipoproteins. The protein contains one glycosylation site and six cysteine residues and forms three intramolecular disulfide bonds. Plasma concentrations range from 0.02 to 0.15 mg/mL, and the protein is primarily associated with HDL [55] (estimated to be present on ~5% of HDL particles [56] but can also associate with LDL). Coprecipitation studies in human plasma demonstrate that apoM can reside on particles with numerous other proteins, including apoA-I, apoA-II, and the apoCs [56]. Mouse studies indicate important roles for apoM in HDL metabolism and size distribution [57].

2.1.14. Apolipoprotein O (apoO)

A more recently discovered apolipoprotein, apoO, is 198 aa's long and contains a 22-residue α-helical repeat region, similar to apoA-I, A-IV, and E [58]. This region is likely responsible for apoO's ability to bind and clear lipids on its own. The protein has a predicted MW of 20 kDa but apoO isolated from human plasma is heavily glycosylated, containing chondroitin sulfate glycans and resulting in an apparent MW of 55 kDa. Synthesis occurs in multiple tissues including adrenal and adipose tissue. The majority of apoO in the plasma is either not associated with lipoproteins or is associated with HDL particles, while small amounts are detected on VLDL and LDL [58].

2.2. Lipid Transfer Proteins

2.2.1. Lecithin: Cholesteryl Acyltransferase (LCAT)

LCAT is an enzyme on HDL that is responsible for the esterification of free cholesterol received from peripheral cells forming cholesteryl ester. Esterified cholesterol accumulates in the hydrophobic core of the HDL particle along with triglyceride, giving mature HDL its spherical morphology.

2.2.2. Phospholipid Transfer Protein (PLTP)

PLTP mediates the exchange of phospholipid between different lipoprotein classes. Movement of phospholipid to HDL will generate larger, less dense particles.

2.2.3. Cholesteryl Ester Transfer Protein (CETP)

The primary function of CETP is as a transporter of cholesteryl esters from HDL to lower density apoB-containing lipoproteins. At the same time, triglyceride is transferred in the opposite direction to HDL.

2.3. Protease Inhibitors

2.3.1. Inter-Alpha-Trypsin Inhibitor Heavy Chains (ITIH2 and 4)

Inter-alpha-trypsin inhibitor family proteins are secreted into the blood primarily by the liver (blood concentrations ~0.15–0.5 mg/mL). Heavy chains represent the main structural unit of the inter-alpha-trypsin inhibitor complex. There are six closely related heavy-chain isoforms (H1–H6), and two of these are commonly found on HDL H2 and H4. Interestingly, H2 and H4 act as acute phase proteins that are down- and upregulated, respectively, during an inflammatory response [59]. Very little is known about the functional properties of these proteins other than that they may play roles in extracellular matrix stabilization.

2.3.2. Plasma Protease C1 Inhibitor (SERPIN1)

This is one of several members of the serine protease inhibitor (SERPIN) family that associates with HDL (some others include alpha-1-antitrypsin and antithrombin). C1 inhibitor is a heavily glycosylated 478-aa protein with a calculated MW of 76 kDa. In the plasma this protein is present at about 0.25 mg/mL, although it is unknown how much of this is likely to be bound to HDL at any given time. Synthesis occurs in hepatocytes, endothelial cells, fibroblasts, and monocytes and macrophages [60,61] and is upregulated during acute phase responses [62]. C1 inhibitor plays many physiological roles in the plasma acting as an important regulator of the complement pathway, contact phase, coagulation, and fibrinolytic systems [63].

2.4. Complement Proteins

2.4.1. Complement C3 (C3)

Another acute phase protein that associates with HDL is C3. Although primarily produced by the liver, C3 can also be secreted by activated macrophages and adipocytes [64]. C3 is a very large protein (1663 aa's) that is cleaved into two active fragments (C3a and C3b) by C3 convertase during its primary physiological role in the complement cascade. It is not clear whether it is the intact protein or one or more fragments that are commonly identified on HDL, but a better understanding of this may help to explain how C3 is functioning on HDL particles. While the roles of C3 in the complement cascade are well defined, recent studies have connected C3 more directly to lipid metabolism [65] and metabolic and cardiovascular disorders [66].

2.4.2. Complement C4b (C4b)

When complement component C4 is cleaved by upstream components of the complement cascade (one of which is activated C1s, also detected on HDL), it forms two fragments, C4a and C4b. C4b is a 193-kDa protein which forms a complex with C2b and this complex has C3 convertase activity, cleaving C3 into C3a and C3b and thus further propagating the complement cascade. It has been reported that patients with mutations leading to low copy number of C4b have an increased incidence of myocardial infarction and stroke [67]. The link between this observation and C4b's roles in complement cascade or its association with HDL are not yet clear.

2.4.3. Complement C9 (C9)

Involved in the final stage of the complement cascade, C9 plays a key role in the formation of the membrane attack complex, forming pores in invading pathogens to disrupt osmotic balance and ultimately resulting in death of the pathogen. Deficiency in humans is very rare but can lead to recurrent meningitis [68]. C9 is also a positive acute phase protein.

2.5. Other Important HDL-Associated Proteins

2.5.1. Serum Amyloid A1 and A2 (SAA1 and SAA2)

These are acute phase proteins produced by the liver and secreted into the blood as part of the inflammatory response. SAA1 has been shown to associate with HDL, displacing apoA-I and inhibiting several of the cardioprotective features of HDL.

2.5.2. Serum Amyloid A4 (SAA4)

In contrast to SAA1 and SAA2, SAA4 is constitutively expressed by the liver, not just during inflammation. SAA4 is found bound to HDL in normal individuals and has been shown to form phospholipid discs on its own similar to many apolipoproteins [69].

2.5.3. Paraoxonase-1 (PON1)

PON1 was discovered in the 1950s [70] but its roles in the prevention of cardiovascular disease were not realized until almost 40 years later. Now the antioxidative properties of this protein and its association with HDL are booming topics of research. PON1 is a 355-aa (47-kDa) enzyme secreted primarily from the liver with a retained leader sequence that directs its binding to HDL in the plasma. Because of PON1's potent protective properties it is one of the most heavily studied proteins on HDL today. The enzyme has the capacity to metabolize several products that are harmful to the body including organophosphates (often in pesticides), nerve gases, and aromatic esters (lactones). Interest in PON1 on HDL started when Mackness and colleagues found that PON1 was the protein component responsible for the majority of HDL's antioxidative capacity [71]. When on HDL, PON1 prevents the accumulation of oxidized lipids on the HDL particle [72], but it can also inhibit the oxidation of lipid on LDL [71]. The prevention of LDL oxidation is thought to be the primary mechanism by which HDL–associated PON1 protects against cardiovascular disease.

3. PROTEOMIC SUBSPECIATION OF HDL

In the preceding section we learned about the vast complexity of the HDL proteome. Now we are left wondering, how can all of these proteins fit on an HDL particle that is only 7 to 12 nm in diameter? The answer is that they cannot. It would be physically impossible for all of the known HDL-associated proteins to be present on every single HDL particle. There is simply not enough surface area on the particle to accommodate more than a few proteins at a time. Additionally, many of these proteins are present in low copy number in the plasma such that there are not enough copies to account for the number of HDL particles. So, how do these HDL-associated proteins interact with HDL? To better understand this, it helps to think about the environment of the blood as a whole. Circulating blood acts as a connective tissue linking all of the body's tissues to each other. The vessels that carry the blood can be thought of as a highway system facilitating the movement of absorbed nutrients to peripheral tissues and the movement of metabolic waste products from peripheral tissues to the liver or kidneys for excretion from the body. The blood is a complex environment containing cellular components (red and white blood cells), sugars (glucose), mineral ions (sodium, potassium, calcium, etc.), lipid-free proteins, and lipoproteins. All of these components play different roles in vital physiological processes. Red blood cells deliver the oxygen necessary for respiration while white blood cells regulate inflammatory processes and protect the body from invading pathogens. Ions maintain proper electrochemical gradients necessary for nerve signal transduction and cell-signaling processes. Lipid-free proteins maintain osmotic balance while performing any number of unique specialized functions. The lipoprotein system allows the body to easily transport lipids such as phospholipid, triglyceride, and cholesterol between organs or tissues. All of these blood components and the tissues that they infuse contribute to the environment of the blood. And this environment is ever changing and adapting based on factors such as age, genetics, diet, state of feeding, overall health, and location within the body. As nascent HDL particles are introduced into the circulation, they begin to interact with the environment by acquiring lipids and proteins, and the manner in which this occurs is likely dependent on the current/local environment of the blood.

The protein environment of blood is extremely complex. Combined datasets from numerous proteomics experiments on plasma have found more than 1900 distinct proteins [73]. Of these, 188 have been detected on HDL in at least one study. That is almost 10%. The HDL proteome does not appear to be just a random sampling of blood proteins; it is clear that there is some driving factor for the association of proteins with HDL particles. There are two possible mechanisms for the association of a protein with any lipoprotein particle: affinity for lipid or affinity for another protein that has affinity for lipid. Several of the "core" apolipoproteins (apoA-I, apoA-IV, apoE, etc.) have defined amphipathic α-helix domains that allow them to associate with lipids and form particles on their own. In the plasma, these proteins are in contact with lipid most of the time. The majority of the other minor HDL-associated proteins interact with the particle more transiently, either by weaker hydrophobic interactions or by interactions with one or more of the core HDL proteins. In this sense, HDL acts as a scaffold where protein components can associate or disassociate depending on the local environment. Therefore, the total pool of HDL in an individual at any given time is composed of numerous proteomic subpopulations or "subspecies" [74]. Current evidence suggests that this dynamic nature of HDL is intended to facilitate the coming together of two or more proteins with cooperative function to drive important physiological processes. This is already evidenced by examples such as the TLF, and the further potential for this is apparent if we look at the structural and functional properties of all of the HDL-associated proteins. By reading over the brief summaries in the previous section, we find that many of these proteins share structural similarities such as retention of hydrophobic signal sequences that direct binding to lipoproteins or similarities in glycation patterns. Additionally, many members of the HDL proteome contribute to similar physiological processes [75] (Figure 12.2); in fact, several are already known to physically interact with each other. The functional likeness of these proteins makes it easy to develop new hypotheses for why these proteins might be coming together on HDL particles, and the broad diversity of their physiological roles makes it exciting to imagine the possibilities for novel and

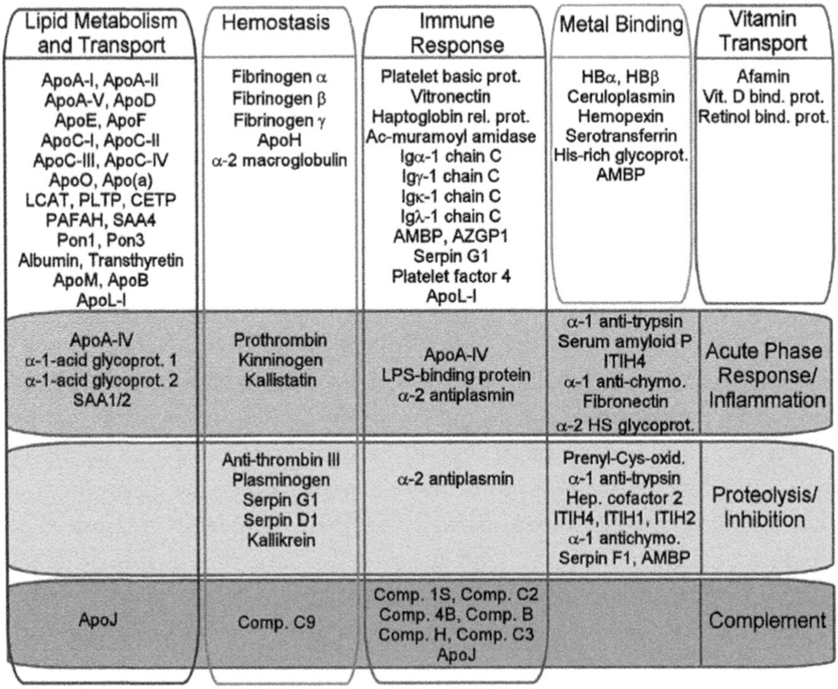

Lipid Metabolism and Transport	Hemostasis	Immune Response	Metal Binding	Vitamin Transport
ApoA-I, ApoA-II ApoA-V, ApoD ApoE, ApoF ApoC-I, ApoC-II ApoC-III, ApoC-IV ApoO, Apo(a) LCAT, PLTP, CETP PAFAH, SAA4 Pon1, Pon3 Albumin, Transthyretin ApoM, ApoB ApoL-I	Fibrinogen α Fibrinogen β Fibrinogen γ ApoH α-2 macroglobulin	Platelet basic prot. Vitronectin Haptoglobin rel. prot. Ac-muramoyl amidase Igα-1 chain C Igγ-1 chain C Igκ-1 chain C Igλ-1 chain C AMBP, AZGP1 Serpin G1 Platelet factor 4 ApoL-I	HBα, HBβ Ceruloplasmin Hemopexin Serotransferrin His-rich glycoprot. AMBP	Afamin Vit. D bind. prot. Retinol bind. prot.
ApoA-IV α-1-acid glycoprot. 1 α-1-acid glycoprot. 2 SAA1/2	Prothrombin Kinninogen Kallistatin	ApoA-IV LPS-binding protein α-2 antiplasmin	α-1 anti-trypsin Serum amyloid P ITIH4 α-1 anti-chymo. Fibronectin α-2 HS glycoprot.	Acute Phase Response/ Inflammation
	Anti-thrombin III Plasminogen Serpin G1 Serpin D1 Kallikrein	α-2 antiplasmin	Prenyl-Cys-oxid. α-1 anti-trypsin Hep. cofactor 2 ITIH4, ITIH1, ITIH2 α-1 antichymo. Serpin F1, AMBP	Proteolysis/ Inhibition
ApoJ	Comp. C9	Comp. 1S, Comp. C2 Comp. 4B, Comp. B Comp. H, Comp. C3 ApoJ		Complement

Figure 12.2 *Gene Ontology Analysis of HDL Associated Proteins.* Proteins found to associate with HDL were analyzed for known physiological roles. Protein names are encircled by functional groups to which they belong. [*This figure was taken from Shah et al. [75]*].

undiscovered functional roles for HDL subspecies. Perhaps more examples of defined HDL subspecies (such as the TLF) exist and play important physiological roles that are equally or even more interesting. Although we have identified a large number of protein components of HDL, we do not yet fully understand how distinct subspecies with defined protein compositions are formed. In fact, we have very little data describing the compositional details of HDL subspecies (i.e., which proteins may be interacting on a single HDL particle).

The dynamic composition of HDL has made it difficult to come to a consensus categorization of HDL subspecies. The compositional heterogeneity of HDL was realized shortly after it was first isolated. Since then, researchers have used a variety of techniques for the isolation and analysis of HDL subspecies. Before shotgun MS techniques were applied to HDL, Asztalos and colleagues used native two-dimensional gel electrophoresis (2DIGE), first separating by charge and then by size followed by immunoreactive identification to characterize the protein composition of HDL [76]. Using this technique, they identified populations of HDL with different electrophoretic mobilities and protein components, laying the groundwork for the study of HDL subspecies. Later, Karlsson and colleagues analyzed samples of HDL2 and HDL3 by 2DIGE using matrix-assisted laser desorption ionization mass spectrometry (MALDI-MS). This approach confirmed many protein associations with HDL and also identified differentially glycosylated forms of apoA-I and apoA-II [4].

With the development of soft-ionization MS techniques came shotgun proteomic analyses of HDL. This is where HDL is isolated from plasma and the total protein is extracted and then analyzed by MS as a complex mixture. The first studies by Vaisar and colleagues applied this technique to pools of total HDL isolated by ultracentrifugation and identified more than 30 protein components of HDL [8], many more than were previously thought to reside on HDL based on immunochemical studies. Taking a deeper look, Davidson and colleagues used density gradient ultracentrifugation to separate HDL into five subfractions and subsequently analyzed each by electrospray ionization mass spectrometry (ESI-MS) to identify the protein components of each subfraction [9]. This important study showed that some HDL-associated proteins were evenly distributed across density fractions while others would prefer lighter or more dense HDL. Considering the destructive effects of ultracentrifugation on the HDL particle, Gordon and colleagues developed a size-based HDL subfractionation method utilizing size exclusion chromatography to divide the total pool of HDL even further into about 10 fractions under physiological salt and force conditions [10]. Similar size exclusion methods had been used previously to separate lipoproteins from plasma, however, because this method separates by size rather than density, there is a large degree of comigration of large or very abundant plasma proteins such as albumin and immunoglobulins in the lipoprotein fractions, preventing confident MS analyses. To overcome this issue a phospholipid-binding resin was utilized to pull down only those proteins which are associated with phospholipid from the size-fractionated plasma, allowing for confident analysis of only lipid-associated proteins [10]. This advance opened the door for MS analysis of HDL fractions collected by methods other than ultracentrifugation. Using this technique, an additional 14 previously undescribed lipid-associated proteins were identified, and a detailed look at how almost 50 different protein components of HDL distributed across lipoproteins of different sizes was produced (Figure 12.3) [10].

Because of the inherent differences in the techniques used to isolate HDL for studies of protein composition, it can be difficult to compare HDL subspecies identified between them—especially because so far there have been no reports of an isolation technique capable of isolating individual species of HDL. Nevertheless, comparing the results of these different techniques sheds some light on the composition of individual HDL subspecies. The development of methods capable of examining the protein components of individual HDL particles should be a goal for future research.

4. INFLUENCE OF DISEASE ON THE HDL PROTEOME

Earlier we referred to HDL as a scaffold for the association of plasma protein components and suggested that it was the environment of the blood that dictated which proteins might be associated with HDL at any given time. So what happens to the HDL proteome when some event changes the plasma environment? One very common example of such an event is inflammation. An inflammatory response can be triggered by physical injury, as a response to infection by an invading organism, as a result of various chronic inflammatory diseases, or even just by eating a poor diet. An inflammatory response causes a series of systemic events, including the production and secretion of acute phase proteins into the bloodstream by the liver and activation and increased numbers of white blood cells. These events change the environment of the blood and subsequently the protein composition of HDL. The most dramatic effect of inflammation on HDL involves the acute phase protein serum

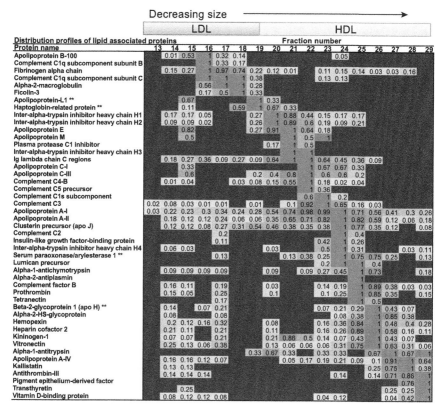

Figure 12.3 *Heat Map of Protein Distribution Across Size-Fractionated Lipoproteins.* The column on the left lists all of the lipid-associated proteins identified by electrospray ionization mass spectrometry across size exclusion chromatography separated plasma fractions. Each row displays the relative distribution of a given protein across fractions as determined by spectral counting with the maximum normalized to 1. From this figure it is clear that the total pool of HDL-associated proteins is differentially distributed across particle sizes such that each fraction (column) contains a unique set of proteins that are the components of the HDL subspecies isolated within that fraction. *[Reprinted with permission from Gordon et al. [10]. Copyright 2010 American Chemical Society].*

amyloid A (SAA). When present in the plasma, SAA can displace apoA-I on HDL and effect a variety of HDL functions. HDL isolated from patients in an acute phase response has decreased capacity to mediate cholesterol efflux [77] and decreased ability to inhibit oxidation of LDL [78], two primary mechanisms by which HDL protects against cardiovascular disease.

Several studies have used mass spectrometry to analyze the HDL proteome in patients with cardiovascular disease and various inflammatory conditions. These studies found that while the majority of HDL-associated proteins do not change dramatically, several key protein components do and some are consistent across numerous inflammatory conditions. HDL-associated SAA increases in acute coronary syndrome (ACS), chronic kidney disease (CKD), rheumatoid arthritis (RA), and psoriasis [11–14]. Several of these studies also noted decreased apoA-I, supporting the earlier observation that SAA displaces apoA-I on HDL. Another protein increased on HDL in all of these groups is complement component C3, a protein involved in innate immunity through the complement cascade. Other proteins change differently depending on the specific disease state, suggesting more specific roles than just being a result of general inflammation. Patients with coronary artery disease (CAD) do not have elevated SAA but do have increased apoE and apoC-IV associated with their HDL [8], an observation that was not common to patients with ACS, CKD, RA, or psoriasis.

Some have used the proteomic signature of HDL as a diagnostic tool to predict CAD. Vaisar and colleagues analyzed HDL_2 from patients using MALDI-MS and pattern-recognition analysis to successfully differentiate between healthy individuals and those with CAD [79]. Targeted MS/MS identified key differentiating features between groups that included oxidized apoA-I and elevated apoC-III. Based on these results the authors devised an analysis model based on specific peptide signals from apo(a), apoC-I, apoC-III, and apoA-I which could accurately discriminate between healthy and CAD patients.

The evidence supporting changes to the HDL proteome in patients with cardiovascular and other inflammatory disease is convincing. However, this is still a very young area of investigation with a lot of room for expansion and further validation. The current data suggest the possibility of disease-specific changes to the HDL proteome. It is also unclear exactly what is driving these variable changes in the HDL proteome and whether they are a result of disease or a contributing factor to disease development. A deeper understanding of the influence of these changes on HDL function will also help to drive the development of targeted therapeutic approaches.

5. HDL PROTEINS AS THERAPEUTIC TARGETS

5.1. Influence of HDL-Raising Therapies on the HDL Proteome

One's risk for cardiovascular disease is most often evaluated based on a balance between pro- and anti-atherogenic lipoprotein classes, that is, LDL and

HDL, respectively. With current therapies we can often achieve target LDL levels through prescription of statins, a class of compounds that inhibit a critical step in cholesterol synthesis and effectively lower LDL-C [80]. However, we have few tools for the manipulation of plasma HDL levels and function. Although HDL-C has long been known to show an inverse correlation with cardiovascular outcomes, therapies that successfully raise plasma HDL-C levels by 30% to more than 100% have been largely unsuccessful in preventing disease [81]. One explanation for this lack of clinical effect is that the currently available therapies such as niacin and CETP inhibitors (CETP is discussed in Chapter 8) were designed to raise HDL-C levels without regard for the type or subspecies of HDL that would be increased and their functional properties. In this chapter, we have explored the evidence indicating that HDL exists as a heterogeneous pool of particles and that different species are likely to have different functional properties based on their different protein and lipid compositions. Based on this, it is clear that some HDL subspecies are likely to be more protective than others, and some can even have pro-inflammatory properties promoting disease [82]. It may be that a broad raising of the cholesterol component of HDL is not necessarily clinically useful and more care should be given to understanding the composition and functionality of the HDL subspecies we are trying to increase.

Some have acknowledged this concept. Green and colleagues studied the effects of combined statin and niacin therapy on the HDL proteome in an attempt to understand how the association of proteins with HDL particles changes during treatment [83]. Looking at the HDL proteome of patients before and 1 year after combination treatment with atorvastatin (10 to 20 mg daily) and extended-release niacin (2 g daily), they noted that along with a 34% increase in plasma HDL-C, levels of three common HDL proteins showed changes in relative abundance determined by spectral counting and extracted ion chromatogram analysis. ApoF and PLTP levels increased by 85% and 247% respectively and apoE decreased by 41%. Previous studies by this group indicated that apoE levels were increased in CAD patients [8], suggesting that the combination therapy used here resulted in at least a partial reversal of this prior observation. However, it is unclear whether this effect was due to the statin or the niacin. At this point it is also unclear whether increased apoE on HDL in CAD patients is a causative factor for CAD development or simply a biomarker of an unhealthy environment in the blood that can predispose one to CAD. In addition to changes in the HDL proteome, niacin has been shown to modulate HDL function by increasing cholesterol efflux and anti-inflammatory functions in vitro [84]. These functional effects of niacin were determined to be a result of increased HDL concentrations and not necessarily improvement of particle function, therefore, it is likely that the observed changes in the protein composition by niacin therapy may influence other functional properties of the particle not yet examined.

The influence of CETP inhibitors on the HDL proteome has also been examined. Krauss and colleagues demonstrated that healthy individuals receiving anacetrapib (20 or 150 mg/day) in addition to having increased levels of large CE-enriched HDL (82% increase in total HDL-C), as would be expected by inhibition of CETP, also showed alterations in their HDL-associated proteins [85]. Using enzyme-linked immunosorbent assays (ELISA) to quantitatively measure apoA-I, apoA-II, apoC-III, and apoE in several density subfractions of HDL, they found that anacetrapib increased the amount of apoA-I and apoC-III on HDL but no differences were detected in apoA-II or apoE. This study indicates that CETP inhibitors are likely influencing the HDL proteome, but the limited ability of immuno-logical techniques to measure large numbers of proteins at once, as can be done with shotgun MS–based approaches, leaves the possibility that many protein alterations may have gone undetected in this study and invites further characterization of CETP-induced modifications of the HDL proteome. HDL from patients treated with anacetrapib showed increased cholesterol efflux in vitro even at matched total HDL concentrations, suggesting increased particle function with this treatment [84].

5.2. Apolipoprotein Mimetic Peptides

Alternative modes of therapy have included the utilization of functional fragments of HDL-associated proteins. They are called apolipoprotein mimetic peptides and are intended to mimic the function of the full-length proteins. Several peptides that imitate the structural properties of apoA-I and apoE have been developed and have shown significant potential for therapeutic use in cardiovascular disease and other chronic inflammatory diseases [86]. A primary structural characteristic of many of these peptides is the amphipathic α-helix, a distinguishing characteristic of the apolipo-protein family. This feature allows them to bind lipids in the plasma, asso-ciating with lipoproteins or forming particles on their own. One of the most studied apoA-I mimetics, the 5A peptide, possesses many functional attributes of full-length apoA-I including cholesterol efflux, inhibition of LDL oxidation, and suppression of inflammation [87]. Additionally, the 5A peptide can reduce atherosclerosis [88] and prevent the induction of asthma in mouse models [89]. Similarly an apoE mimetic peptide representing the LDL receptor-binding region of full-length apoE (aa's 130–149) reduced airway hyperreactivity and goblet cell hyperplasia in house dust mite–challenged mice [90].

The use of small peptide-based therapies offers several advantages over the administration of full-length proteins, including increased stability and significantly lower manufacturing costs. The binding of mimetic peptides to endogenous lipoprotein particles is likely to influence the association of other proteins, however this has not yet been studied. A more complete discussion of HDL mimetic peptides can be found in Chapter 9.

6. CHALLENGES OF MS-BASED ANALYSES OF THE HDL PROTEOME

There are still many challenges that complicate detailed analysis of the HDL proteome. The three most pressing challenges are related to isolation of HDL, quantification of associated proteins, and complications due to post-translational modifications to HDL-associated proteins.

Throughout this chapter we described proteomic studies of HDL that have been isolated from blood by four different techniques: ultracentrifugation, 2DIGE, size exclusion chromatography, and immunoaffinity purification. Each of these techniques has inherent advantages and disadvantages when it comes to subsequent MS analysis. Ultracentrifugation involves high salt concentrations and high g-forces which have been shown to disrupt protein interactions with HDL [91]. Electrophoresis-based separation of HDL has limited resolution and requires inefficient gel-extraction procedures. Size exclusion chromatography methods have a high potential for coseparation of large or very abundant plasma proteins, although this obstacle has been partially overcome by the use of PL-binding resin prior to MS [10]. Finally, isolation of HDL by immunocapture with anti–apoA-I antibodies can be discriminatory toward particles that have exposed apoA-I epitopes on their surface. Although the majority of HDL particles do contain apoA-I, it is likely that a small percentage of lipoprotein particles that fall within the size and density range of HDL do not. On other particles containing apoA-I, the binding of peripheral HDL proteins could potentially shield apoA-I epitopes from capture. In addition to complicating a thorough MS-based analysis of HDL, the broad use of various isolation techniques complicates the study of lipoproteins by making it difficult to compare different studies, because each technique isolates a different population of HDL or introduces its own artifacts.

Methods for accurate quantitative analyses of proteins by MS have been in development for many years. The most common technique for estimating protein abundance used in analyses of the HDL proteome is spectral counting [8-10]. This is essentially a count of the total number of peptides identified from a given protein in the sample being analyzed. Because the number of peptides generated by a protein is not only dependent on the concentration of that protein in the sample but also the number of fragments produced by a digest of that protein and the ionization properties of those fragments, spectral counting cannot be used to determine an absolute quantitative value for a protein. However, spectral counting can be used to determine the relative abundance of a protein across samples prepared similarly and when analyzed at the same time.

Another area of investigation related to the HDL proteome is interested in examining modifications that occur on proteins under various conditions and their effects on HDL function. For example, several modifications to apoA-I have been characterized, including oxidation of specific tryptophan and methionine residues [92,93] and nitration and chlorination of multiple sites [92,94]. Detecting these modifications is not difficult if you already know what you are looking for. Many protein database search engines allow you to choose optional variable modification settings that will account for the change in mass that occurs if some of these simple modifications are present, allowing accurate identification of peptides. Other protein modifications such as glycosylation are common on many apolipoproteins and can complicate MS/MS identification, shifting peptide mass in a sometimes unpredictable manner. Additionally, it is likely that previously uncharacterized and unsuspected modifications are occurring on HDL proteins in vivo. Such modifications would prevent MS/MS identification of peptides and if using spectral counting strategies could result in an apparent but false decrease in relative protein abundance between, for example, a disease group with the modification and a healthy group without.

7. CONCLUSION

Studies of the HDL proteome have generated a long and diverse list of proteins, at least longer and definitely more diverse than most lipoprotein biologists would have guessed prior to the application of MS-based technologies. While this list may be a little daunting, it is also very exciting for the field of HDL research and may represent an extremely fruitful road ahead as we try to tease out and understand what all of these proteins are doing on HDL. It is still unclear exactly what influence the HDL proteome has on HDL function and what impact manipulation of these proteins can have. An important aim of future research should be to carefully characterize the protein compositions of individual HDL subspecies and examine their functional properties. This will require significant advances in HDL isolation techniques but would answer a lot of important questions about why so many proteins are attracted to HDL and what they do while in contact with the particle. With the recent discovery of so many proteins with functional ties outside of lipid transport and metabolism, we will undoubtedly discover new physiological roles for HDL.

REFERENCES

[1] Rifkin MR. Identification of the trypanocidal factor in normal human serum: high density lipoprotein. Proc Natl Acad Sci USA 1978;75(7):3450–4.

[2] Hajduk SL, Moore DR, Vasudevacharya J, Siqueira H, Torri AF, Tytler EM, et al. Lysis of *Trypanosoma brucei* by a toxic subspecies of human high density lipoprotein. J Biol Chem 1989;264(9):5210–7.

[3] Pays E, Vanhollebeke B. Human innate immunity against African trypanosomes. Curr Opin Immunol 2009;21(5):493–8.

[4] Karlsson H, Leanderson P, Tagesson C, Lindahl M. Lipoproteomics II: mapping of proteins in high-density lipoprotein using two-dimensional gel electrophoresis and mass spectrometry. Proteomics 2005;5(5):1431–45.

[5] Heller M, Stalder D, Schlappritzi E, Hayn G, Matter U, Haeberli A. Mass spectrometry-based analytical tools for the molecular protein characterization of human plasma lipoproteins. Proteomics 2005;5(10):2619–30.

[6] Hortin GL, Shen RF, Martin BM, Remaley AT. Diverse range of small peptides associated with high-density lipoprotein. Biochem Biophys Res Commun 2006;340(3):909–15.

[7] Rezaee F, Casetta B, Levels JH, Speijer D, Meijers JC. Proteomic analysis of high-density lipoprotein. Proteomics 2006;6(2):721–30.

[8] Vaisar T, Pennathur S, Green PS, Gharib SA, Hoofnagle AN, Cheung MC, et al. Shotgun proteomics implicates protease inhibition and complement activation in the anti-inflammatory properties of HDL. J Clin Invest 2007;117(3):746–56.

[9] Davidson WS, Silva RA, Chantepie S, Lagor WR, Chapman MJ, Kontush A. Proteomic analysis of defined HDL subpopulations reveals particle-specific protein clusters: relevance to antioxidative function. Arterioscler Thromb Vasc Biol 2009;29(6):870–6.

[10] Gordon SM, Deng J, Lu LJ, Davidson WS. Proteomic characterization of human plasma high density lipoprotein fractionated by gel filtration chromatography. J Proteome Res 2010;9(10):5239–49.

[11] Alwaili K, Bailey D, Awan Z, Bailey SD, Ruel I, Hafiane A, et al. The HDL proteome in acute coronary syndromes shifts to an inflammatory profile. Biochim Biophys Acta 2012;1821(3):405–15.

[12] Watanabe J, Charles-Schoeman C, Miao Y, Elashoff D, Lee YY, Katselis G, et al. Proteomic profiling following immunoaffinity capture of high-density lipoprotein: association of acute-phase proteins and complement factors with proinflammatory high-density lipoprotein in rheumatoid arthritis. Arthritis Rheum 2012;64(6):1828–37.

[13] Holzer M, Birner-Gruenberger R, Stojakovic T, El-Gamal D, Binder V, Wadsack C, et al. Uremia alters HDL composition and function. J Am Soc Nephrol 2011;22(9): 1631–41.

[14] Holzer M, Wolf P, Curcic S, Birner-Gruenberger R, Weger W, Inzinger M, et al. Psoriasis alters HDL composition and cholesterol efflux capacity. J Lipid Res 2012;53(8):1618–24.

[15] Weichhart T, Kopecky C, Kubicek M, Haidinger M, Doller D, Katholnig K, et al. Serum amyloid A in uremic HDL promotes inflammation. J Am Soc Nephrol 2012;23(5):934–47.

[16] Mange A, Goux A, Badiou S, Patrier L, Canaud B, Maudelonde T, et al. HDL proteome in hemodialysis patients: a quantitative nanoflow liquid chromatography-tandem mass spectrometry approach. PLoS One 2012;7(3):e34107.

[17] Weisgraber KH, Mahley RW. Apoprotein (E–A-II) complex of human plasma lipoproteins. I. Characterization of this mixed disulfide and its identification in a high density lipoprotein subfraction. J Biol Chem 1978;253(17):6281–8.

[18] Blanco-Vaca F, Via DP, Yang CY, Massey JB, Pownall HJ. Characterization of disulfide-linked heterodimers containing apolipoprotein D in human plasma lipoproteins. J Lipid Res 1992;33(12):1785–96.

[19] Lagrost L, Persegol L, Lallemant C, Gambert P. Influence of apolipoprotein composition of high density lipoprotein particles on cholesteryl ester transfer protein activity. Particles containing various proportions of apolipoproteins AI and AII. J Biol Chem 1994;269(5):3189–97.

[20] Albers JJ, Wolfbauer G, Cheung MC, Day JR, Ching AF, Lok S, et al. Functional expression of human and mouse plasma phospholipid transfer protein: effect of recombinant and plasma PLTP on HDL subspecies. Biochim Biophys Acta 1995;1258(1): 27–34.

[21] Durbin DM, Jonas A. Lipid-free apolipoproteins A-I and A-II promote remodeling of reconstituted high density lipoproteins and alter their reactivity with lecithin:cholesterol acyltransferase. J Lipid Res 1999;40(12):2293–302.

[22] Thuren T. Hepatic lipase and HDL metabolism. Curr Opin Lipidol 2000;11(3): 277–83.

[23] Castellani LW, Navab M, Van Lenten BJ, Hedrick CC, Hama SY, Goto AM, et al. Overexpression of apolipoprotein AII in transgenic mice converts high density lipoproteins to proinflammatory particles. J Clin Invest 1997;100(2):464–74.

[24] Castro G, Nihoul LP, Dengremont C, de Geitere C, Delfly B, Tailleux A, et al. Cholesterol efflux, lecithin-cholesterol acyltransferase activity, and pre-beta particle formation by serum from human apolipoprotein A-I and apolipoprotein A-I/apolipoprotein A-II transgenic mice consistent with the latter being less effective for reverse cholesterol transport. Biochemistry 1997;36(8):2243–9.

[25] Tso P, Liu M, Kalogeris TJ. The role of apolipoprotein A-IV in food intake regulation. J Nutr 1999;129(8):1503–6.

[26] Qin X, Swertfeger DK, Zheng S, Hui DY, Tso P, Apolipoprotein AIV. a potent endogenous inhibitor of lipid oxidation. Am J Physiol 1998;274(5 Pt 2):H1836–40.

[27] Schaap FG, Rensen PC, Voshol PJ, Vrins C, van der Vliet HN, Chamuleau RA, et al. ApoAV reduces plasma triglycerides by inhibiting very low density lipoprotein-triglyceride (VLDL-TG) production and stimulating lipoprotein lipase-mediated VLDL-TG hydrolysis. J Biol Chem 2004;279(27):27941–7.

[28] Jong MC, Hofker MH, Havekes LM. Role of ApoCs in lipoprotein metabolism: functional differences between ApoC1, ApoC2, and ApoC3. Arterioscler Thromb Vasc Biol 1999;19(3):472–84.

[29] Drayna D, Fielding C, McLean J, Baer B, Castro G, Chen E, et al. Cloning and expression of human apolipoprotein D cDNA. J Biol Chem 1986;261(35):16535–9.

[30] Yang CY, Gu ZW, Blanco-Vaca F, Gaskell SJ, Yang M, Massey JB, et al. Structure of human apolipoprotein D: locations of the intermolecular and intramolecular disulfide links. Biochemistry 1994;33(41):12451–5.

[31] Peitsch MC, Boguski MS. Is apolipoprotein D a mammalian bilin-binding protein? New Biol 1990;2(2):197–206.

[32] Weech PK, Camato R, Milne RW, Marcel YL. Apolipoprotein D and cross-reacting human plasma apolipoproteins identified using monoclonal antibodies. J Biol Chem 1986;261(17):7941–51.

[33] Camato R, Marcel YL, Milne RW, Lussier-Cacan S, Weech PK. Protein polymorphism of a human plasma apolipoprotein D antigenic epitope. J Lipid Res 1989;30(6):865–75.

[34] Steyrer E, Kostner GM. Activation of lecithin-cholesterol acyltransferase by apolipoprotein D: comparison of proteoliposomes containing apolipoprotein D, A-I or C-I. Biochim Biophys Acta 1988;958(3):484–91.

[35] Westerlund JA, Weisgraber KH. Discrete carboxyl-terminal segments of apolipoprotein E mediate lipoprotein association and protein oligomerization. J Biol Chem 1993;268(21):15745–50.

[36] Vedhachalam C, Narayanaswami V, Neto N, Forte TM, Phillips MC, Lund-Katz S, et al. The C-terminal lipid-binding domain of apolipoprotein E is a highly efficient mediator of ABCA1-dependent cholesterol efflux that promotes the assembly of high-density lipoproteins. Biochemistry 2007;46(10):2583–93.

[37] Plump AS, Smith JD, Hayek T, Aalto-Setala K, Walsh A, Verstuyft JG, et al. Severe hypercholesterolemia and atherosclerosis in apolipoprotein E-deficient mice created by homologous recombination in ES cells. Cell 1992;71(2):343–53.

[38] Riddell DR, Graham A, Owen JS. Apolipoprotein E inhibits platelet aggregation through the L-arginine:nitric oxide pathway. Implications for vascular disease. J Biol Chem 1997;272(1):89–95.

[39] Kim J, Basak JM, Holtzman DM. The role of apolipoprotein E in Alzheimer's disease. Neuron 2009;63(3):287–303.

[40] He Y, Greene DJ, Kinter M, Morton RE. Control of cholesteryl ester transfer protein activity by sequestration of lipid transfer inhibitor protein in an inactive complex. J Lipid Res 2008;49(7):1529–37.

[41] Lagor WR, Fields DW, Khetarpal SA, Kumaravel A, Lin W, Weintraub N, et al. The effects of apolipoprotein F deficiency on high density lipoprotein cholesterol metabolism in mice. PLoS One 2012;7(2):e31616.

[42] Lagor WR, Brown RJ, Toh SA, Millar JS, Fuki IV, de la Llera-Moya M, et al. Overexpression of apolipoprotein F reduces HDL cholesterol levels in vivo. Arterioscler Thromb Vasc Biol 2009;29(1):40–6.

[43] Agar C, de Groot PG, Levels JH, Marquart JA, Meijers JC. Beta2-glycoprotein I is incorrectly named apolipoprotein H. J Thromb Haemost 2009;7(1):235–6.

[44] Nakaya Y, Schaefer EJ, Brewer Jr HB. Activation of human post heparin lipoprotein lipase by apolipoprotein H (beta 2-glycoprotein I). Biochem Biophys Res Commun 1980;95(3):1168–72.

[45] Nimpf J, Wurm H, Kostner GM. Interaction of beta 2-glycoprotein-I with human blood platelets: influence upon the ADP-induced aggregation. Thromb Haemost 1985;54(2):397–401.

[46] Chonn A, Semple SC, Cullis PR. Beta 2 glycoprotein I is a major protein associated with very rapidly cleared liposomes in vivo, suggesting a significant role in the immune clearance of "non-self" particles. J Biol Chem 1995;270(43):25845–9.

[47] Cassader M, Ruiu G, Gambino R, Veglia F, Pagano G. Apolipoprotein H levels in diabetic subjects: correlation with cholesterol levels. Metabolism 1997;46(5):522–5.

[48] de Silva HV, Stuart WD, Park YB, Mao SJ, Gil CM, Wetterau JR, et al. Purification and characterization of apolipoprotein J. J Biol Chem 1990;25;265(24):14292–7.

[49] Carver JA, Rekas A, Thorn DC, Wilson MR. Small heat-shock proteins and clusterin: intra- and extracellular molecular chaperones with a common mechanism of action and function? IUBMB Life 2003;55(12):661–8.

[50] Klock G, Baiersdorfer M, Koch-Brandt C. Chapter 7: Cell protective functions of secretory Clusterin (sCLU). Adv Cancer Res 2009;104:115–38.

[51] Zoubeidi A, Gleave M. Small heat shock proteins in cancer therapy and prognosis. Int J Biochem Cell Biol 2012;44(10):1646–56.

[52] Charnay Y, Imhof A, Vallet PG, Kovari E, Bouras C, Giannakopoulos P. Clusterin in neurological disorders: molecular perspectives and clinical relevance. Brain Res Bull 2012;88(5):434–43.

[53] Duchateau PN, Movsesyan I, Yamashita S, Sakai N, Hirano K, Schoenhaus SA, et al. Plasma apolipoprotein L concentrations correlate with plasma triglycerides and cholesterol levels in normolipidemic, hyperlipidemic, and diabetic subjects. J Lipid Res 2000;41(8):1231–6.

[54] Zhang XY, Dong X, Zheng L, Luo GH, Liu YH, Ekstrom U, et al. Specific tissue expression and cellular localization of human apolipoprotein M as determined by in situ hybridization. Acta Histochem 2003;105(1):67–72.

[55] Xu N, Dahlback B. A novel human apolipoprotein (apoM). J Biol Chem 1999;274(44):31286–90.

[56] Christoffersen C, Nielsen LB, Axler O, Andersson A, Johnsen AH, Dahlback B. Isolation and characterization of human apolipoprotein M-containing lipoproteins. J Lipid Res 2006;47(8):1833–43.

[57] Wolfrum C, Poy MN, Stoffel M. Apolipoprotein M is required for prebeta-HDL formation and cholesterol efflux to HDL and protects against atherosclerosis. Nat Med 2005;11(4):418–22.

[58] Lamant M, Smih F, Harmancey R, Philip-Couderc P, Pathak A, Roncalli J, et al. ApoO, a novel apolipoprotein, is an original glycoprotein up-regulated by diabetes in human heart. J Biol Chem 2006;281(47):36289–302.

[59] Zhuo L, Kimata K. Structure and function of inter-alpha-trypsin inhibitor heavy chains. Connect Tissue Res 2008;49(5):311–20.

[60] Zuraw BL, Lotz M. Regulation of the hepatic synthesis of C1 inhibitor by the hepatocyte stimulating factors interleukin 6 and interferon gamma. J Biol Chem 1990;265(21):12664–70.

[61] Yeung Laiwah AC, Jones L, Hamilton AO, Whaley K. Complement-subcomponent-C1-inhibitor synthesis by human monocytes. Biochem J 1985;226(1):199–205.

[62] Kalter ES, Daha MR, ten Cate JW, Verhoef J, Bouma BN. Activation and inhibition of Hageman factor-dependent pathways and the complement system in uncomplicated bacteremia or bacterial shock. J Infect Dis 1985;151(6):1019–27.

[63] Zeerleder S. C1-inhibitor: more than a serine protease inhibitor. Semin Thromb Hemost 2011;37(4):362–74.

[64] Alper CA, Johnson AM, Birtch AG, Moore FD. Human C'3: evidence for the liver as the primary site of synthesis. Science 1969;163(3864):286–8.

[65] Engstrom G, Hedblad B, Eriksson KF, Janzon L, Lindgarde F. Complement C3 is a risk factor for the development of diabetes: a population-based cohort study. Diabetes 2005;54(2):570–5.

[66] Hernandez-Mijares A, Jarabo-Bueno MM, Lopez-Ruiz A, Sola-Izquierdo E, Morillas-Arino C, Martinez-Triguero ML. Levels of C3 in patients with severe, morbid and extreme obesity: its relationship to insulin resistance and different cardiovascular risk factors. Int J Obes (Lond) 2007;31(6):927–32.

[67] Szilagyi A, Fust G. Diseases associated with the low copy number of the C4B gene encoding C4, the fourth component of complement. Cytogenet Genome Res 2008;123(1-4):118–30.

[68] Zoppi M, Weiss M, Nydegger UE, Hess T, Spath PJ. Recurrent meningitis in a patient with congenital deficiency of the C9 component of complement. First case of C9 deficiency in Europe. Arch Intern Med 1990;150(11):2395–9.

[69] Bausserman LL, Herbert PN, Forte T, Klausner RD, McAdam KP, Osborne Jr JC, et al. Interaction of the serum amyloid A proteins with phospholipid. J Biol Chem 1983;258(17):10681–8.

[70] Aldridge WN. Serum esterases. II. An enzyme hydrolysing diethyl p-nitrophenyl phosphate (E600) and its identity with the A-esterase of mammalian sera. Biochem J 1953;53(1):117–24.

[71] Mackness MI, Arrol S, Abbott C, Durrington PN. Protection of low-density lipoprotein against oxidative modification by high-density lipoprotein associated paraoxonase. Atherosclerosis 1993;104(1–2):129–35.

[72] Aviram M, Rosenblat M, Bisgaier CL, Newton RS, Primo-Parmo SL, La Du BN. Paraoxonase inhibits high-density lipoprotein oxidation and preserves its functions. A possible peroxidative role for paraoxonase. J Clin Invest 1998;101(8):1581–90.

[73] Farrah T, Deutsch EW, Omenn GS, Campbell DS, Sun Z, Bletz JA, et al. A high-confidence human plasma proteome reference set with estimated concentrations in PeptideAtlas. Mol Cell Proteomics 2011;10(9); M110 006353.

[74] Gordon S, Durairaj A, Lu JL, Davidson WS. High-density lipoprotein proteomics: identifying new drug targets and biomarkers by understanding functionality. Curr Cardiovasc Risk Rep 2010;4(1):1–8.

[75] Shah AS, Tan L, Lu Long J, Davidson WS. The proteomic diversity of high density lipoproteins: our emerging understanding of its importance in lipid transport and beyond. J Lipid Res 2013 Feb 24; [Epub ahead of print].

[76] Asztalos BF, Sloop CH, Wong L, Roheim PS. Two-dimensional electrophoresis of plasma lipoproteins: recognition of new apo A-I-containing subpopulations. Biochim Biophys Acta 1993;1169(3):291–300.

[77] Artl A, Marsche G, Lestavel S, Sattler W, Malle E. Role of serum amyloid A during metabolism of acute-phase HDL by macrophages. Arterioscler Thromb Vasc Biol 2000;20(3):763–72.

[78] Van Lenten BJ, Hama SY, de Beer FC, Stafforini DM, McIntyre TM, Prescott SM, et al. Anti-inflammatory HDL becomes pro-inflammatory during the acute phase response. Loss of protective effect of HDL against LDL oxidation in aortic wall cell cocultures. J Clin Invest 1995;96(6):2758–67.

[79] Vaisar T, Mayer P, Nilsson E, Zhao XQ, Knopp R, Prazen BJ. HDL in humans with cardiovascular disease exhibits a proteomic signature. Clin Chim Acta 2010;411(13-14):972–9.

[80] Rader DJ. Therapy to reduce risk of coronary heart disease. Clin Cardiol 2003;26(1):2–8.

[81] Joy T, Hegele RA. Is raising HDL a futile strategy for atheroprotection? Nat Rev Drug Discov 2008;7(2):143–55.

[82] Navab M, Reddy ST, Van Lenten BJ, Anantharamaiah GM, Fogelman AM. The role of dysfunctional HDL in atherosclerosis. J Lipid Res 2009;50(Suppl.):S145–9.

[83] Green PS, Vaisar T, Pennathur S, Kulstad JJ, Moore AB, Marcovina S, et al. Combined statin and niacin therapy remodels the high-density lipoprotein proteome. Circulation 2008;118(12):1259–67.

[84] Yvan-Charvet L, Kling J, Pagler T, Li H, Hubbard B, Fisher T, et al. Cholesterol efflux potential and antiinflammatory properties of high-density lipoprotein after treatment with niacin or anacetrapib. Arterioscler Thromb Vasc Biol 2010;30(7):1430–8.

[85] Krauss RM, Wojnooski K, Orr J, Geaney JC, Pinto CA, Liu Y, et al. Changes in lipoprotein subfraction concentration and composition in healthy individuals treated with the CETP inhibitor anacetrapib. J Lipid Res 2012;53(3):540–7.

[86] Gordon SM, Davidson WS. Apolipoprotein A-I mimetics and high-density lipoprotein function. Curr Opin Endocrinol Diabetes Obes 2012;19(2):109–14.

[87] D'Souza W, Stonik JA, Murphy A, Demosky SJ, Sethi AA, Moore XL, et al. Structure/function relationships of apolipoprotein a-I mimetic peptides: implications for antiatherogenic activities of high-density lipoprotein. Circ Res 2010;107(2):217–27.

[88] Amar MJ, D'Souza W, Turner S, Demosky S, Sviridov D, Stonik J, et al. 5A apolipoprotein mimetic peptide promotes cholesterol efflux and reduces atherosclerosis in mice. J Pharmacol Exp Ther 2010;334(2):634–41.

[89] Yao X, Dai C, Fredriksson K, Dagur PK, McCoy JP, Qu X, et al. 5A, an apolipoprotein A-I mimetic peptide, attenuates the induction of house dust mite-induced asthma. J Immunol 2011;186(1):576–83.

[90] Yao X, Fredriksson K, Yu ZX, Xu X, Raghavachari N, Keeran KJ, et al. Apolipoprotein E negatively regulates house dust mite-induced asthma via a low-density lipoprotein receptor-mediated pathway. Am J Respir Crit Care Med 2010;182(10):1228–38.

[91] van't Hooft F, Havel RJ. Metabolism of apolipoprotein E in plasma high density lipoproteins from normal and cholesterol-fed rats. J Biol Chem 1982;257(18):10996–1001.

[92] Shao B, Oda MN, Bergt C, Fu X, Green PS, Brot N, et al. Myeloperoxidase impairs ABCA1-dependent cholesterol efflux through methionine oxidation and site-specific tyrosine chlorination of apolipoprotein A-I. J Biol Chem 2006;281(14):9001–4.

[93] Hadfield KA, Pattison DI, Brown BE, Hou L, Rye KA, Davies MJ, et al. Myeloperoxidase-derived oxidants modify apolipoprotein A-I and generate dysfunctional high-density lipoproteins: comparison of hypothiocyanous acid (HOSCN) with hypochlorous acid (HOCl). Biochem J 2013;449(2):531–42.

[94] Zheng L, Settle M, Brubaker G, Schmitt D, Hazen SL, Smith JD, et al. Localization of nitration and chlorination sites on apolipoprotein A-I catalyzed by myeloperoxidase in human atheroma and associated oxidative impairment in ABCA1-dependent cholesterol efflux from macrophages. J Biol Chem 2005;280(1):38–47.

INDEX

Note: Page numbers with "f" denote figures; "t" tables; "b" boxes.

Printed and bound by CPI Group (UK) Ltd, Croydon, CR0 4YY

03/10/2024

01040425-0010